The Icon Critical Dictionary of the New Cosmology

The Icon Critical Dictionary of the New Cosmology

Edited by

Peter Coles

ICON BOOKS

Published in 1998 by Icon Books Ltd.,
Grange Road, Duxford, Cambridge CB2 4QF
e-mail: icon@mistral.co.uk

Reprinted 1999

Distributed in the UK, Europe, Canada, South Africa and Asia by
the Penguin Group:
Penguin Books Ltd., 27 Wrights Lane, London W8 5TZ

Published in Australia in 1998 by Allen & Unwin Pty. Ltd.,
PO Box 8500, 9 Atchison Street, St. Leonards, NSW 2065

Cover illustration by Andrzej Klimowski
Design and layout by Christos Kondeatis
Managing editor John Woodruff
Typesetting by Hands Fotoset, Leicester

ISBN 1 874166 64 1

Printed and bound in Great Britain by
Mackays plc., Chatham, Kent

CONTENTS

Foreword

On the cosmic scale, things have fallen out rather well for cosmologists here on planet Earth. We could have found ourselves ensconced on an overcast planet whose weather forecast was all scattered showers and no sunny intervals. Our Solar System might easily have been located in one of the dustier parts of the Milky Way, through which visible light from neither stars nor galaxies could penetrate. Both outcomes would have left astronomers down on their luck. We would have been thousands of years behind in our knowledge of planetary motions and the laws that govern the changing of the sky. Our deep understanding of Nature's laws of motion and gravity, gleaned primarily from our study of the magisterial celestial motions, would have been stunted; our view of ourselves and the immensity of the Universe around us smaller and more parochial.

Instead, Nature has been kind to us, allowing us to see enough to begin to build up something of the big picture of the expanding Universe in which we live. Nearly four hundred years ago, observations made with the unaided eye were first supplemented by telescopes that magnified images in visible light, but the 20th century has witnessed the unveiling of the Universe across the whole electromagnetic spectrum. We have telescopes and receivers that eavesdrop on the Universe in the radio, millimetre, ultraviolet, infrared and X-ray wavebands. Each paints a different portrait of the Universe, clearly revealing things that once could be seen only through a glass darkly.

Technical developments have played a key role. Physicists and engineers have provided instruments whose sensitivity was undreamt of just twenty-five years ago. Where once we had to build bigger and bigger mirrors if our telescopes were to see farther and fainter than ever before, we can now advance more rapidly, more cheaply and more flexibly, by improving the quality of the receivers that register photons from billions of light years away. And, most dramatically of all, we have at last placed a telescope in space. High above the twinkling of the Earth's atmosphere, the Hubble Space Telescope has provided us with breathtaking images of the Universe. They have a sharpness and unsuspected intricacy that has turned them into some of our most instantly recognisable natural works of art.

During the past fifteen years our theoretical understanding of the Universe's structure and past history has grown at least as fast as these observational capabilities. The growth in our understanding of the behaviour of matter and the forces that govern its most elementary constituents at very high energies has enabled us to reconstruct the earliest moments of the expanding Universe, with dramatic successes. In the next few years those theories will be subjected to critical observational tests. But calculation and observation are not the only ways of learning about the Universe: there is now a third way. Large supercomputers allow us to build simulations of the complicated sequences of events that led to the formation of stars and galaxies. Different scenarios can be explored experimentally to select those which produce patterns of light similar to those that we see. As we look to the future, these triple fronts of theoretical investigation, computer simulation and multi-wavelength observation promise to converge upon a single picture of how the large-scale panorama of galaxies arose, what the Universe was like in its first moments, and of what its material contents consist.

Already, these advances have created an outpouring of popular books, journalism, TV specials and Internet sites. Cosmology has become one of the highest-profile sciences. Even the money markets had to have a 'big bang' (maybe they will come to discover that there can also be a 'big crunch'). It is therefore an ideal time to provide a thoroughly up-to-date dictionary of the science behind the news. *The Icon Critical Dictionary of the New Cosmology* does this and much more besides. First, the reader is treated to six lucid overviews of different aspects of modern cosmology by leading researchers who are actively involved in the work they describe. In a little over a hundred pages, readers will receive a fast-track education in cosmology that will enable them to follow the great advances expected in the next few years. There follows a detailed dictionary that includes all the important ideas and personalities of cosmology. This will be a lasting and authoritative work of reference. The entries are substantial, with an emphasis on explanation tempered by elegant economies of lexicography. Whether you are a student, a journalist, a cosmologist or just curious to know what it's all about, from *absorption lines* to *Fritz Zwicky*, it's all here.

John D. Barrow
Professor of Astronomy, University of Sussex
January 1998

Editor's Preface

Cosmology is a large and complex subject. It is also one that holds a fascination for a wide range of individuals, from ordinary people to professional research scientists. The idea behind this book is to provide a guide to the latest cosmological developments in such a way that it will prove useful to as many different categories of reader as possible.

The book is divided into two main sections. The first consists of a set of six introductory essays in which experts give overviews of the latest developments in both observation and theory. The second is constructed more along the lines of a dictionary, and gives in-depth explanations of specific concepts or terms. This section also contains brief biographical notes on some of the most important cosmologists and physicists, past and present.

The essays and the dictionary entries contain cross-references in **bold** so that related words and concepts can be explored. There is also an index at the end of the book as an aid to locating terms which do not have their own entries. Many such terms appear in other entries in *italics*. Having an index in a dictionary may seem strange, but the alphabetical entries are much longer than are normally found in a technical dictionary, and related topics are frequently discussed under one heading rather than being spread over a number of entries.

Feel free to use this book however you like. If you want an overview of a particular area of cosmology, head first for the appropriate essay, then follow the cross-references into the main dictionary. If you need an explanation of a specific term, then look for a dictionary entry. If it has no entry, try the index and you may discover that your target term is dealt with in another entry. If you just want to browse through entries at random, you can do that too: they are all relatively self-contained.

It is in the nature of cosmology that some aspects of it are inherently mathematical, or involve subtle physical concepts that are difficult to explain in simple language. Some of the dictionary entries are quite technical, as they need to be if they are to be of use to those intending to study cosmology. Such entries are flagged by one or two stars. Unstarred entries, which comprise the bulk of the dictionary, should be intelligible to well-informed lay readers with no specific training beyond high-school science and mathematics. One-star entries require a knowledge of concepts from mathematics or science that might be encountered in first-year undergraduate courses in British universities. Two-star entries present advanced concepts and mathematical treatments. They are not for the faint-hearted, but I have wherever possible started these entries with straightforward outlines of the key ideas before launching into the technical descriptions.

Where appropriate, suggestions for further reading are also given. These range from popular texts which require no special knowledge, to citations of research papers in which major discoveries were first reported. Standard textbooks are referred to where it has not been possible to provide a self-contained explanation of all the required physics. All the references are also gathered into a classified listing at the end of the book.

I hope that the result is a book which can be used in many ways and by many different kinds of reader.

Peter Coles
Queen Mary & Westfield College
January 1998

ACKNOWLEDGEMENTS

All pictures taken by the Hubble Space Telescope are reproduced by courtesy of NASA and the Space Telescope Science Institute.

We are grateful to EUROPACE 2000 for permission to use diagrams from their multimedia course Topics of Modern Cosmology.

All other diagrams and figures are in the public domain.

Contributors

Professor **Peter Coles** (Editor, and Essay 1), formerly of Queen Mary & Westfield College, University of London, is Professor of Astrophysics at the University of Nottingham (UK). He studied Natural Sciences (Physics) at Cambridge and then moved to the University of Sussex where he obtained a D.Phil. in 1989. As well as scores of journal articles, he has co-written two textbooks on cosmology.

Dr **Julianne Dalcanton** (Essay 4) studied physics as an undergraduate at the Massachusetts Institute of Technology (MIT), and then went to graduate school at Princeton to study astrophysical sciences and obtained her Ph.D. there. She was then awarded a prestigious Hubble Fellowship, which she took at the Observatories of the Carnegie Institution of Washington at Pasadena, California. She is now a professor elect at the University of Washington in Seattle. Her main research area is extragalactic observational astrophysics, including studies of quasars and galaxy evolution.

Professor **Carlos Frenk** (Essay 2) holds a chair in the department of physics at the University of Durham (UK). He comes originally from Mexico, and obtained his Ph.D. from the University of Cambridge (UK). He was one of the pioneers of the use of massive computers in the study of cosmological structure formation, and was one of the architects of the cold dark matter theory.

Dr **Andrew R. Liddle** (Essay 3) gained his Ph.D. from the University of Glasgow, then moved to the Astronomy Centre at the University of Sussex where he is currently a Royal Society University Research Fellow. His main research interests concern the observational consequences of the physics of the early Universe, especially the connection with observed structures in the galaxy distribution and the cosmic microwave background.

Dr **Charles H. Lineweaver** (Essay 5) received a Ph.D. in physics from the University of California at Berkeley in 1994. He was a member of the

Cosmic Background Explorer 'COBE-DMR' team which discovered the famous 'ripples' – temperature variations in the cosmic microwave background – in 1992. He is an author of more than thirty articles dealing with the cosmic microwave background, and an editor of the recently published *The Cosmic Microwave Background* (Kluwer, Dordrecht, 1997). He is currently a research fellow at the University of New South Wales in Sydney, Australia.

Dr **Priyamvada Natarajan** (Essay 6) read physics and mathematics at the Massachusetts Institute of Technology, Cambridge (USA) and has recently completed her Ph.D. at the Institute of Astronomy at Cambridge University (UK). She is currently a research fellow at Trinity College, Cambridge, and a postdoctoral fellow at the Canadian Institute for Theoretical Astrophysics (CITA) in Toronto. Her research interests lie primarily in cosmology, especially astrophysical applications of gravitational lensing.

I

Modern Cosmology: Observation and Experiment

1

Foundations of the New Cosmology

Peter Coles

Introduction

Cosmology is the study of the origin and evolution of the **Universe** as a whole. Nowadays, this is a subject with immense popular appeal. Hardly a day seems to go by without the media announcing a new discovery by astronomers using one of the bewildering array of high-tech instruments now at their disposal. This popular appeal has at least partly to do with the universal desire to understand where we came from, what the Universe is all about and why we are here. These are questions traditionally addressed by religions, and it may be that the amazing growth of interest in cosmology is related in some way to the decline of the religious tradition, at least in the Western world. But in any case, the task of unravelling the nature of the Universe using both sensitive observations of impossibly distant objects and complex, obscure mathematical theories is an ambitious goal indeed. And even those who do not understand the technicalities of the work being done can hardly fail to be impressed by the achievements of the 1990s. Cosmologists themselves often describe the current era as the 'Golden Age' of cosmology, with developments in instrumental technology making possible observations that could scarcely have been imagined in the previous decade. And significant breakthroughs in fundamental physics have led to important changes in the way we think about the Universe and how we interpret the new observational data.

Both the observational and the theoretical sides of the subject continue not only to fascinate the general public, but also to occupy some of the world's most talented professional scientists. Part of the attraction is that cosmology lies at the intersection of many scientific disciplines. The subject therefore requires many seemingly disparate branches of physics, astronomy and astrophysics to be mastered.

Some scientists are interested in cosmology primarily as a branch of astronomy, and seek to understand how the various constituents of the

Universe, from stars to galaxies and giant clusters of galaxies, came into being and evolved. This requires an understanding of how the Universe at large is constructed, and how its properties change with time. Others have an interest in more fundamental physical properties of the Universe. For example, the field of *astro-particle physics* involves taking present-day observations and turning back the clock in an attempt to understand the behaviour of the Universe in the very early stages of its evolution, tiny fractions of a second after the initial Big Bang, when the energies were way beyond anything that can be reached in a terrestrial laboratory. Yet others see cosmology as an application of **general relativity**, and exploit the cosmological setting as a testing-ground for Albert **Einstein**'s beautiful but mathematically challenging theory.

Cosmology is, by nature, a very peculiar subject that is set apart from other branches of physics and astronomy. The Universe is, by definition, unique. We cannot prepare an ensemble of universes with slightly different properties and look for differences or correlations in their behaviour. In many branches of physical science it is this kind of experimentation that often leads to the formulation of empirical laws, which give rise to models and subsequently to theories. Cosmology is different. We have only one Universe from which to extract the empirical laws we then try to explain by theory, as well as the experimental evidence we use to test the theories we have formulated. Although the distinction between them is not clear-cut, it is fair to say that physics is characterised by experiment and theory, and cosmology by observation and modelling. Subtle influences of personal philosophy, and of cultural and, in some cases, religious background, can lead to different choices of model (or 'paradigm') in many branches of science, but this tendency is particularly noticeable in cosmology. For example, the **expansion of the Universe**, which is now regarded as one of the 20th century's most important scientific discoveries, could have been predicted on the basis of Newtonian physics as early as the 17th century. However, a philosophical predisposition in Western societies towards an unchanging, regular cosmos apparently prevented scientists from drawing this conclusion until it was forced upon them by observations made in the 20th century.

The nature of cosmology, from its beginnings in mythology to the present 'Golden Age' of frontier science, has undergone many radical changes. Sometimes these changes have been in response to changing social circumstances (as in the Industrial Revolution), and sometimes they have been brought about by changes in philosophical outlook among contemporary thinkers. Now the subject is undergoing another upheaval: the dawn of the 'new cosmology' of the 21st century. This essay attempts to put these developments in context by charting the evolution of cosmology into a branch of physical science.

The cosmology of ancient Greece

One can learn much about what cosmology actually means from its history. Since prehistoric times, humans have sought to make sense of their existence, and that of the external world, in the context of some kind of theoretical framework. The first such theories, not recognisable as 'science' in the modern sense of the word, were mythological and fall more within the province of anthropology than cosmology. Cosmology emerged as a recognisably modern scientific discipline with the Greeks, first with Thales (625–547 BC) and Anaximander (610–540 BC), and then with the Pythagoreans of the 6th century BC, who regarded numbers as the basis of all natural things.

The most important early step on the road to modern cosmology was taken by Plato (427–348 BC). In the tradition of the Greek mythologists, his description takes the form of a creation story, narrated by a fictional philosopher named Timaeus of Locris, who explains how the whole of nature is initiated by a divine creator called the Demiurge ('craftsman' in Greek). The Demiurge seeks, as far as he is able, to replicate through physical copies the ideal, perfect structures of true being which exist in the world of 'ideas'. What is created is the domain of things that can change. Birth, growth, alteration and death are then parts of the physical world. But the Demiurge merely prepares the model for the world: he does not carry out its construction or play a role in its day-to-day maintenance; these tasks he delegates to a set of divine subordinates. These 'gods', usually in human form, control the physical world. For this reason, the whole cosmic system is described in terms of the behaviour of humans. All phenomena of nature are represented as an interplay of two fundamental forces: reason and necessity. Plato takes reason to represent a kind of 'world soul'; the material medium (the four elements of earth, air, water and fire) represents the domain of necessity.

The logical successor to Plato was one of his pupils, Aristotle (384–322 BC). It was Aristotle's ideas that would dominate Western thought in the Middle Ages and pave the way for what was to come during the Renaissance. In some ways Aristotle's ideas are similar to modern scientific reasoning, but there are also important differences. For example, his *De caelo* contains a discussion of the basic properties of motion. According to Aristotle, all motion is either straight or circular (or a combination of the two). All bodies are either simple (i.e. composed of a single element, such as fire or earth) or are compounds. The element fire and bodies composed of it have a natural tendency to upward movement, while bodies composed of earth move downwards (i.e. towards the centre of the Universe, which is the Earth). Circular movement is natural for substances other than the four elements. It is considered more 'divine' than straight-line motion, and substances that move in a circular way are consequently considered more divine than those that move in straight lines.

In Aristotle's cosmology the Universe is spherical, and divided into two: a changing region, which extends as far as the Moon and at whose centre sits the Earth surrounded by the other elements; and an unchanging region, in which the heavenly bodies perform stately circular motions. There is a separate set of physical laws for each of the two regions, since they are composed of different types of matter. Aristotle argues that the Universe is not infinite because it moves in a circle (as we can see with our eyes if we watch the stars). If the Universe were infinite, it would be moving though an infinite distance in a finite time, which is impossible. He also claimed that there is only one world. If there were more than one world – each with a centre as the natural place for earthy material to move towards, and a circumference for fire to move towards – then the Earth could move towards any of the centres and fire could move towards any of the circumferences. Chaos would ensue. Since we observe order instead of chaos, then there must be only one world. Aristotle also showed that the Earth is spherical (since it casts a circular shadow on the Moon during a lunar eclipse, and different stars are seen from different parts of the Earth), and held that it was stationary and at the centre of the heavenly sphere, which rotated around it.

In these arguments, Aristotle's use of observation is in stark contrast to Plato's dictum that nothing can be learnt by using the senses. This is seen by some historians as a turning point in science, marking the beginning of extensive empirical investigations. But, while he did place a new emphasis on the value of observation, Aristotle's method still differs in important ways from modern scientific practice. His observations, for example, are used more to persuade his readers of the truth of his conclusions than as an aid to arriving at those conclusions. Moreover, it would never have occurred to him to test his conclusions by conducting experiments: he regarded the laws of nature as being self-evident. The strength of his arguments lies largely in their commonsense nature and his ability to marshal disparate phenomena into a single overarching scheme of things.

Although intellectually appealing, Aristotle's view of perfectly circular motions did not stand up to detailed scrutiny in the light of astronomical observations. The body of accumulated empirical knowledge of the motions of the planets increased as the Greek empire expanded to the east under Alexander the Great, and the vast archives of astronomical data assembled by the Babylonians and Egyptians were discovered. These observations made it clear that the planets did not move in circular orbits around the Earth, as Aristotle had asserted. The culmination of this new interplay between theory and observation was the *Almagest*, compiled by Ptolemy in the 2nd century AD. This magnificent book lays down complex mathematical and geometrical formulae for calculating the positions of the planets: the first complete, quantitative and empirically tested mathematical model for the Universe.

Towards the Renaissance

Much of the knowledge of ancient Greece was lost to Christian culture during the dark ages. It did not disappear entirely, however, because it formed the basis of Islamic astronomy, which made enormous progress during this period. Compared with the sophistication of the *Almagest*, the knowledge of astronomy in medieval Europe was extremely limited. Thomas Aquinas (1225–74) seized upon Aristotle's ideas (which were available in Latin translation at the time, whereas Ptolemy's were not) and forged a synthesis of the Christian view of creation with the pagan cosmology of Aristotle. Western astronomical thought was dominated by these ideas until the 16th and early 17th centuries.

The dismantling of the Aristotelian world-view is usually credited to Nicolaus **Copernicus**. He was unhappy that Ptolemy's theory of the Solar System essentially treated each planet separately. The *Almagest* gave formulae for predicting where the planets should be at particular times, but these formulae were very different for different planets: the scheme lacked any concept of universal behaviour. As set out in his *De revolutionibus* (published in 1543, the year of his death), Copernicus had come to the conclusion that the old *geocentric* model with the Earth at the centre of the Solar System was unsatisfactory, and that the orbits of the planets could be better explained by the *heliocentric* model, in which the Sun, not the Earth, lay at the centre of the cosmos. The *Copernican principle*, the notion that we (on the Earth) do not inhabit a special place in the Universe (a forerunner of the modern **cosmological principle**), was symptomatic of the philosophical and religious changes that took place during the Renaissance. The philosophical impact of this work was, however, lessened by the insertion of a disclaimer (without Copernicus's knowledge) at the front of the book. This preface, written by the German theologian Andreas Osiander (1498–1552), claimed that Copernicus was not arguing that nature was really like this, merely that it provided a convenient way to calculate planetary positions. For many years, Copernicus's true message thus remained obscured. **Galileo** championed Copernicus's cause. In 1609, after acquiring one of the first telescopes, he was able to show that the planet Jupiter appeared to have satellites orbiting around it. If this were so, why then could not the Earth and the other planets be orbiting the Sun? There thus began a long struggle between Galileo and the Vatican, which was still wedded to an Aristotelian world-view.

In any case, it is not really fair to say that Copernicus himself overthrew the view of the Earth at the centre of the Universe. His model for the planetary motions did not actually fit the observational data very well, and was certainly not as successful in this regard as the Ptolemaic system, though it was indeed much simpler. Johannes **Kepler**, working with detailed and highly accurate observations made by his late employer, Tycho Brahe (1546–1601), changed the Copernican model to incorporate elliptical rather than circular orbits. His new model fitted the available observations perfectly (within the limits of contemporary observational accuracy), but the price that had to be paid was the complete rejection of Aristotle's view of the divine circular motions of the heavenly bodies. Interestingly, Galileo did little to propagate Kepler's theory; he appears to have lacked the patience to struggle through Kepler's difficult books.

Newton and after

In Kepler's time, the idea that the planets were moving in elliptical orbits must have seemed rather ugly. The grand symmetry of a sphere was much more aesthetically appealing. It was about eighty years after the publication of Kepler's new theory in 1619 that Isaac **Newton** demonstrated (in the *Principia*, first published in 1687) that these odd motions could be explained by a universal law of gravitation which was itself simple and elegant. This is perhaps the first instance of an idea which is now common in modern physics: that a symmetrical law can have asymmetrical outcomes (see, for example, **spontaneous symmetry-breaking**).

Newton's law of **gravity** is still used by physicists today, as it is a good approximation in many circumstances to the more complete theory of general relativity on which modern cosmology is based. But Newton's famous laws of motion also initiated a change in philosophy: it ushered in the *mechanistic* view of the Universe as a kind of giant clockwork device, a view which began to take hold with the emergence of mathematical physics and the first stirrings of technological development. The dawn of theoretical physics also brought with it a new approach to cosmology based on the idea of universal mathematical laws. Not just Newton, but also René Descartes (1596–1650), Immanuel **Kant** and Pierre-Simon de Laplace (1749–1827) attempted to apply the concept of universal laws to the Universe as a whole.

Kant, for example, constructed one of the first reasonably complete models of a scientific view of the Universe. His cosmology was thoroughly mechanistic and materialistic, but it makes clear that every cosmology must begin with the perception of a 'systematic constitution' that could be viewed as evidence of some sort of 'grand design'. Although most of Kant's main tenets were mistaken, his work was of unprecedented scope, made detailed use of physical theory and contained a number of fundamental insights. His cosmological explanation takes the form of showing how the 'systematic constitution' arose, by way of Newton's laws of motion and the law of universal gravitation, from a primaeval state of chaos. The chaos consisted of atoms or particles of matter spread throughout an infinite space. According to Kant, this chaos was unstable: the denser particles began at once to attract the more tenuous. This is the explanation of the origin of motion, and of the formation of bodies and, eventually, of the planets. Despite its ambitions, though, Kant's cosmology was largely sketchy and qualitative.

The dominant view of scientists at the time was of a mechanistic, deterministic Universe performing its show on the eternal stage furnished by Newton's absolute space and time. The culmination of this view, in the spirit of the burgeoning Industrial Revolution, was the notion of the Universe as a gigantic engine. In the late 19th century, physicists became preoccupied with the relationship between cosmology and **thermodynamics**, the theory of energy and heat. In particular there was the widespread belief that the **heat death of the Universe**, a consequence of the second law of thermodynamics, would lead to the eventual failure of the cosmic machine. The development of the mechanistic view, into the idea that the Universe might be controlled by 'timeless' physical laws but could nevertheless itself be changing with time, would prove a vital step towards the construction of the **Big Bang theory**.

The birth of the Big Bang: the 20th century

Physical science underwent a major upheaval in the early 20th century as a result of twin developments in, first, atomic theory and then **quantum theory** and, second, the theory of relativity (both **special relativity** and general relativity). Out went the picture of a deterministic world-system, because quantum physics embodies a fundamental indeterminacy. And with it went the idea of absolute space and **time**, because relativity shows that time depends on who measures it.

The full implications of quantum theory for cosmology have yet to be elucidated, but the incorporation of ideas from relativity theory was enough in itself to revolutionise early 20th-century approaches to cosmology: modern relativistic cosmology emerged in this period. Special relativity had already shattered the illusion of absolute space and time. In 1915, Einstein advanced his theory of general relativity, in which space is not only relative, it is also curved. When he applied the theory to cosmology he was startled to find that the resulting 'field equations' (see **Einstein equations**) said the Universe should be evolving. Einstein thought he must have made a mistake, and promptly modified the equations to give a static cosmological solution by introducing the infamous **cosmological constant**. It was not until after 1929 and the work of Edwin **Hubble** that the astronomical community became convinced that the Universe was actually expanding after all.

As well the new physics, the rapid development of observational astronomy, in both telescope design and detector technology, played an important role in shaping modern cosmology. Alexander **Friedmann**, Georges **Lemaître**, Willem **de Sitter** and others constructed other viable cosmological solutions to the Einstein equations which could be tested only by observation. Eventually, the turmoil of experiment and theory resolved itself into two rival camps: on one side stood the supporters of the **steady state theory**, which described an eternal, infinite Universe in which matter is continuously created; ranged against them were the proponents of the Big Bang theory, in which the entire Universe is created in one fell swoop.

The respective advocates of these two world-views began a long and acrimonious debate about which was correct, the legacy of which lingers still. For many cosmologists this debate was resolved by the discovery in 1965 of the **cosmic microwave background radiation**, which was immediately perceived as evidence in favour of an evolving Universe that was hotter and denser in the past. It is reasonable to regard this discovery as marking the beginning of *physical cosmology*. Counts of distant galaxies had already begun to show evidence of evolution in the properties of these objects, and the first calculations had been made, notably by Ralph **Alpher** and Robert **Herman** in the late 1940s, of the relative proportions of different chemical elements (see **light element abundances**) expected to be produced by nuclear reactions in the early stages of the Big Bang. These and other considerations left the Big Bang model as the clear victor over its steady state rival.

THE STANDARD MODEL

At this point we should give a brief overview of what the Big Bang theory is and, perhaps more importantly, what it is not. To begin with, the theory requires a self-consistent mathematical description of the large-scale properties of the Universe. The most important step towards constructing such a description is the realisation that the **fundamental interaction** that is most prominent on the large scales relevant to cosmology is gravity; and the most complete theory of gravity presently available is Einstein's theory of general relativity. This theory has three components:

A description of the spacetime geometry (see **curvature of spacetime**);

Equations describing the action of gravity;

A description of the bulk properties of matter.

The fundamental principle on which most cosmological models are based is the so-called cosmological principle, which states that the Universe is, at least on large scales, homogeneous and isotropic. That is to say, we occupy no special place within it (an extension of the Copernican principle) and it appears much the same to observers wherever in the Universe they may be. This assumption makes the description of the geometry in cosmological models a much simpler task than in many other situations in which general relativity is employed; but it is by no means obvious why the Universe should have these simple properties. (For further discussion see **horizon problem** and **inflationary Universe**; for now we shall just assume that the cosmological principle holds, and provides a satisfactory starting point for Big Bang models.)

The first thing to do is to describe the geometrical properties of **spacetime** compatible with the cosmological principle. It turns out that all mathematical spacetimes can be described in terms of the **Robertson–Walker metric**, a mathematical function which describes a geometry that can represent a **flat universe**, a **closed universe** or an **open universe**. It does not tell us to which of these three the real Universe approximates. The Robertson–Walker metric has a simple form because there is a preferred time coordinate in our smoothly expanding Universe. Observers everywhere can set their clocks according to the local density of matter, which is the same at all places at a given time. Aside from the geometry, the evolution of the Universe is simply described by its overall size: the cosmos at different times keeps the same geometry, so that snapshots taken at different times look like different-size blow-ups of earlier snapshots. This is the **expansion of the Universe**, as predicted by general relativity. An important consequence of the curved geometry and expansion is that light signals are affected by both as they propagate with finite speed from a source to an observer. The finite speed of light means that we are always looking at the Universe as it was in the past, rather than as it is in the present. This means that we have to be very careful about how we interpret observations, but it does mean that in principle we can study cosmic history (what the Universe was like in the past) as well as cosmic geography (what it looks like now).

The dynamics of the Big Bang model are determined by the gravitational Einstein equations of general relativity. In the general case, this theory involves a complicated **tensor** formulation (essentially, it consists of ten independent nonlinear partial differential equations) which is extremely difficult to understand, let alone solve. However, with the simplifying assumption of the geometry afforded by the Robertson–Walker metric, the Einstein equations simplify considerably, and we end up with a single equation describing the entire evolution of the Universe: the *Friedmann equation*. The family of mathematical solutions to this equation are called the **Friedman models**, and they provide the foundations of the Big Bang theory.

It is a property of the homogeneous and isotropic expansion of the Universe around every point that all such models reproduce **Hubble's law**, which states that the velocity at which a galaxy or any other distant object appears to be receding from us is proportional to its distance. The constant of proportionality in Hubble's law is called the Hubble parameter, or **Hubble constant**, and is usually given the symbol H_0. The actual value of H_0 is not known to any great accuracy at present; the observational problems involved in determining the **extragalactic distance scale**, which is what H_0 represents, are formidable. But it is a very important quantity because it determines, for example, the **age of the Universe** and the scale of our observable **horizon**.

The important factor that determines the long-term evolution of a Friedman universe is the **density parameter**, Ω. This is simply the ratio of the actual density of the Universe to a critical value that is required to make it halt its expansion and start to recollapse (see **closed universe**). If Ω is greater than 1, the curvature of spacetime has a positive value and the Universe will recollapse into a **singularity**, a process sometimes known as the *Big Crunch*. If Ω is less than 1, then the curvature is negative and the Universe will expand for ever with ever-decreasing density. Poised between these two alternatives is the flat universe, corresponding to $\Omega = 1$. The precise value of Ω is not known at present, and the best we can do is to say that it probably lies between about 0.1 and, say, 2 or 3. It is not predicted in the standard Friedmann models – it has the role of a parameter, and must be determined by observational investigations, particularly by searches for **dark matter** in whatever form it exists.

When we apply the formalism of general relativity to the study of cosmology, we necessarily enter mathematical territory. But cosmology is not just mathematics: it is a branch of physical science, and as such it should be capable of making predictions that can be tested against observations. The Big Bang theory is therefore more than the Friedmann equations and the values of Ω and H_0. These are just part of the toolkit that cosmologists use to study the physical processes that have operated at various stages of the **thermal history of the Universe**. Extrapolating back into the past would be foolhardy if there were no empirical evidence that the basic picture outlined by the Friedmann models is correct.

The first major piece of supporting evidence is the expansion of the Universe itself, as embodied in **Hubble's law**, which gives the relationship between **redshift** and distance for relatively nearby sources. Hubble was actually rather lucky, because his sample of galaxies was very small, and the statistical correlation between redshift and distance was rather weak. But in recent years Hubble's law ('relation' is perhaps a better way of describing it) has been convincingly demonstrated to hold out to rather large distances, so we can be sure that what is sometimes called the *Hubble expansion* is observationally secure. On the other hand, it is not exactly true to say that the Big Bang explains the Hubble expansion, because there is nothing in it that explicitly requires the Universe to be expanding rather than contracting. The Big Bang provides a sort of half-explanation.

The next piece of evidence for the Big Bang, and probably the most compelling, is the existence of the cosmic microwave background radiation. The unmistakable **black-body** signature of this radiation, as shown by the spectrum obtained by the **Cosmic Background Explorer** (COBE) satellite, proves beyond all reasonable doubt that, wherever it came from, it was produced in **thermal equilibrium** with matter. In the Big Bang models this is accounted for by taking the present radiation background and winding back the clock to when the Universe was about one-thousandth of its present size. Under these conditions matter would be fully ionised, and scattering of the background photons by free electrons is expected to have maintained equilibrium in the required way. The background radiation is therefore taken to be a relic of the first few hundred thousand years of the Universe's evolution in a Big Bang model. Indeed, it is very difficult to see how the microwave background radiation could have been generated with a black-body spectrum unless the cosmology is very much like the Big Bang model.

The third main success of the Big Bang theory is the accurate prediction of the observed light element abundances: the proportions of helium, deuterium, lithium and beryllium present in the Universe. In the Big Bang model, these elements are produced by **nucleosynthesis** in the first few seconds of the Universe's existence, when the conditions resembled those in the explosion of a thermonuclear device. These light element abundances are calculated under the assumption that the early Universe was in thermal equilibrium. Now, the abundances of the light nuclei depend very sensitively on the total density of matter that exists in a form in which it is capable of participating in nuclear reactions. If the predictions are to be matched with observations, then a strong constraint emerges on the amount of baryonic matter (i.e. matter composed of, for the most part, protons and neutrons) in the Universe.

So we can now summarise the content of the standard Big Bang model. It incorporates the expansion of the Universe from a hot state of thermal equilibrium where nucleosynthesis of the light elements took place, giving rise to the cosmic microwave background radiation. This model describes the overall properties of the Universe we observe today, about 15 billion years after the Big Bang, all the way back to about the first millionth of a second (see Essay 3) or so after the creation event, which is assumed to mark the origin of time. It is important to recognise that the Big Bang is an incomplete theory which leaves many questions unanswered (and indeed many questions unasked), but it is nevertheless the best model we have, and it forms the basic framework within which virtually all observational data are interpreted.

TOWARDS THE NEW COSMOLOGY

The story told by the standard Big Bang theory is accepted by most cosmologists as being basically true, even if we do not know the values of the cosmological parameters that would fine-tune it. Now that the basic framework of the Big Bang theory appears to be in place, future explorations will be aimed at filling in the gaps, and extending it into areas that are not penetrated by the standard model. Two such areas have been the goal of much recent research.

The first is **structure formation**. The standard model is globally homogeneous, but we know that the Universe is rather inhomogeneous, at least on the relatively small scales on which we can observe it directly. We also know, from the near isotropy of the cosmic microwave background radiation, that the Universe was extremely smooth when this radiation was last scattered, at a time when the Universe was very young. So the problem is to explain how the structure and complexity we see around us today can have evolved from such an apparently simple initial state. There is a standard theory for how this happened, and it is based on the idea of gravitational instability. The details of how it works are not known at the present, but only a few years ago this field was almost purely theoretical, since there were virtually no data against which to test the various models. A 'standard' picture of structure formation emerged in the 1970s, based on a phenomenon called the **Jeans instability**. Since gravity is an attractive force, a region of the Universe which is slightly denser than average will gradually accrete material from its surroundings. In so doing the original, slightly denser regions get denser still and therefore accrete even more material. Eventually this region becomes a strongly bound 'lump' of matter surrounded by a region of comparatively low density.

After two decades, gravitational instability continues to form the basis of the standard theory for structure formation, though the basic idea has undergone several refinements. In the 1980s, for example, there was a standard model of structure formation called the *cold dark matter model*, in which the gravitational Jeans instability of a fluid of **weakly interacting massive particles** (WIMPs) is taken to be the origin of structure in the Universe. This model was accepted by many in the 1980s as being the 'right' answer. In the 1990s, however, the picture has changed enormously, with observations taking the driving seat and theorists struggling to find a model that explains them. The cold dark matter model is now thought to be excluded, in particular by observations of large-scale galaxy clustering and by the cosmic microwave background anisotropies detected by the COBE satellite. As new data continue to accumulate it seems likely that theory will lag behind observations in this area for many years to come. The details of how structures of the form we observe today were produced are, however, still far from completely understood.

The 1980s saw another important theoretical development: the idea that the Universe may have undergone a period of inflation, during which its expansion rate accelerated and any initial inhomogeneities were smoothed out (see **inflationary Universe**). Inflation provides a model which can, at least in principle, explain how such homogeneities might have arisen and which does not require the introduction of the cosmological principle *ab initio*. While creating an observable patch of the Universe which is predominantly smooth and isotropic, inflation also guarantees the existence of small fluctuations in the cosmological density which may be the initial perturbations needed to feed the gravitational instability thought to be the origin of galaxies and other structures (see Essay 2).

The history of cosmology in the latter part of the 20th century is marked by an interesting interplay of opposites. For example, in the development of structure formation theories we can see a strong element of *continuity* (such as the survival of the idea of gravitational instability), but also a tendency towards *change* (the incorporation of WIMPs into the picture). The standard cosmological models have an expansion rate which is decelerating because of the *attractive* nature of gravity. In inflationary models (or those with a cosmological constant) the expansion is accelerated by virtue of the fact that gravity effectively becomes *repulsive* for some period. The cosmological principle asserts a kind of large-scale *order*, while inflation allows this to be achieved locally within a Universe characterised by large-scale *disorder*. The confrontation between the steady state and Big Bang models highlights the distinction between *stationarity* and *evolution*. Some inflationary variants of the Big Bang model posit a *metauniverse* within which *miniuniverses* of the size of our observable patch are continually being formed. The appearance of miniuniverses also emphasises the contrast between *whole* and *part*: is our observable Universe all there is, or even representative of all there is? Or is it just an atypical 'bubble' which happens to have the properties required for life to evolve within it? This is the territory of the **anthropic principle**, which emphasises the *special* nature of the conditions necessary to create observers, as opposed to the *general* homogeneity implied by the cosmological principle in its traditional form.

A related set of cosmological problems concerns the amount of matter in the Universe, as well as its nature. Cosmologists want to know the value of Ω, and to do so they need to determine the masses of large astronomical objects. This has led to overwhelming evidence for the existence of large amounts of cosmic dark matter. But it is also important to know what kind of material this is: if nucleosynthesis theory is correct, it cannot be in the form with which we are familiar: atoms made of electrons, protons and neutrons. It seems likely that the dark matter will turn out to be some form of exotic **elementary particle** produced by the fundamental interactions that operated in the early Universe. If this is so, then particle cosmologists will be able to test theories of ultra-high-energy physics, such as **grand unified theories**, using the early Universe as their laboratory.

This leads us to another direction in which cosmologists have sought to extend the Big Bang theory: into the period well before the first microsecond. One particularly important set of theoretical ideas has emerged. In the theory of the **inflationary Universe**, a **phase transition** that the Universe underwent as it cooled initiated a rapid acceleration in the expansion for a very brief period of time. This caused the Universe today to be very much bigger than is predicted by a standard Friedmann model. To put it another way, our observable patch of the Universe grew from a much smaller initial patch in the inflationary Universe than it would do in a standard Friedmann model. Inflation explains some of the properties of our Universe which are just taken for granted in the standard Friedmann models. In particular, it suggests that Ω should be very close to 1. Another thing inflation does is to generate very small 'quantum fluctuations' in the density of the Universe, which could be the **primordial density fluctuations** upon which gravitational instability acted to produce structure. These considerations may also explain away some of the mystery surrounding the cosmological constant, and whether it should appear in the Einstein equations.

One issue is of fundamental concern in any attempt to extend the model to earlier times. If we extend a Friedmann model (based on classical relativity theory and the behaviour of forms of matter with which we are familiar) back to $t = 0$, we invariably find a singularity, in which the density of matter increases to an infinite value. This breakdown of the **laws of physics** at the creation event means that the standard models just cannot be complete. At times before the **Planck time** and energies above the **Planck energy**, the effects of **quantum gravity** must have been important; this may or may not provide an explanation of what happened in the very earliest moment of the Big Bang.

An interesting characteristic of cosmology is the distinction, which is often blurred, between what one might call cosmology and *metacosmology*. We take cosmology to mean the scientific study of the cosmos as a whole, an essential part of which is the testing of theoretical constructions against observations. Metacosmology is a term for those elements of a theoretical construction, or paradigm, which are not amenable to observational testing. As the subject has developed, various aspects of cosmology have moved from the realm of metacosmology into that of cosmology proper. The cosmic microwave background radiation, whose existence was postulated as early as the 1940s but which was not observable by means of the technology available at that time, became part of cosmology proper in 1965. It has been argued by some that the inflationary metacosmology has now become part of scientific cosmology because of the COBE discovery of **ripples** in the temperature of the cosmic microwave background radiation. This claim is probably premature, though things are clearly moving in the right direction for this transfer to take place at some time in the future. Some ideas may remain for ever in the metacosmological realm, either because of the technical difficulty of observing their consequences or because they are not testable even in principle. An example of the latter difficulty may be furnished by Andrei Linde's chaotic inflationary picture of eternally creating miniuniverses, which lie beyond the radius of our observable Universe (see **baby universes**).

Despite these complexities and idiosyncrasies, the new cosmology presents us with clear challenges. On the purely theoretical side, we require a full integration of particle physics into the Big Bang model, and a theory which treats gravitational physics at the quantum level. We need to know what kinds of elementary particle could have been produced in the early Universe, and how structure formation happened. Many observational targets have also been set: the detection of candidate dark matter in the form of weakly interacting massive particles in the galactic halo; **gravitational waves**; more detailed observations of the temperature fluctuations in the cosmic microwave background; larger **redshift surveys** of galaxies and measurements of **peculiar motions**; and the elucidation of how the properties of galaxies have evolved with cosmic time. Above all, cosmology is a field in which many fundamental questions remain unanswered and where there is plenty of scope for new ideas.

The early years of the new millennium promise to be a period of intense excitement, with experiments set to probe the microwave background in finer detail, and powerful optical telescopes mapping the distribution of galaxies out to greater distances. Who can say what theoretical ideas will be advanced in the light of new observations? Will the theoretical ideas described in this book turn out to be correct, or will we have to throw them all away and go back to the drawing board?

FURTHER READING

Crowe, M. J., *Modern Theories of the Universe from Herschel to Hubble* (Dover, New York, 1994).

Hetherington, N. S., *Encyclopedia of Cosmology* (Garland, New York, 1993).

Hoskin, M. (editor), *The Cambridge Illustrated History of Astronomy* (Cambridge University Press, Cambridge, 1997).

Kline, M., *Mathematics in Western Culture* (Penguin, London, 1987).

North, J., *The Fontana History of Astronomy and Cosmology* (Fontana, London, 1994).

Overbye, D., *Lonely Hearts of the Cosmos: The Story of the Scientific Quest for the Secret of the Universe* (HarperCollins, New York, 1991).

2

The Emergence of Cosmic Structure

Carlos Frenk

Introduction

The Universe is thought to have begun with a great explosion – the Big Bang. At an early time, about 15 billion years ago, all the mass in the **Universe** was contained in a tiny region which was very dense and hot. Since then the Universe has been steadily expanding, cooling as it has done so and creating the conditions for the formation of stars, galaxies, planets and, eventually, life and people. This essay covers some of the most important events that have occurred during the Universe's life history. In particular, it focuses on the physical phenomena through which the Universe evolved from its primaeval amorphous state to its present, highly structured complexity. Several fundamental aspects are fairly well understood, for others we have only some tentative ideas, and for many more we have had no more than a glimpse of what might have happened. Two specific issues are addressed:

What is the Universe made of?

What physical processes gave birth to the galaxies and other structures?

A grand tour of the cosmos

From the perspective of a cosmologist, the basic building blocks of the Universe are the **galaxies**. A galaxy is an assembly of stars – ranging from a few million to several hundred billion of them – held together by gravitational forces. The Sun belongs to the Milky Way (or the Galaxy, with a capital G), a medium-sized galaxy, typical of those we call *spirals*. This name comes from their prominent spiral arms, which are generated by the revolution of huge gas clouds in circular orbits about the galactic centre. In these clouds, new stars are continually being formed. A typical galaxy is about 60,000 light years across. (Compare this with the distance to the nearest star, Proxima Centauri, which is just over 4 light years away.)

Galaxies like the company of other galaxies. This gives rise to a characteristic pattern of galaxy clustering which forms the **large-scale structure** of the Universe. The smallest assemblages of galaxies are groups with a handful of members. The Milky Way is part of the Local Group of about thirty galaxies, mostly dwarfs, but containing another large spiral, the Andromeda Galaxy. This is over 2 million light years away and is quite similar to the Milky Way Galaxy. The largest galaxy clusters contain thousands of bright members; the nearest is the Virgo Cluster, about twenty times farther away from us than the Andromeda Galaxy. On even larger scales, the galaxies are arranged in gigantic structures known as *superclusters* which contain several thousand bright galaxies. (Our own Galaxy is part of the 'Local Supercluster'.) These are the largest structures that have been identified to date. They tend to have capricious, elongated shapes, with typical dimensions exceeding 100 million light years and a mass of about a hundred thousand galaxy masses (or one hundred million billion Suns).

There are lots of large numbers in cosmology. These large numbers simply reflect our choice of units, which are naturally based on human experience. We find such large numbers in cosmology because a human lifespan is very small compared with the age of the Universe: typically, we live for only one hundred-millionth of the age of the Universe. However, it is complexity, not sheer size, that makes things difficult to understand. Biological processes occur on a small scale, but are harder to understand than galaxies.

We have seen that the galaxies are arranged in a hierarchy of ever-increasing size: groups, clusters, superclusters. How far does this structuring go? Large-scale **redshift surveys** seem to suggest that there is a scale, encompassing a few superclusters, on which the Universe appears to be fairly homogeneous in a broad sense. That is, if you were to draw a circle whose diameter matched this scale on a map of galaxies and counted the galaxies that fell within it, the number would not vary too much from place to place. On these very large scales – a few hundred million light years – the Universe is nearly amorphous or homogeneous. Thus, although highly complex on small scales, on very large scales the Universe appears rather well organised. It is only because of this overall large-scale simplicity that we can make any progress at all in understanding the evolution of the Universe. One of the basic tenets on which cosmological theory is based is the *Copernican principle* – the assumption that we do not occupy a privileged position in the Universe and that any other (hypothetical) observer would see pretty much the same picture that we ourselves see. The Copernican principle in a cosmological setting is usually phrased in terms of the **cosmological principle**: that the Universe is isotropic (i.e. it looks the same in all directions) and homogeneous (i.e. it looks the same in all places).

On large scales, the galaxies exhibit an amazing collective behaviour. This was discovered in the 1920s, and it revolutionised our view of the entire Universe. Edwin **Hubble**, using the largest telescope in the world at the time (the 100-inch (2.5 m) telescope at Mount Wilson in California), realised that all galaxies are moving at formidable speeds away from us. He had discovered the **expansion of the Universe**. For this, he made use of a simple phenomenon of physics with which we are familiar in everyday life, the **Doppler shift**. Hubble observed that the spectral lines of galaxies are all shifted towards the red end of the spectrum, indicating that the galaxies are all receding from us, and that this **redshift** increases in proportion to the distance of the galaxy. That is, the farther away the galaxy is, the faster it is receding from us, and the velocity of recession is directly proportional to the distance. This last property – a uniform expansion – is very important for it tells us that the observed expansion is not exclusive to our viewpoint. In a uniform expansion every observer sees a similar situation. To understand this, imagine the surface of a balloon that is being inflated. If you paint dots on the balloon, then as the balloon expands the distance between any two dots increases. An observer located on any one dot would see all the other dots moving away from it: in a uniform expansion, every observer sees exactly the same phenomenon – just as the Copernican principle leads us to expect.

If galaxies are all moving away from us today, then they must have been closer together in the past. In the very remote past, they would have been very close indeed. In fact, at a very early time the entire Universe would have been concentrated into a very dense, hot state. And it was not just the matter, but also space and time that were compressed into this very dense, hot state. To describe this state we need to resort to the theory of **general relativity**, according to which it is the whole of space that stretches as the Universe ages. The initial state from which the expansion began was the Big Bang. It was not just unimaginably dense, it was also unimaginably hot. As it expanded, the Universe cooled down in much the same way as a compressed gas cools as it expands. Since the Universe has been around for rather a long time, we would expect its present-day temperature to be rather low. In fact, the temperature to which an initially hot Universe would have cooled by the present was calculated in the 1940s by George **Gamow**, but unfortunately his calculation was ignored. In 1965 two physicists, Arno **Penzias** and Robert **Wilson**, discovered the residual heat left over from the hot early phase in the life of the Universe, in the form of a uniform sea of microwave radiation (see Essay 5).

This **cosmic microwave background radiation** has recently been remeasured with exquisite accuracy by the **Cosmic Background Explorer** (COBE) satellite. Its properties are exactly what we would expect from an early hot phase in the Universe's history. Its emission spectrum has the **black-body** shape characteristic of a hot body, and it is almost uniform across the whole of space. It has a temperature of only 2.73 degrees above absolute zero (2.73 K). Its uniformity tells us that it does not come from our Galaxy and that it must therefore be of cosmological origin; the black-body spectrum tells us that it comes from a time when the Universe was hot and opaque. The cosmic microwave background radiation has been propagating freely since the Universe was about 100,000 years old (when atoms first formed and the fog of the original fireball lifted). It provides direct evidence that the Universe was once very small and very hot – direct evidence in favour of the **Big Bang theory**. In 1990, at a meeting of the American Astronomical Society where the COBE spectrum was shown for the first time, the audience of several thousand astronomers gave the presenters a standing ovation.

The universal expansion of the galaxies and the microwave background radiation are two important pieces of empirical evidence in support of the Big Bang theory. But there is a third, equally important one: the chemical composition of the Universe. At very early times, when the Universe was about 1 second old, its mass–energy was in the form of a 'cosmic soup' of **electromagnetic radiation** and **elementary particles**: protons and electrons. No other type of particle could exist in the midst of the tremendous heat. However, by the time the Universe was about 10 seconds old the temperature had dropped sufficiently (to about 10 billion degrees) to allow protons and electrons to combine into neutrons. Neutrons and protons are the raw material from which atomic nuclei are formed. When the Universe was about 100 seconds old, protons began to fuse with neutrons, first to make deuterium (sometimes called heavy hydrogen), and later helium, lithium and beryllium, by a process called **nucleosynthesis**. These nuclear reactions lasted about a minute, and stopped when the expansion of the Universe had driven the particles too far apart to collide with sufficient energy. (The formation of carbon and heavier elements had to wait until the formation of stars, about 5 billion years later – it is stars that form all the heavy atoms, including those of which we are made.) Thus, after about 3 minutes the primordial chemical composition of the Universe had been established. From our knowledge of atomic physics gained in the laboratory, it is possible to predict quite accurately the chemical composition that must have emerged from the Big Bang. The prediction is that matter in the Universe should consist of about 75% hydrogen and 23% helium by mass, with trace amounts of other elements. When astronomers measure the chemical composition of primordial gas clouds (clouds unpolluted by stars), they measure exactly 75% hydrogen, 23% helium! This is a great triumph of modern science in general and of the Big Bang theory in particular. (For further details, see **light element abundances**.)

The geometry of space

Since the Universe is expanding today, we might wonder what its long-term fate will be. Is it destined to continue expanding for ever, or will the expansion eventually come to a halt and perhaps even reverse? Again, for an answer we must resort first to general relativity and then to observations. Qualitatively, the answer is intuitively obvious: the fate of the expanding Universe, and indeed its present rate of expansion, are determined by a single quantity – the mean density of matter. If the density is high enough then **gravity**, the major cosmic player on large scales, will win in the end. The self-gravity of matter will eventually arrest the expansion; the Universe will stop for an instant and then begin gradually to fall back on itself, reversing the initial expansion. (All sorts of fascinating physics would occur then, but that is beyond the scope of this essay.) The alternative is that there is not enough density to arrest the expansion, which would then continue unabated *ad infinitum*.

Mathematically, there is an intermediate state between these two: a particular density, the so-called 'critical' density, which is just sufficient to hold the Universe at the borderline between these two extremes. (Formally, a critical Universe has just the right density to continue expanding for ever.) As we shall see later, this critical state is the one that many cosmologists believe our Universe is in. Although it is enough to eventually turn the whole Universe around, the critical density is laughably small by Earth standards: only three hydrogen atoms per cubic metre. In general relativity, gravity and geometry are one and the same thing; the density of the Universe then determines its geometrical structure. A universe that expands for ever has an 'open' geometry like that of a saddle; a universe that recollapses has a 'closed' geometry, like that of a sphere; and a universe with the critical density has a 'flat geometry', like that of a sheet. The mean density (or geometry) of the Universe is usually expressed in terms of a parameter called the **density parameter** and usually given the symbol Ω. An **open universe** has Ω less than 1, a **closed universe** has Ω greater than 1, and a critical-density **flat universe** has $\Omega = 1$ exactly.

General relativity lays down these three alternatives for the dynamical behaviour of the Universe, but it does not tell us which of the three is the one that applies to our Universe. To find this out, we need to consider different kinds of theoretical idea, and also observations. I shall look at some new theoretical developments, and then discuss how we go about measuring the mean density of matter in the Universe. Around 1980, Alan **Guth**, while worrying about a problem in particle physics, stumbled upon a very elegant idea which not only solved his particle physics problem, but may also solve the riddle of the cosmic geometry. This idea, which bears the inelegant name of *inflation*, goes back to the physics of the very early Universe (see Essay 3; see also **inflationary Universe**). A remarkable prediction of these new theories is that the Universe should have almost exactly the critical density. But is this really so?

The stuff of the Universe

We might naïvely think that determining the mean density of matter in the Universe is relatively straightforward. After all, we can count how many galaxies there are in a given volume. The density is then just the ratio of the mass of the galaxies divided by the volume. But when we do this calculation, we find that galaxies contribute only about 1% of the critical density predicted by inflation. Can we be sure that all we see is all we get? We cannot, and, in fact, we already know that there is much more to the Universe than meets the eye – or even the most powerful telescope. Evidence for vast amounts of invisible matter, the so-called **dark matter**, has been accumulating and is now overwhelmingly persuasive. The tell-tale sign of dark matter is the gravitational force. All stable structures in the Universe result from a balance between gravity and some other force. In a spiral galaxy, for example, the self-gravity of the stars in it is balanced by the centrifugal force that arises from their circular motion. We can measure the speed at which stars in a galaxy rotate (from the Doppler shifts in the spectral lines of stars in the disc of the galaxy) and hence the centrifugal force (see **rotation curves**).

It turns out that if the mass in a galaxy were all in the form of the stars we can see, there would not be enough gravitational force to hold the galaxy together against centrifugal forces. Since the galaxies we observe appear to be perfectly stable, we conclude that there must be matter in the galaxy in addition to that contributed by the stars. This invisible matter is arranged in a roughly spherical *halo*. A recent analysis of the motions of satellite galaxies – small companion galaxies orbiting larger galaxies similar to the Milky Way – shows that these haloes extend well beyond the regions occupied by stars. A similar argument can be made for elliptical galaxies and even clusters of galaxies: the visible matter does not provide enough gravitational force, so more material has to be present. When we measure the mass of a galaxy cluster, simply by requiring that gravity should balance the force produced by the motions of the galaxies in the cluster, we find that this mass contributes about 20% of the critical density – about twenty times the density contributed by the visible stars.

Most of the mass in galaxy clusters is dark matter. Could there be yet more of this stuff hiding away in the vast regions between the clusters? Only very recently has it become possible to attempt to answer this question reliably by direct measurement. And again, the basic ingredient is gravitational physics. One of the most difficult things in astronomy is to measure distances to other galaxies directly. However, redshifts are easy to measure. Thanks to the expansion law of the Universe – **Hubble's law**, which relates the distance to the velocity – we can infer the distances to galaxies simply by measuring their redshifts. In 1988 I was involved in a consortium of universities (three British and one Canadian) which undertook one of the largest ever programmes to measure a large number of galaxy redshifts. The resulting survey, known as the *QDOT survey* (after the initials of the participating institutions: Queen Mary & Westfield, Durham, Oxford and Toronto) allowed us to construct the deepest map so far of the distribution of galaxies around us, in three dimensions.

This map allows us not only to determine the cosmography of our local Universe, but also to measure the mean density of the whole Universe. The lumps of galaxies visible in the map produce gravitational accelerations on nearby galaxies and cause their paths to deviate slightly from the overall universal expansion. These so-called **peculiar motions** depend on the mean density of matter. Thus, by comparing predictions of the way galaxies should move as a result of the lumpiness of the QDOT map with actual measured peculiar motions, we can determine the mean density of matter on very large scales. The result is immensely rewarding: to explain the motions of galaxies, the density has to have the critical value, with one proviso – that the lumpiness in the distribution of galaxies be similar to the lumpiness in the distribution of invisible matter. This is something we cannot be sure of because it is only the galaxies that shine, not, of course, the dark matter.

I have already mentioned that the cosmic microwave background radiation has properties which are extremely uniform across the sky. In one specific direction, however, it appears slightly hotter than everywhere else, and in exactly the opposite direction it appears slightly cooler. This is just what the Doppler effect would predict if the source or, equivalently, the observer (in this case us) is moving. The very small deviation from uniformity (it amounts to only about one part in a thousand) is the signature of the motion of the Milky Way in the Universe. We are moving at about 600 km/s in a direction pointing roughly towards the Virgo Cluster of galaxies. This motion of our galaxy is typical of the peculiar motions of galaxies. Such motions are induced by the gravitational pull of the surrounding matter, over and above the uniform expansion of the Universe. Apart from this slight *anisotropy*, the cosmic microwave background radiation is very uniform indeed. As we shall see, this has important consequences for the process of galaxy formation.

If we have a map that shows how galaxies are distributed in space, we can calculate the net gravitational force acting on our Galaxy (or, for that matter, on any other nearby galaxy) caused by all the material around it. We can thus 'predict' the speed with which our Galaxy should be moving. In making this calculation we must take into account the fact that the Universe is expanding; the predicted speed then depends on the mean cosmic density of matter. We know that our Galaxy is moving at about 600 km/s. It turns out that, for the matter traced in our map to be capable of inducing such a speed, the mean cosmic density must have the critical value! This means that 99% of the mass of the Universe must be dark. (There might, however, be some subtle effects that have been overlooked, so this result should not be regarded as definite.)

So, there are indications that we live in a critical-density Universe, just as predicted by inflation – a Universe that will expand for ever, but only just. However, we have seen that only 1% of the mass of the Universe is in the form of visible galaxies. This means that 99% of the mass of the Universe is in some dark, invisible form. What can this dark matter be? A crucial clue to the identity of the dark matter is provided by the theory of Big Bang nucleosynthesis, discussed earlier. One of the great triumphs of Big Bang theory is its ability to predict the relative abundances of the light elements: hydrogen, helium, deuterium, lithium and beryllium. The exact amounts of the light elements that were produced in the Big Bang depend sensitively on the density of the protons and neutrons at the time of nucleosynthesis, 3 minutes after the Big Bang. Protons and neutrons (the particles that make up the bulk of ordinary matter) are collectively known as *baryons*. It turns out that, for Big Bang nucleosynthesis to work, the maximum allowed density of baryons must be only about 10% of the critical density. Yet, we have seen how recent measurements imply that the Universe has the critical density. The inescapable conclusion is that the bulk of the cosmic mass exists not as baryons or ordinary matter, but in some more exotic form.

In recent years, particle physicists have come up with new theories of the fundamental structure of matter. Some of these theories (which have grandiose names such as **grand unified theories** or **supersymmetry**) require the existence of exotic elementary particles with names such as axions, photinos and neutralinos. These theories are still controversial, and the predicted particles have yet to be detected in particle accelerators. Nevertheless, these exotic particles are prime candidates for the dark matter. It is a sobering thought that not only do we humans not occupy a privileged position at the centre of the Universe, but we may not even be made of the same stuff that makes up most of its mass! (Some have called this 'the demise of particle chauvinism'.) One particular type of exotic dark matter – that made up of supersymmetric particles or axions – is known as *cold dark matter*. The cold dark matter theory has had a profound influence in cosmology since it was developed during the 1980s. Before examining it, however, I must emphasise that, exciting as these ideas are, they are still rather tentative. Our train of thought follows logically only if we believe that the Universe has the critical density.

In 1993, a number of colleagues and I published a paper in the journal *Nature* which seems to contradict our previous critical density result from the QDOT survey. Our argument was based on the observed properties of rich galaxy clusters which form part of the pattern of the large-scale structure. These clusters contain, in addition to galaxies and dark matter, large amounts of hot gas at a temperature of 100 million degrees. This gas emits X-rays, and the properties of this emission had recently been measured very accurately by the German/US/UK ROSAT satellite. From these data we calculated the fraction of the total mass in a typical cluster that is in the form of baryons. The surprising result is that this fraction is about 15%, half as large again as the 10% we would have expected from Big Bang nucleosynthesis and the assumption that the Universe has the critical density. In our paper we showed that the baryon fraction in clusters should be representative of that of the Universe as a whole. Our unpalatable conclusion was that either the Universe does not have the critical density – it is an open universe – or that there is something slightly wrong with the standard Big Bang nucleosynthesis argument. I emphasise the 'slightly' – our results did not imply that there is anything fundamentally wrong with the principles of Big Bang nucleosynthesis, but simply that if $\Omega = 1$, either Big Bang nucleosynthesis must have been more complex than previously thought, or some of the observational data must be wrong. Note that if we accept the Big Bang nucleosynthesis result, then the implied value of Ω is 0.3. Since this value is larger than the 0.1 allowed for baryonic matter, the conclusion that most of the dark matter must be non-baryonic still holds. (For more information, see **baryon catastrophe**.)

The jury is still out on the issue of whether or not we live in a Universe with the critical density. There are strong theoretical and observational arguments in favour of this view, but there is this nagging issue of the baryon fraction in clusters. One is reminded of the words of the biologist Francis Crick (one of the discoverers of the DNA double helix), who said that a theory which agreed with all the experiments had to be wrong because, at any given time, at least some of the experiments are wrong. An open universe may turn out to be T. H. Huxley's 'great tragedy of science: the slaying of a beautiful hypothesis by an ugly fact'.

For those who feel nervous about the discrepancy mentioned above and would rather have a boring Universe with nothing other than baryons (thus ignoring the weight of evidence), there are a few baryonic dark matter candidates which have not yet been excluded by the extensive searches which astronomers have carried out over the past few years. One possibility is black-hole remnants of old massive stars. These, however, seem very unlikely to have been produced in the required abundance, for they would have led to a brighter Universe containing higher proportions of heavy elements than we observe today. Another possibility is the existence of objects of Jupiter's mass (known as brown dwarfs). These are in effect failed stars: objects whose mass is too low (less than a tenth the mass of the Sun) to ignite the nuclear reactions that make stars shine. They are also sometimes given the name of MACHOs (**massive compact halo objects**) to distinguish them from the WIMPs (**weakly interacting massive particles**) that make up the non-baryonic dark matter. Searches for MACHOs using **gravitational lensing** are described in Essay 6.

This, then, is the second unsolved problem in cosmology today: the identity of the dark matter. Whatever this dark stuff is, one thing is clear: identifying the main constituent of our Universe is the most pressing problem of contemporary physics. Today, many researchers are pursuing this holy grail. For example, the UK is one of the world leaders in attempts to develop very sensitive detectors to capture dark matter particles from the halo of our Galaxy. (There should be several hundred thousand per cubic metre everywhere, including in our own bodies.) This experiment is being carried in the deepest underground mine in Europe.

Dead stars?

Let us now take a step back to our own neighbourhood. The Solar System is the only direct evidence we have so far for the existence of planets and **life in the Universe**. The Sun, however, is a typical star, so it is likely that there are many other similar systems. One of the things we do know quite a lot about is the life cycle of stars. Stars are born when clouds of interstellar gas and dust collapse under their own gravity. When the density and temperature in the inner regions are high enough, the interior of a star turns into a thermonuclear fusion reactor, transmuting hydrogen, the simplest and commonest element in the Universe, into helium. This happens when the central temperature reaches about 15 million degrees. The energy thus liberated gives rise to a force that opposes the all-pervasive pull of gravity. The nuclear reaction rate adjusts itself so as to balance gravity exactly. When this equilibrium is reached, a new star is born.

Since it is nuclear reactions that balance gravity, as the star uses up more and more of its nuclear fuel it evolves as the balance shifts. This evolution is quite well understood and can be calculated quite precisely using a large computer. The star goes through a whole chain of fusion reactions, slowly contracting and expanding as different types of fuel are consumed. A star like the Sun has enough fuel to last about 10 billion years. Since it is now about 4.5 billion years old, it is a middle-aged star.

Eventually, the nuclear fuel will all be used up and the star will come to the end of its life. For a star like the Sun, the end will be foretold by a huge expansion. The outer regions of the Sun itself will expand to engulf the Earth, and shortly after the outermost layers will be expelled as a shell of hot gas (creating what is misleadingly called a planetary nebula). The now inert core will turn into a *white dwarf*, a ball of gas in which gravity is balanced by quantum forces acting on the electrons. Its temperature is now only about 3000 degrees. However, since there is no more energy generation, the white dwarf gradually cools and becomes a *black dwarf*. The death throes of a star, from its expansion to its collapse and transformation into a white dwarf, occupy no more than a few thousand years.

Stars more massive than the Sun end their lives in a more spectacular fashion. The more massive a star, the greater the central temperature and density, and the faster the consumption of nuclear fuel. Such stars therefore live their lives faster, and are also brighter than stars like the Sun. They can also sustain nuclear reactions that produce heavier elements than can be manufactured in Sun-like stars: they are capable of burning all elements up to iron. Iron is the most tightly bound atom in the Universe. It is not possible to extract energy by fusing two iron atoms; on the contrary, such a reaction is possible only by injecting energy. Once the central part of a star is made of iron, after only a few million years from birth, there is no source of energy that can stand up to gravity. The core of the star implodes catastrophically, sending a shock wave that rips the outer parts of the star apart: it blows up as a **supernova**. In the last few instants before the explosion, the tremendous pressures that are achieved are capable of squeezing protons into existing atomic nuclei, and this produces many of the elements which, like uranium, are heavier than iron. The enormous amount of energy liberated in these reactions makes the supernova explosion incredibly bright – it can outshine its entire home galaxy.

The supernova explosion ejects processed chemical elements. These wander about in interstellar space and eventually find their way to a cloud of gas and dust destined to become a new star. In this way, material is recycled and the new star (with any attendant planets) incorporates the elements processed in an earlier generation of stars. This is the origin of the atoms of which we are made: they were once cooked in the nuclear furnace of a now dead star. We are all made of stardust.

After the explosion the stellar core that remains may become a *neutron star*, one in which gravity is balanced by another quantum force, in this case due to protons rather than electrons. However, if the mass of the core is large enough, not even this force can arrest the power of gravity. The core then continues to shrink inexorably and becomes a **black hole**. Black holes are among the most fascinating objects in nature. Matter in them is packed so densely that their gravitational pull does not allow even light to escape. Inside a black hole, spacetime becomes distorted and all sorts of strange phenomena, understood only in terms of general relativity, can happen.

The existence of dark matter in itself should not come as a shock. After all, stars shine only because they are capable of sustaining nuclear reactions in their interiors. But they cannot do this for ever and, indeed, all stars are doomed to fade in due course. It turns out, however, that even the afterglow of a dead star can be detected with the most sensitive modern instruments.

The emergence of structure

Intimately linked to the identity of the dark matter is the mystery of when and how the galaxies formed. Once again, gravity takes the leading role, hence the importance of the dark matter – the main source of gravity. Since the 1940s, it had been conjectured that the origin of galaxies, clusters and other large-scale structure should be sought in the early Universe. If, at early times, the Universe was not completely smooth but instead contained small irregularities, these irregularities would grow. This is because an irregularity, or 'clump', represents an excess of gravitational attraction over the surrounding material. Some of the matter which would otherwise be expanding away with the Universe is attracted to the clump and is eventually accreted by it. In this way the clump steadily grows and eventually becomes so heavy that it collapses into a gravitationally bound structure, a galaxy or a galaxy cluster. This theory had been worked out in great detail, but remained essentially a conjecture until one of the most remarkable discoveries of recent times was announced.

On 22 April 1992, I was woken at 6.00 a.m. by a telephone call from a newspaper reporter in the USA seeking my views on an announcement that had just been made in Washington. This was the announcement by the COBE team of the discovery of **ripples** in the cosmic microwave background radiation. This radiation is the relic of the primaeval fireball which has been propagating through space since the Universe was about 100,000 years old, and a thousand times smaller than its present size. The ripples that the COBE satellite had detected were tiny irregularities in the 2.73 K radiation, with an amplitude of only 1 part in 100,000. These irregularities in the radiation are caused by equally tiny irregularities in the matter distribution at an early time, precisely those irregularities which, it had been conjectured, were required for galaxies to form. These are the **primordial density fluctuations**. Thus COBE discovered the fossil remnants of the progenitors of galaxies, the missing link between the early regular Universe and the present-day structured Universe. The discovery made headline news all over the world.

The discovery of ripples in the cosmic microwave background radiation confirmed the basic theory for cosmological **structure formation** – the so-called gravitational **Jeans instability** theory whereby small ripples in the expanding Universe are amplified by their own gravity. But it does more than that: it gives us yet another clue to the identity of the dark matter. For the ripples to be as small as the COBE measurements indicate, the primordial clumps must be made of non-baryonic dark matter. This is because clumps made entirely of baryons, if of the size indicated by the COBE measurements, would not have had enough time to recollapse and form galaxies by the present day. Before the emission of the cosmic microwave background radiation, the baryons would have been continuously pushed around by the radiation, and prevented from growing. By contrast, clumps of non-baryonic dark matter would have been unaffected by the radiation, and could have started to grow even before the fog lifted, before the cosmic microwave background radiation was emitted. They would have had a head start over clumps of baryons, and had enough time to collapse by the present day to form the dark halos of galaxies. The baryons, once freed from the radiation, simply fell into the pre-existing clumps of non-baryonic dark matter and eventually became converted into stars, giving rise to the luminous parts of galaxies.

COBE, then, revealed the presence of small irregularities – the progenitors of galaxies – present when the Universe was 100,000 years old. But where did these ripples come from in the first place? Yet again, the answer may lie in the physics of the very early Universe, in the epoch of inflation itself. The inflationary phase was triggered by quantum processes: a **phase transition** associated with the breaking of the original symmetry in the forces of nature. One of the most important properties of quantum processes is the generation of irregularities, called quantum fluctuations, associated with Heisenberg's uncertainty principle (see **quantum physics**). This principle allows small quanta of energy to appear out of nothing, as it were, provided they then disappear very quickly. Whenever there are quantum fields there are always quantum fluctuations. In the normal course of events these fluctuations come and go very rapidly, but when the Universe is inflating, a quantum fluctuation gets carried away with the wild expansion and is blown up to macroscopic scales. When this happens, the fluctuation is no longer subject to quantum effects and becomes established as a genuine ripple in the fabric of spacetime – a ripple in the energy density of the Universe. Cosmologists can calculate the evolution of ripples formed this way, and can also derive their 'spectrum' – the relative amplitudes of ripples of different physical size. The astonishing fact is that the spectrum of the ripples measured by COBE is exactly the same as the spectrum that was derived by assuming that the ripples arose from quantum processes during inflation. This agreement suggests a truly amazing connection between the physics of the microscopic – the world of subatomic particles – and the physics of the macroscopic – the cosmos as a whole. It is difficult to conceive of any other synthesis which rivals this in power or beauty. Particle physics and cosmology are two of the frontiers of modern science. What we seem to be uncovering is a very deep connection between the two.

We now have a theory for the origin of the ripples seen by COBE – the progenitors of galaxies. How do these tiny ripples – departures from the mean density of only one part in a hundred thousand – develop into the majestic structures that dominate the Universe today, galaxies and clusters? I have already discussed the idea of gravitational instability, whereby these ripples are amplified by their own gravity as the Universe expands. In the 1990s it became possible to build computer simulations of the growth of primordial clumps and the process of galaxy formation. These simulations can follow the motions of dark matter, and also of baryons in the expanding Universe from early times to the present. The nature and evolution of primordial clumps is very closely linked to the identity of the dark matter. Different assumptions lead to different model universes. There are two classes of non-baryonic dark-matter particle candidates: 'hot' and 'cold' dark matter. These names come from the different temperatures or speeds of these particles in the early Universe. Light particles, the classic example of which is a neutrino, move fast and are therefore hot; heavy particles, like those predicted in supersymmetric theories of elementary particles, move slowly and are therefore cold. Computer simulations have shown that the growth of structure in a universe containing hot dark matter would not have led to a galaxy distribution like that observed. Cold dark matter, on the other hand, proved much more successful and became established at as the standard model of cosmogony.

The cold dark matter (CDM) cosmogony – based on the assumptions that the dark matter is made of cold, non-baryonic elementary particles (WIMPs), that the Universe has the critical density, and that primordial fluctuations were of the type produced during inflation – has had a profound influence on cosmological thinking. The theory was extensively explored throughout the 1980s using large computer simulations, which showed that it could account for many of the detailed properties of galaxies and galaxy clusters. As the observational data became more plentiful and more precise, it gradually became apparent that, successful as the theory is on the scales of galaxies and clusters, it may well not be the last word. The QDOT survey, for example, showed that superclusters in the real Universe appear to be larger than the superclusters that form in the computer models of a CDM universe. Similarly, the ripples found by COBE seem to have twice the amplitude than those predicted in the standard CDM theory. There is a growing feeling that the CDM theory may be incomplete, and that it needs to be modified if it is to provide an accurate explanation of the origin of structure in the Universe. What these modifications will turn out to be are currently the topic of active research. Nevertheless, it is very impressive that the simplest and most elegant theory, CDM, comes so close – to within a factor of two – to the observations.

As for the future, there is much to look forward to. Major new surveys of galaxies are now under way. A consortium of British universities is currently using the 'two-degree field' (2dF) on the Anglo-Australian Telescope to carry out a major new survey to measure 250,000 galaxy redshifts – a hundred times as many as in the QDOT survey – and thus to map the cosmic structure to unprecedented depths and accuracy. At the same time, other research groups around the world are measuring and mapping the ripples in the cosmic microwave background radiation, again to unprecedented accuracy, thus sharpening the view provided by this unique window on the early Universe. On Earth, experiments aimed at detecting the elementary particles that may constitute the dark matter will hopefully reach the required sensitivity levels early in the next millennium.

Conclusions

These are very exciting times for cosmologists. We seem to be on the verge of a number of major breakthroughs: unravelling the identity and the amount of dark matter, and understanding the physical processes by which our Universe became organised into stars and galaxies. All the indications are that these breakthroughs will reveal a deep-rooted connection between particle physics and cosmology, raising the enthralling possibility that the behaviour of the cosmos as a whole may be understood in terms of the properties of matter on the subatomic scale. If these ideas turn out to be correct, and only experiments will decide, they will represent one of the most comprehensive syntheses in the history of the physical sciences. I end with a quotation from Thomas Wright of Durham (1711–86), the English philosopher who, way ahead of his time, proposed that the Milky Way is a flattened disk, with the Sun located away from its centre, and that the nebulae were distant objects beyond it:

> Which of these theories is most probable, I shall leave undetermined, and must acknowledge at the same time that my notions here are so imperfect I hardly dare conjecture.

FURTHER READING

Riordan, M. and Schramm, D., *The Shadows of Creation: Dark Matter and the Structure of the Universe* (Oxford University Press, Oxford, 1993).

Rowan-Robinson, M. *et al.*, 'A sparse-sampled redshift survey of IRAS Galaxies: I. The convergence of the IRAS dipole and the origin of our motion with respect to the microwave background', *Monthly Notices of the Royal Astronomical Society*, 1990, **247**, 1.

Saunders, W. *et al.*, 'The density field of the local Universe', *Nature*, 1991, **349**, 32.

Silk, J., *The Big Bang*, revised and updated edition (W. H. Freeman, New York, 1989).

White, S. D. M. *et al.*, 'The baryon content of galaxy clusters: A challenge to cosmological orthodoxy', *Nature*, 1993, **366**, 429.

3
The Very Early Universe

Andrew R. Liddle

Themes of the very early Universe

For much of the early 20th century, **cosmology** barely existed as a scientific discipline. For instance, what we now recognise as distant **galaxies** very much like our own were thought to be unusual objects, nebulae, contained within our own Milky Way. In the light of our present knowledge, the progress of modern cosmology can be regarded as one of the great scientific endeavours of the 20th century, and as we reach the millennium the standard model of the **Big Bang theory**, described in Essay 1, is so well established that it is almost impossible to find a cosmologist willing to take issue with it. The successful predictions of the **expansion of the Universe**, of the **cosmic microwave background radiation**, of the **light element abundances** (the proportions of e.g. hydrogen, deuterium and helium present in the Universe) and of the **age of the Universe** are dramatic indeed.

Emboldened by these successes, cosmologists are now taking on the ambitious task of pushing our knowledge back to earlier and earlier moments in the Universe's history. As the light elements are believed to have originated when the Universe was but a second old, we are now talking about probing back to minuscule fractions of a second after the Universe began. As we shall see shortly, this requires an understanding of the physics of the very very small: we need to look at the fundamental constituents of matter. It is fascinating that a study of the very large – the Universe itself – should lead us in this direction.

Cosmology is all about big questions. The hot Big Bang model lets us ask such questions as 'How old is the Universe?' and 'Will it expand for ever, or recollapse in a Big Crunch?', and to attempt to answer them by studying the Universe around us. In Essay 2, Carlos Frenk shows that definitive answers are at the moment just tantalisingly out of reach, but there are many reasons to be optimistic that accurate answers will soon be within our grasp (see e.g. Essay 5). While those questions can be addressed in the standard Big Bang model, over the years it has become clear that other questions require us to go back to the very earliest moments in the history of the Universe if we are to find an explanation. Such questions are:

Why is the Universe full of matter but not antimatter?

Is it possible for the Universe to contain material that we cannot see directly?

Why is the material in the Universe grouped into structures such as stars, galaxies, clusters of galaxies and other **large-scale structure**?

THE EARLIER IT IS, THE HOTTER IT IS

One of the intriguing properties of the present Universe is that it is so very cold. To be precise, its temperature is 2.73 K (degrees kelvin), less than three degrees above absolute zero. We know this because we have measured the radiation emitted from it, the cosmic microwave background (CMB) radiation. Think first of a star like our Sun. It is fiercely hot (several thousand degrees at its surface) and emits light mostly in the visible part of the electromagnetic spectrum, corresponding to quite short wavelengths. The central heating radiator in your room is much cooler, a few tens of degrees Celsius (about 300 K); you cannot see its radiation because it is of a much longer wavelength, mostly infrared, though you can feel the heat energy on your skin. The CMB radiation is of an even longer wavelength, in the microwave part of the electromagnetic spectrum, so it must correspond to something much cooler again. In fact it has the very low temperature given above. So, the shorter the wavelength of light, the more energetic it is and the higher a temperature it corresponds to. Now, because the Universe is expanding, light travelling through it will have its wavelength stretched, as shown in Figure 1, and so it becomes cooler. Turning this reasoning on its head, we know that the Universe is cool today, and that it has been expanding for a very long time and cooling in the process. Therefore, in the distant past the Universe must have been much hotter than it is now, and indeed the further back we go in the Universe's history the hotter it must have been.

It turns out that the temperature of the Universe (in degrees kelvin) is inversely proportional to its size. For example, when the Universe was half its present size its temperature must have been 2×2.73 K = 5.46 K,

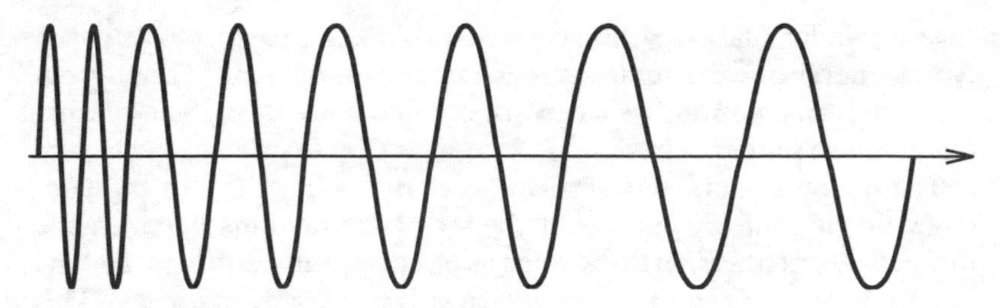

Figure 1 As the Universe expands, the wavelength of light is stretched and the Universe cools down. The Universe was much hotter in the past.

still not far from absolute zero. But when it was one-thousandth of its present size, its temperature would have been nearly 3000 K. Instead of the puny microwaves we have today, at that time the Universe would have been filled with radiation as intense as if you were hovering just above the surface of the Sun. Although such conditions undoubtedly constitute a hostile environment, we understand them rather well because we routinely recreate them in laboratories on Earth in order to test materials to their limits. It was this understanding that made it possible to predict the CMB, which has played such a central role in the acceptance of the Big Bang model. The CMB was formed at a temperature of a few thousand degrees; before this the radiation in the Universe was hot enough to interact strongly with the material in it, whereas afterwards radiation was too weak to interact with matter and travelled freely, just as radiation of a few thousand degrees is able to escape the surface of the Sun.

As we look further and further back into the Universe's history, where it becomes ever smaller and ever hotter, we begin to encounter conditions which it is presently impossible to recreate in the laboratory. However, all is not lost because we can rely on information gleaned by particle physicists, working with their huge particle accelerators. To understand the connection between the early Universe and particle physics, we must look to the physicists' standard model of a gas, in which the gas is treated as a collection of particles. As the temperature of the gas increases, the individual particles gain energy, so they move faster and undergo more violent collisions. We can talk theoretically about temperatures of a million million (10^{12}) degrees, but we lack the technology to heat a sizeable amount of gas up to that sort of temperature: by using magnetic confinement in nuclear fusion devices, we can only (!) get up to 10^8 K or so. But the big particle accelerators at, for example, CERN in Switzerland and Fermilab in America are able to smash together individual **elementary particles** with energies as high as the typical energy of particles in a gas at a temperature of 10^{15} K, and from studying such collisions we can work out how a gas of particles at that temperature should behave.

The many results from particle accelerators have led to the construction of a highly successful theory of interactions between the different elementary particles (such as electrons and neutrinos, and the quarks that join together in threes to make protons and neutrons). This theory is known as the *standard model* (see **fundamental interactions**). The standard model describes three of the fundamental forces of nature: the electromagnetic, strong and weak interactions. Electromagnetism is familiar to us all in the form of electricity, magnetism and light (see **electromagnetic radiation**). The other two forces do not impinge directly on our consciousness, but are crucial for our existence. The weak interaction, or *weak nuclear force*, is responsible for radioactivity, and in particular the processes in the Sun that govern the nuclear reactions which ultimately provide all the energy for life on Earth. The strong interaction, or *strong nuclear force*, is what makes quarks 'stick together' to form protons and neutrons; without it our bodies would simply fall apart.

One of the great successes at CERN was the discovery that, though electromagnetism and the weak force seem very different at familiar temperatures, once we reach an energy corresponding to a temperature of 10^{15} K, their character changes and they become different aspects of a single force, known as the *electroweak force*. This theory has been very accurately verified, and one of the main preoccupations of particle physicists is further unifications of the other fundamental interactions into a single framework – a **theory of everything**. The strong force, also included in the standard model though not fully unified with the other two, appears quite amenable to this sort of treatment, but the fourth and final force, gravity, has so far proved an enormous stumbling block and has yet to be incorporated into the standard model.

So far, the standard model has stood up to every test thrown at it, and it gives us a framework of physical laws which enables us to describe the Universe up to that fantastic temperature of 10^{15} K or so. As a hot Universe has an expansion rate proportional to the square root of time, and the Universe is presently about ten billion (10^{10}) years old and at a temperature of 2.73 K, the Universe would have been at this prodigiously high temperature when its age was about

$$10^{10} \text{ years} \times (3 \text{ K}/10^{15} \text{ K})^2 \approx 10^{-19} \text{ years} \approx 10^{-12} \text{ seconds}$$

So we have in our hands the physical tools necessary to describe the Universe to within that instant – one picosecond – of its birth! The formation of light elements (see **nucleosynthesis**), when the Universe was about one second old, and the formation of the microwave background when it was about a hundred thousand years old, are thus easily accessible to us.

But still we want to go on, to understand earlier and earlier moments, because the Big Bang cosmology, allied to the standard model of particle interactions, has proved unable to answer the questions we raised above. Unfortunately, the temperatures before 10^{-12} seconds would have been too high for us to replicate them in experiments on Earth. In this uncharted territory, physical certainty based on experimental verification is replaced by theoretical speculation about what kind of physics might apply. Might the forces of nature become further unified to include the strong force, in a **grand unified theory**? Might the mathematical symmetry that describes the low-energy world be replaced at high energies by a form of **supersymmetry**? Might the fundamental constituents of matter turn out to be not points, but instead one-dimensional 'strings', as suggested by an idea known as superstring theory (see **string theory**)? Might Einstein's theory of relativity have to be replaced by some new theory of gravity, just as Newton's gravitational theory was supplanted by general relativity as means of describing very powerful gravitational forces? All these ideas have dramatic consequences for the processes that may have been at work in the first 10^{-12} seconds of the Universe's existence.

Figure 2 (overleaf) shows a possible history of the different epochs of the Universe as it cools. Before 10^{-12} seconds we are relying on speculation, and several possible ideas are listed for what might happen at those times. These ideas and others are explored in the rest of this essay. We hope that by extrapolating from what we currently know about physics we shall discover something about the Universe all the way back to a time of just 10^{-43} seconds after the Big Bang – an inconceivably short period of time, and an extraordinarily bold claim for modern physics. For still earlier times it is believed that a proper description would be possible only by merging the theories of **gravity** (Einstein's general relativity) and **quantum mechanics**. At the moment no one has any idea how this could be achieved, so speculation stops here! The study of the early Universe thus falls in the period between 10^{-43} seconds and 10^{-12} seconds after the Big Bang, during which time the temperature fell from 10^{31} K to 10^{15} K. As we shall now see, there are many ideas as to what might have taken place during this interval.

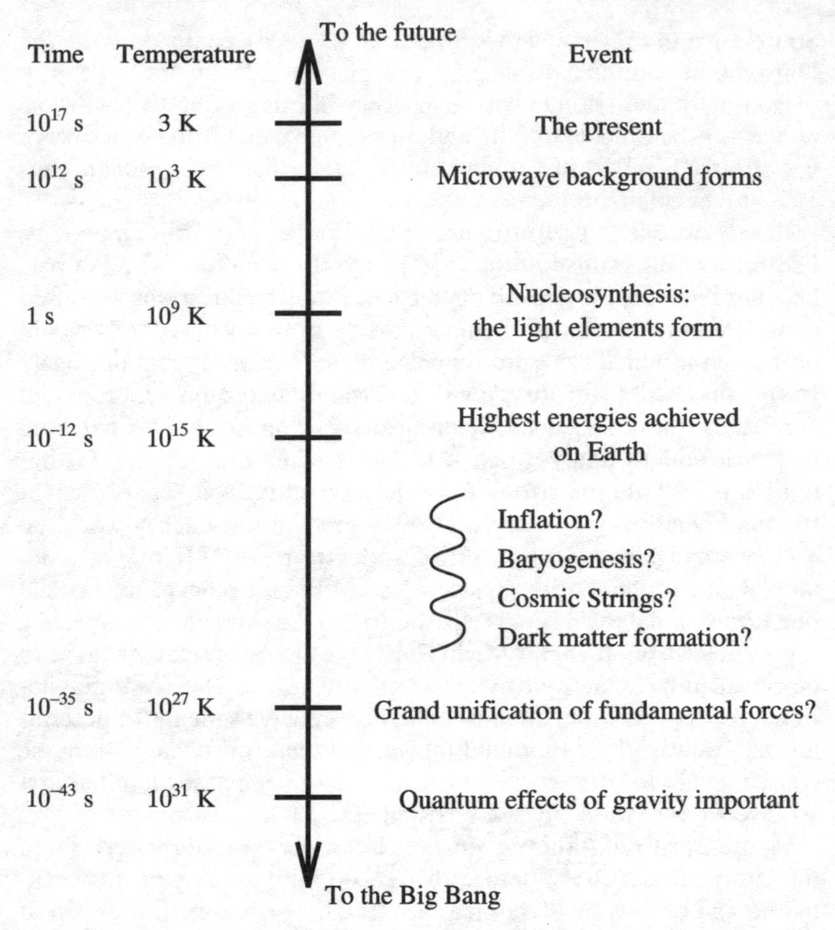

Figure 2 A chronology of the Universe. Studies of the early Universe ask what might have happened in the first 10^{-12} seconds.

THE STUFF WE'RE MADE OF

The Universe is full of matter, not **antimatter**. This is quite a strange state of affairs, for every particle has an *antiparticle* – the *antiproton* for the proton, and the *positron* for the electron, for example – which has the same properties, except that it has the opposite charge. Nothing stops an antiproton and a positron from combining to make an anti-atom, or a large collection of anti-atoms to form an anti-Sun with its own anti-Solar System, and so on. Except that if the matter and antimatter were ever brought together, there would be an astonishingly fierce explosion as they mutually annihilated.

How much matter there is in the Universe, compared with the amount of antimatter, is measured by a quantity known as the *baryon number* which, if you like, is just the number of protons and neutrons minus the number of antiprotons and antineutrons. (A **baryon** is any particle, such as a proton or neutron, that is made from three quarks.) As the Universe is full of protons and neutrons, but not their opposites, its baryon number is positive. The intriguing thing is that baryon number is conserved (i.e. it does not change) in the standard model of particle physics. For example, radioactive decay can change a neutron into a proton, but there are no decays that change a proton into, say, an electron. So if the baryon number is positive now (a fact to which we owe our very existence, since we are made mostly of protons and neutrons), and the standard model is all there is, then the Universe must always have had the same baryon number. There would be no physical explanation of where the material of which we are made comes from because there would be no way of creating it.

But is the standard model all there is? We know that it provides a good description of all the physical phenomena experienced on Earth, including the results of the most powerful particle accelerator experiments. The energies reached in the most energetic of these processes correspond to temperatures of around 10^{15} K. This was the temperature of the Universe itself, some 10^{-12} seconds after the Big Bang. But what happened before then, when the temperature was even higher? It turns out that there is good reason to believe that at extraordinarily high temperatures baryon number is no longer conserved. The **grand unified theories** which bring together the different forces predict that protons can decay into electrons, changing the baryon number. So these theories of unification predict that in the very early stages of the Universe it may have been possible to create a baryon number where formerly none existed. This process goes under the name of **baryogenesis**.

It may have happened as follows. Imagine that, at some very early stage, the Universe was perfectly symmetric between matter and antimatter, thus with zero baryon number. At this very early time the Universe contained far more protons, antiprotons, and so on (strictly, the constituent quarks and antiquarks that would later combine to form the protons and antiprotons) than at present because of the enormously high temperature and energy prevailing, perhaps a billion times as many. As the Universe cooled, grand unified interactions violated baryon number conservation and created what seems a tiny imbalance: for every billion antiprotons, there were a billion and one protons. The Universe cooled further, and grand unified interactions ceased to become important. The standard model now applies, and the baryon number became conserved. Finally, matter–antimatter annihilation took place; the billion antiprotons annihilated with a billion of the protons. But the solitary extra proton had nothing with which to annihilate, and it was left behind in a Universe which had no antimatter but which was filled with the left-over protons.

These one-in-a-billion survivors make up everything we see in the Universe today, including all the stars, the galaxies, the planets and ourselves. Figure 3 illustrates the process. This scenario gives us a qualitative picture of where the atoms from which we are made could have come from. But since we know next to nothing about the correct theory of grand unification, or even if grand unification as a concept is correct, we are still a long way from the quantitative success of nucleosynthesis, the theory of the formation of the light elements which is a central plank of the Big Bang model. Also, more recently it was

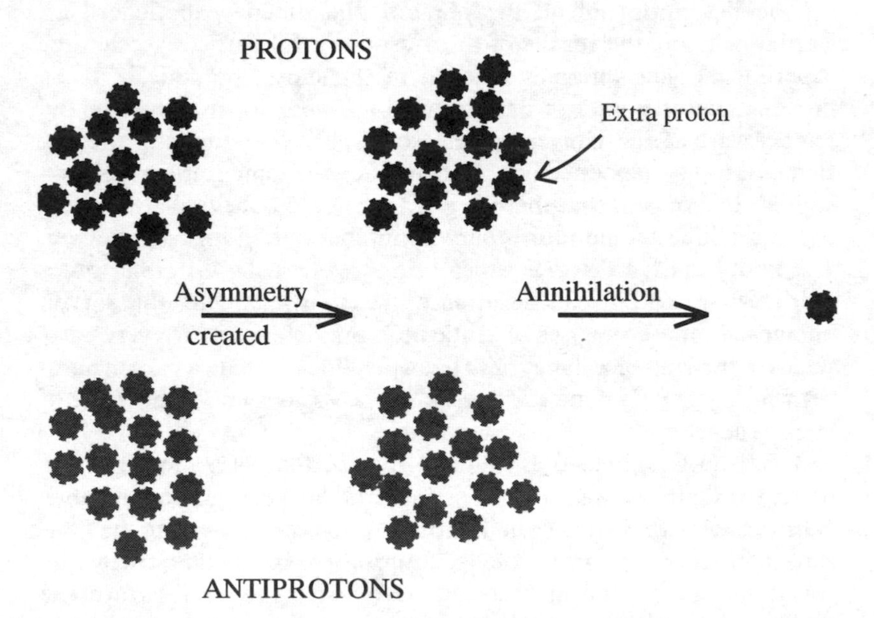

Figure 3 The creation of matter over antimatter, illustrated with fifteen protons (rather than a billion!). An extra proton is created by processes which violate baryon number conservation, and after annihilation the extra protons are all that's left.

discovered that there may after all be baryon number changing processes in the standard model, provided the temperatures are very high, so perhaps grand unified theories are not necessary for baryogenesis after all. It looks as if it may be some time before this topic enters the mainstream of cosmology.

THE CURIOUS CASE OF THE DARK MATTER

So we cannot claim to understand the formation of the material – the protons, neutrons and electrons – from which we are made. That is hardly encouraging when we try to come to terms with the strong likelihood that most of the Universe is made up of some other kind of matter, which we can as yet neither see nor feel. This is the famous **dark matter**.

There is a lot of circumstantial evidence that the Universe is filled with material that we cannot see, but which nevertheless exerts a gravitational force (some of this evidence is reviewed in Essays 2 and 6). The extra gravitational force from invisible matter enables galaxies to spin faster than they otherwise would, and also pulls galaxies towards one another more quickly than would be expected if what we see is all there is (see **peculiar motions**).

The usual assumption is that the dark matter is in the form of elementary particles which interact only very weakly with anything else. The standard model of particle physics throws up just a single candidate, the neutrino, whose existence was first postulated by Wolfgang **Pauli** in 1930 to explain an apparent energy deficit in radioactive decay processes. It was some time before the neutrino was detected directly, because it interacts so weakly with everything else (a famous statistic is that if you fire a neutrino through a slab of lead, the slab has to be 200 light years thick for the neutrino to be likely to interact). There are now believed to be three types of neutrino, each with a corresponding antineutrino.

In the standard model, neutrinos are supposed to have no mass and so cannot be the dark matter. But because they are so hard to detect, there are only fairly weak limits on any mass they might have, and it turns out to be perfectly possible that they could account for all of the missing mass. Indeed, for the two less common types of neutrino, the requirement that they should not collectively account for more than the total missing mass is a stronger constraint than any Earth-based one involving their direct detection. From time to time, it has in fact been claimed that a mass for the neutrino has been detected in radioactive decay or accelerator experiments (indeed, such a claim was circulating as this essay was being written), but none of these claims has yet been supported by subsequent experimental results.

Since the neutrino is the only standard model candidate, the only way that we can have other candidates is to accept some extension beyond the standard model. Particle physicists have not been shy of suggesting such extensions, which carry such names as **supersymmetry**, grand unified theories and superstring theories. Typically, these new theories are stacked full of new particles which have not been seen at the energies available on Earth, but which might have been copiously produced in the early Universe. What is perhaps the favourite candidate has been given the cumbersome name of the 'lightest supersymmetric particle'.

At present, the indirect astrophysical arguments for exotic dark matter, based on its gravitational effects, are compelling, but it is by no means clear whether particles with the required properties can be produced naturally within the theoretical framework of present-day particle physics. The resulting tension between astronomers and theoretical physicists will probably be resolved only if a dark-matter particle is detected *directly* as a result of it interacting with matter. The candidate particles are supposed to be all around us, but interacting so weakly that we are unaware of their presence. Even so, several experiments have been designed and built, mostly in huge underground tanks, in an attempt to capture sufficient numbers of these elusive species for the theoretical models to be substantiated by experimental proof. If these experiments succeed, we shall have learnt that not only do we live in an uninspiring region of an average-looking galaxy, but that we are not even composed of the material that comprises most of the Universe.

INFLATION: THE UNIVERSE GETS BIG

One of the key ideas in early Universe cosmology is known as cosmological inflation, and was introduced by Alan **Guth** in 1981. The **inflationary Universe** is a response to some observations which defy explanation within the context of the standard Big Bang model, such as:

Why is the Universe so big and flat?

Why is the microwave background temperature so similar coming from opposite sides of the sky?

Why is the material in the Universe so evenly distributed?

Let us go through these one at a time. If you balance a screwdriver on its tip, go away for a year and then come back, you would be expecting the screwdriver to be lying flat on the ground rather than still be balanced. Intuitively, we might expect the Universe to be something like that. If it contained too much matter, the extra gravitational pull would very quickly cause it to recollapse in a Big Crunch; if it contained too little, gravity would not be sufficient to stop everything from flying apart. In practice, our Universe is observed to be delicately balanced between the two: gravity is pulling everything back, but not so strongly as to cause a recollapse. The Universe has performed a remarkable balancing act in remaining poised between these two regimes for as long as it has. This special circumstance is known as a **flat universe**, where the normal laws of geometry apply. In the other two cases space itself is curved, much in the way that the surface of the Earth is. So what we are saying is that the curvature of space ought to be very noticeable – but it isn't. This is known as the **flatness problem**.

The second question is called the **horizon problem**. Nothing travels faster than light, but of course light itself does manage to go that fast and the microwaves we see as the cosmic microwave background radiation are no exception. Those coming from, say, our right have spent almost the entire history of the Universe coming towards us as fast as it is possible to go, and have just got here. So it seems a safe bet that nothing could have got from their point of origin all the way across the Universe past us to the point of origin of the microwaves we see coming from our left, which have also been travelling for most of the history of the Universe. So no interactions could have occurred between these two regions – in effect, they are unaware of each other's existence. Yet despite that, they appear to be in **thermal equilibrium**: at the same temperature, to an extraordinary degree of accuracy.

Related to this is the smoothness with which material in the Universe is distributed. While locally we see very pronounced structures such as planets, stars and galaxies, if we take a big enough box (say a hundred million light years across), wherever in the Universe we choose to put it we shall find that it contains pretty much the same number of galaxies. On big enough scales, then, the Universe is basically the same wherever we look. It would be nice to able to say that even if the Universe started out in a hideous mess, it would be possible for physical processes to smooth it out and make it nice and even. But unfortunately the argument we used just now tells us that they cannot – there simply has not been enough time for material to move sufficiently far to have smoothed out the initial irregularities.

The inflationary Universe model solves these problems by postulating that, at some point in its earliest stages, the Universe underwent a stage of dramatically rapid expansion (in fact, the expansion accelerated for a certain period of time). For this to happen, the gravitational force must in effect have become repulsive – a kind of antigravity. It turns out that this can be achieved if matter existed not in the form familiar to us today, but as what is called a **scalar field**. This is not some new funny form of matter devised specially with inflation in mind; scalar fields have been long known to particle physicists, and play an essential role in understanding fundamental particles and their interactions. The most famous, though still hypothetical, example is the *Higgs field*, which it is hoped will be discovered in the next generation of experiments at CERN and which is believed to be responsible for the electromagnetic interaction and the weak interaction having their separate identities.

The inflationary expansion has to be dramatic (see Figure 4). It must increase the size of the Universe by a factor of at least 10^{30}, all within the first 10^{-20} seconds or so of its existence. Fortunately the Universe starts out very small, so this is not as difficult as it would be today when the Universe is huge. For cosmologists, the problem with inflation is not that it is hard to make it happen, but rather that we know so many ways of making it happen that we have trouble deciding which, if any, is the right one!

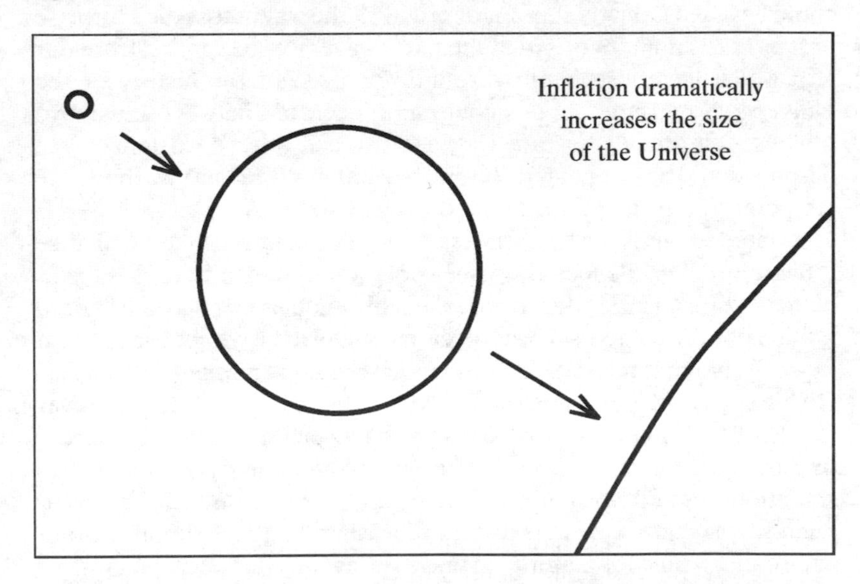

Figure 4 Inflation takes a small portion of the Universe, and makes it very big indeed.

The cosmological problems discussed above are resolved by the inflationary expansion, because it implies that what we see today as the observable Universe – the region from which light has had time to reach us – is but a tiny region of the actual Universe. Moreover, that light came from a region sufficiently small for interactions to have been able to establish thermal equilibrium, making physical conditions the same throughout the small patch within which our Universe resides (see Figure 5). The observed flatness of the Universe is also explained. If you take, say, a balloon (rather obviously a curved surface), and expand it 10^{30} times, it will become a sphere somewhat bigger than our entire observable Universe. If you stood on its surface you would not be able to detect its curvature (it is quite hard to tell that the surface of the Earth, which is much much smaller, is curved). What inflation does, in fact, is to destroy any memory of the way the Universe started out. However complicated the Big Bang might have been, inflation takes a region small enough to be smoothly distributed, and blows it up into a large flat region inside which our observable Universe resides. Anything undesirable is carried away to huge distances. It is the cosmic equivalent of sweeping things under the carpet!

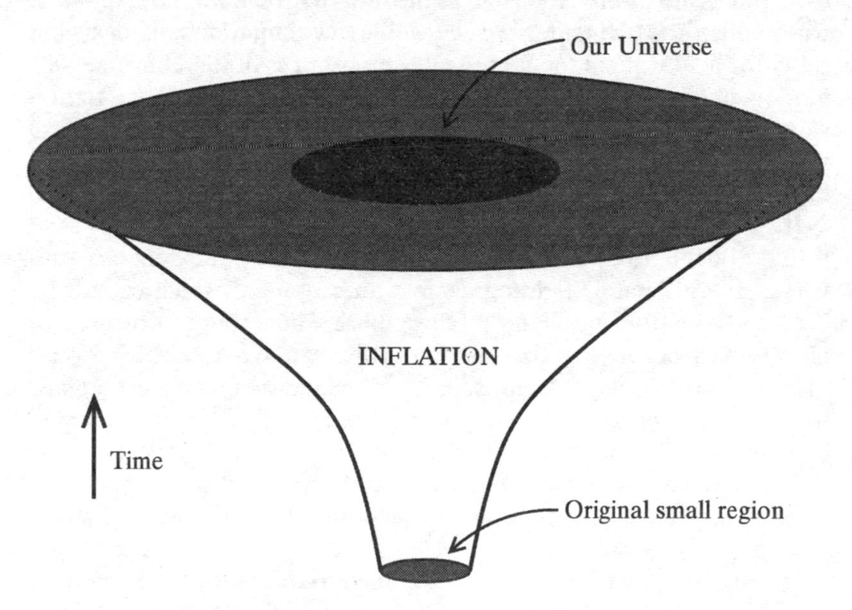

Figure 5 Although it seems big to us, our Universe lies in a very small region blown up to a huge size by inflation. The original region is small enough to have been able to attain a smooth equilibrium.

WHERE DO THE STRUCTURES COME FROM?

One of the most important topics in modern cosmology is the development of structure in the Universe, such as galaxies and clusters of galaxies, and also irregularities in the cosmic microwave background. Essays 2 and 5 examine how structures develop, and what this means for cosmology.

The central idea here is the gravitational **Jeans instability**. This simply says that if one region has more material than another region, then it will pull with a stronger gravitational force and so will be better at assembling yet more material. So the dense regions get denser, and the underendowed regions get emptier. As time goes by, then, gravity will increase the amount of structure in the Universe. Exactly how it does so depends on a number of factors – on how much material there is in the Universe, how rapidly it is expanding, and what form the dark matter takes – and so by studying the evolving structures we can potentially learn about all these questions.

At issue is how it all gets started. Gravitational instability is all very well, but it needs something, however small, to work with. These small somethings are often referred to as the *seeds* of gravitational instability, from which the structures we see will grow. Inflation was designed specifically to explain the large-scale smoothness of the Universe – its homogeneity – and by creating a homogeneous Universe inflation would seem to be acting against our wishes. That is, we desire a Universe which is fairly homogeneous (to explain the large-scale smoothness), but not too homogeneous or structures will never form.

It's a tall order, but remarkably inflation is able to do it. The reason is that, though inflation is trying its best to make a homogeneous Universe, it is ultimately fighting against an opponent which cannot be beaten – quantum mechanics. Heisenberg's uncertainty principle of **quantum theory** tells us that there is always some irreducible amount of uncertainty in any system, sometimes called *quantum fluctuations*, which can never be completely eliminated. Caught up in the rapid expansion of the inflationary epoch and stretched to huge sizes, these quantum fluctuations can become the seed irregularities (the so-called **primordial density fluctuations**) from which all the observed structures grow.

In our present world we are conditioned to thinking of the quantum world as being part of the very small, to do with the electrons in atoms or radioactive decay processes in atomic nuclei. Even a bacterium is way too big to experience quantum mechanics directly. But the inflationary cosmology may turn this view on its head. Its dramatic claim is that everything we see in the Universe, from planets to galaxies to the largest galaxy clusters, may have its origin in quantum fluctuations during an inflationary epoch.

A BOX OF STRING?

Suppose you take a quantity of water molecules at quite a high temperature, say a few hundred degrees Celsius. What you will have is steam. As you let the steam cool down its properties change a little, until you get to 100 degrees Celsius when suddenly there is an abrupt change as the steam condenses into liquid water. Cool this water further, and there will be little change until you reach zero degrees Celsius, whereupon the water suddenly solidifies into a block of ice. The dramatic changes that the water undergoes as it passes from one physical state to another are known as **phase transitions**. The Universe underwent such phase transitions too, and we have already seen some examples. At nucleosynthesis, the Universe changed from being a 'sea' of protons and neutrons to one containing nuclei such as hydrogen, helium and deuterium. At the time when the cosmic microwave background radiation was created, the sea of nuclei and free electrons changed into a sea of atoms, and the Universe rapidly switched from being opaque to being transparent. An example of an earlier phase transition is the so-called *quark–hadron phase transition*, where individual quarks collected together in threes to form protons and neutrons for the first time.

The very early Universe too is expected to have undergone phase transitions, but ones in which the nature of the fundamental forces changed. For example, at temperatures above 10^{15} K the electromagnetic and weak interactions have no separate identity, and as the Universe cooled they would have split off and started to be noticeably different. And at the much higher temperatures of grand unification, the strong interaction would also have been indistinguishable from the other two forces, splitting off only as the Universe cooled.

The most cosmologically interesting aspect of phase transitions is that they tend not to occur perfectly. For example, an ice cube is never a perfect crystal, but has defects and fracture lines within it. When a ferromagnet is cooled, the magnetic spins of the individual atoms want to line up. They tend to split into regions called domains; in each domain all the atoms are aligned, but the magnetisations of the individual domains point in different directions (see Figure 6). Similarly, when the

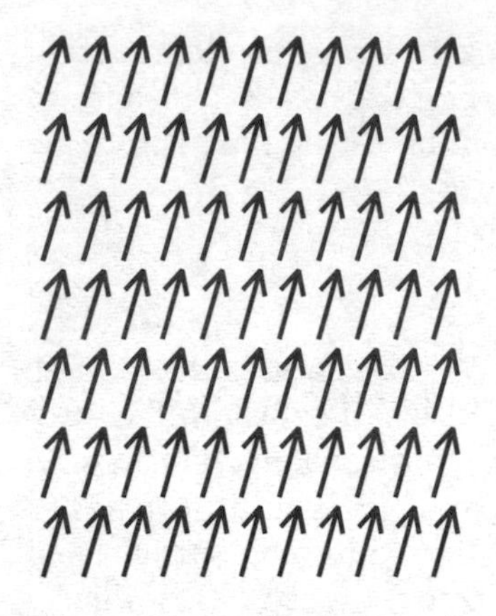

Perfect alignment

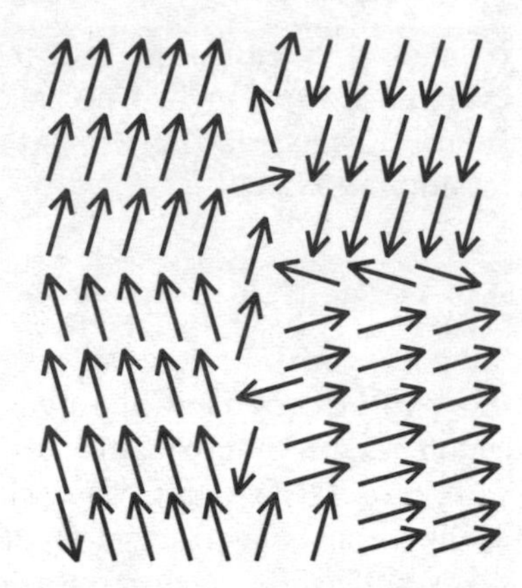

What actually happens

Figure 6 Ideally the individual atoms' magnetic spins (think of them as miniature bar magnets) in a ferromagnet would line up in a perfect structure as the magnet is cooled. In practice, though, domains form where nearby spins line up, but the separate domains do not necessarily line up with one another and the overall structure is disorganised.

Universe cooled its phase transitions may not have occured perfectly, and the phase transition may have left behind what are technically known as **topological defects**.

Depending on the type of phase transition, these defects can be of different types. They may be walls, they may be points or they may be one-dimensional 'lines'. This last possibility has received the most attention (even cropping up in an episode of *Star Trek: The Next Generation*!), and these are known as cosmic strings. Figure 7 shows what cosmic strings might have looked like in the early Universe. They would have tended to thrash around, colliding and interacting with one another.

Cosmic strings (and other topological defects) are one way of making a perfectly homogeneous Universe into an inhomogeneous one, and it has been suggested that they provide an alternative to inflation as a means of seeding structure in the Universe. However, presently the state of affairs in observational cosmology is rather against cosmic strings successfully playing this role; the structures that they can make are not nearly as pronounced as those observed. Indeed, while topological defects are a very elegant idea, it is presently not at all clear what they might be good for!

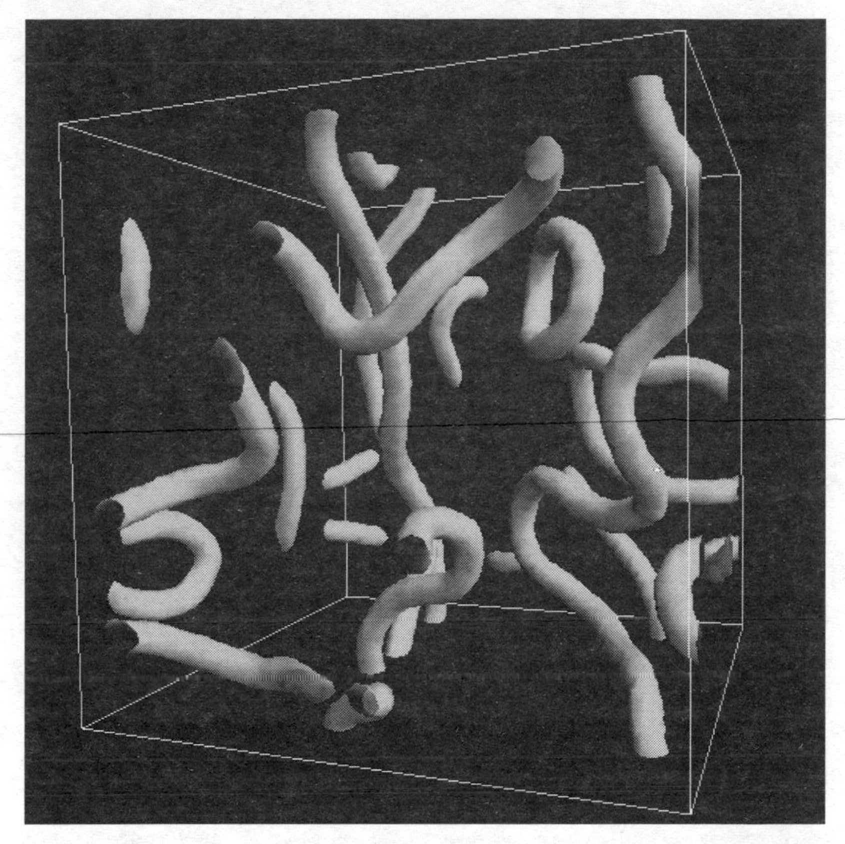

Figure 7 A computer simulation of what cosmic strings might have looked like. The 'ends' are where the strings reach the edge of the simulation box (they re-enter on the opposite face); proper cosmic strings would have had no ends.

Baby black holes

When a star of about ten times the Sun's mass runs out of nuclear fuel to burn and thus reaches the end of its life, one possible fate is for it to collapse into a **black hole**, an object of such density that even light itself is unable to escape from its surface. It may well be that our Galaxy contains many such black holes, and there is also evidence that the centres of other galaxies contain *supermassive black holes* which have grown by swallowing up surrounding material.

One of the more fascinating aspects of black holes is that, whereas in classical physics nothing can escape from them, when we bring **quantum**

mechanics to bear on them we find that they can radiate. The reason is that in quantum mechanics what we call 'empty' space is a seething mass of particles and antiparticles, continually popping in and out of existence and ruled only by Heisenberg's uncertainty principle. As we have seen, this concept has been used to show how inflation can provide the seeds for structure formation. Near a black hole, it may that one of the particle–antiparticle pair falls into the black hole, leaving the other to escape from its environs as radiation. This is shown in Figure 8, and is known as **Hawking radiation**. The trick by which this works is that near black holes there are orbits with negative energy; the negative-energy particle falls into the black hole, thus reducing the black hole's total energy (and hence its mass), while the balancing positive energy is carried away by the escaping antiparticle. Overall, as always, energy is conserved.

Left to its own devices for long enough, a black hole will eventually evaporate away completely by Hawking radiation. But it can take a long time: $10^{10}(M/10^{15})^3$ years, where M is the mass of the black hole in grams. Now, our Sun weighs about 10^{33} grams, so a black hole of that mass would have a lifetime of 10^{64} years! This is so enormous compared with the lifetime of the entire Universe (about 10^{10} years) that Hawking radiation is completely negligible as a cosmological phenomenon. The sort of black holes that are forming today do not exhibit noticeable evaporation.

To get significant evaporation, we need much lighter black holes. These cannot form today, but perhaps they could have formed in the early Universe when the density was vastly higher. A black hole of mass 10^{15} grams – about the mass of a mountain, albeit compressed into a volume smaller than an atomic nucleus – will evaporate in around the lifetime of the present Universe, so if black holes of such mass were formed, we might hope to observe them today. But no evaporating black hole has been positively identified, and quite strong upper limits have been set on how many black holes there could be. During its late stages, black hole evaporation can be rather violent. The formula above shows that the final 10^9 grams or so is released in the last second (remember there are about 3×10^7 seconds in a year). That is about the mass of a swimming pool, all converted into pure energy (a process around a hundred times more efficient than a nuclear weapon of the same size). Rather a dramatic event!

There are several ways in which baby black holes, usually known as *primordial black holes*, might have formed in the early Universe. They may have been caused by large irregularities in the early Universe left by

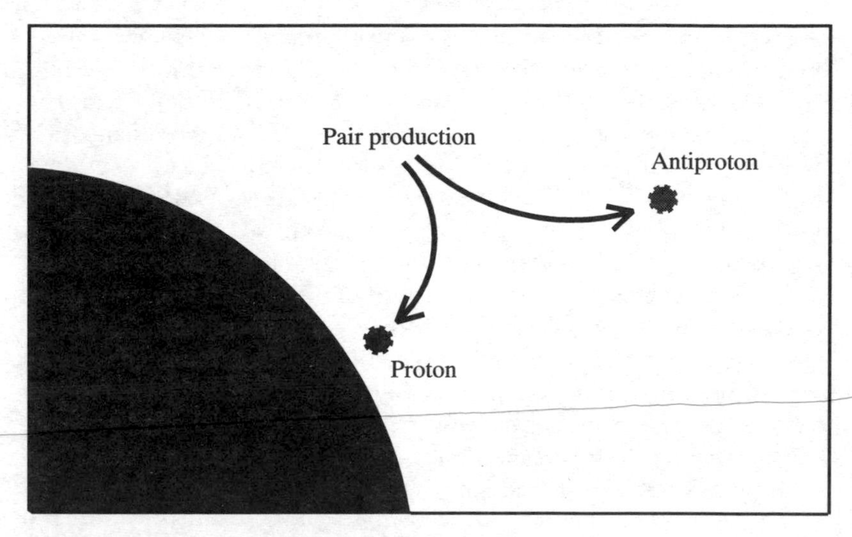

Figure 8 A black hole swallows a proton, and the antiproton escapes.

inflation. Or they may have formed if, for some reason, the pressure fell dramatically in the early Universe and no longer impeded gravitational collapse. They may be the products of turbulence caused at a phase transition. Or they may have formed from collapsing 'loops' of cosmic string. Any identification of evaporating black holes in our present Universe would be a vital clue to its very early stages.

The quest for the 'Bang'

Our wanderings through modern cosmology have taken us back to within a tiny fraction of a second after the Big Bang. But at 10^{-43} seconds we come up against a wall which we cannot yet climb, and further progress requires us to combine quantum mechanics with gravity. It is fortunate for cosmologists that gravity is so weak a force (for example, a magnet the size of a button can easily overcome the gravitational pull of a planet the size of the Earth!), otherwise we would encounter this wall at a much later time.

So does our ignorance stop us getting to the Big Bang itself? At the moment it is fair to say 'yes', though there are some ideas about how we might proceed. One is **quantum cosmology**: the idea that somehow the entire Universe was created by *quantum tunnelling*. Now, quantum tunnelling itself is quite familiar: for example, it is how an alpha particle escapes from a nucleus during radioactive alpha decay. But it is rather bold to suggest that a whole Universe can be created in this way – especially since the only thing it can tunnel from is 'nothing', and that does not just mean empty space, but the complete absence of any such thing as time and space, since those entities are supposed to have come into being only when the Universe was created. 'Tunnelling from nothing' is probably the most speculative idea in all of cosmology.

Inflation may well provide a barrier to us ever learning about the Big Bang itself. Remember that the purpose of inflation was to generate a Universe like our own, regardless of how it started. It achieves this by gathering everything up in a huge expansion so that awkward questions are swept aside. If inflation did indeed occur, it would hide any observational signals created at the time of the Big Bang. Exactly how the Bang itself might have occurred may be concealed from us for ever. How close we can get, though, remains to be seen.

Further reading

Barrow, J. D., *The Origin of the Universe* (Orion, London, 1995).
Guth, A. H., *The Inflationary Universe* (Jonathan Cape, New York, 1996).
Kolb, E. W. and Turner, M. S., *The Early Universe* (Addison-Wesley, Redwood City, CA, 1990).
Narlikar, J. V., *Introduction to Cosmology*, 2nd edition (Cambridge University Press, Cambridge, 1993).
Novikov, I., *Black Holes and the Universe* (Cambridge University Press, Cambridge, 1995).
Overbye, D., *Lonely Hearts of the Cosmos: The Story of the Scientific Quest for the Secret of the Universe* (HarperCollins, New York, 1991).
Roos, M., *Introduction to Cosmology*, 2nd edition (John Wiley, Chichester, 1997).

4

Opening New Windows on the Cosmos

Julianne Dalcanton

Introduction

Imagine being trapped in a room. The room has no windows, and no doors. The only way you can learn about the outside world is to listen. If you listen with your ears, you can perhaps learn about what is happening immediately outside the room. You might hear a violent thunderstorm taking place a few miles away, if it were very loud and you listened very carefully. Now suppose that you could build a receiver which would help your hearing by amplifying distant sounds. You could maybe hear people talking several buildings away, or cars on a distant highway. The better your receiver was, the more you could hear and the more you could learn. Now suppose you are an astronomer, trying to understand the forces that created the stars, the galaxies that contain them and the planets that orbit them. You too are trapped, not in a room but on the Earth. You cannot ask the stars what shaped them, nor can you experiment on a galaxy. All you can do is sit and watch.

The miracle of modern astronomy is that we have managed to learn so much about the way the Universe works, with no means beyond our ability to look at the skies. It is a tribute to human creativity that we have devised so many different ways to observe the heavens, and so many tools to do so. Astronomy has seamlessly joined our understanding of the physical processes that occur on Earth to the limited types of observation we can make, using the former to improve our understanding of the latter. Given the close link between our skill in observing the skies and our understanding of the Universe, astronomy is one of the most technologically driven sciences. Because our understanding is limited only to what we can 'see', any tool that allows us to see more effectively, or in a novel way, opens new vistas before us.

Astronomers have at their disposal just two basic diagnostic aids – images and spectra. We can take pictures of the sky, giving us images, or we can use **spectroscopy** to measure the distribution of energy emitted by stars and galaxies, giving us spectra. Although we are restricted to images and spectra, they do come in many varieties. We are not limited to taking photographs of the sky as our eyes would see it. We can create images using forms of **electromagnetic radiation** which our eyes cannot detect. For example, using large radio dishes we can 'see' the Universe in long-wavelength, low-energy radiation, and study the cool gas from which stars form. With X-ray telescopes orbiting the Earth we can create pictures of the sky at very high energies and short wavelengths, allowing us to map hot plasma stretched between galaxies, or the fires created by gas and stars sinking into black holes. In the same way, we are not limited to spectra in visible light: we can study the energy distributions of astronomical sources over ranges of wavelengths from other regions of the electromagnetic spectrum.

From astronomy to cosmology

Images and spectra each provide us with their own particular perspective on the world. Years of evolution have conferred on us humans a spectacular ability to visually classify the world we see, and with images alone we can begin to categorise the substance of the Universe. As early as the 18th century, astronomers recognised the existence of two very different kinds of object in their images: point-like stars and extended objects (some of which were other **galaxies**) which they called 'nebulae'. That these galaxies are not actually *in* the Milky Way, but are in fact distant collections of stars identical to our own Galaxy, was not scientifically confirmed until the early 1920s, however. And that confirmation was also made with the help of images.

Training a large telescope on some of the largest 'nebulae' in the sky, our closest companions the galaxies NGC 6822, M33 and M31 – the last of these also known as the Andromeda Galaxy, Edwin **Hubble** resolved them into individual stars. (In extragalactic astronomy, 'close' means about 2 million light years away; this is 340,000,000,000 times farther than the smallest separation between Earth and Mars.) Moreover, he identified in them Cepheid variable stars, whose brightness oscillates with time, with well-defined periods. Luminous Cepheids have longer periods than faint ones (the so-called *period–luminosity law*), so Hubble was able to use the rate at which the Cepheids in the Andromeda Galaxy vary to calculate their luminosity. By comparing their luminosity to their observed brightness he made the first determination of the distance to another galaxy, and demonstrated that it was well outside our own Milky Way Galaxy. This landmark from the history of astronomy shows how a wealth of knowledge can spring from only images and human thought. With nothing but well-calibrated photographs of the sky, astronomers discovered not only that the Sun was one of billions of stars in the Milky Way, but that the Universe itself was filled with other 'island universes' like our own.

The second tool in the astronomers' kit is the ability to separate the light of distant stars and galaxies into spectra. To physicists, 'light' is an electromagnetic wave carried by photons, each of which has a unique energy. A photon's energy can be characterised by its wavelength, which is the distance between peaks in the electromagnetic wave with which the photon is associated. Very energetic photons, like X-rays or gamma-rays, have very small wavelengths (less than 1 nm, or about ten times the size of a single atom of hydrogen), while low-energy photons, such as radio waves, have wavelengths measured in metres. The light we can detect with our eyes falls between these extremes in energy, and ranges in wavelength from roughly 400 to 750 nm. Many different energies (and thus many different wavelengths) of light can be emitted simultaneously from a single star or galaxy. A spectrum helps us to untangle these different energies, by separating light according to its wavelength.

Because of the close link between wavelength and energy, spectra can reveal much about the physical processes that operate in the objects we observe. Almost all physical processes produce energy distributions which are characteristic, either in the shape of the distribution or because of the presence of features at very specific energies. These sharp features are known as **emission lines** and **absorption lines**, and correspond respectively to sharp excesses and deficiencies of light at very specific wavelengths. Very hot gases, such as the hot clouds surrounding young stars, tend to produce emission lines, whereas the cooler outer atmospheres of stars produce absorption lines by selectively blocking specific energies of light travelling out from the star's centre. (The absorption lines in the Sun's spectrum are called Fraunhofer lines.) The wavelength at which a line occurs signifies the processes that are operating within the specific atom or molecule responsible for that spectral feature. By looking at the wavelengths of various lines in the spectra of stars or galaxies, we can therefore understand many things. Because specific emission and absorption lines are produced by specific atoms and molecules, we can deduce the chemical elements from which a star is made, simply by identifying the lines in its spectrum.

We can also measure the temperature of stars by looking at their spectra. The light produced by an object with a single temperature has a very characteristic shape, known as a **black-body** curve. The shape changes with temperature, so by looking at the distribution of energies traced by a spectrum we can measure a temperature. These shape changes are manifested as changes in colour, hotter objects being bluer (shorter wavelengths) and cooler objects redder (longer wavelengths). You can see the same effect in the tip of a welder's torch, where the hottest part of the flame is a glowing blue, fading to yellow and red at the cooler edges. The colour of galaxies and stars can also be measured with images, by observing the sky through filters – slabs of coloured glass which can be put in front of an astronomical camera. Galaxies or stars which are very cool and red will be very bright when observed through a red filter, and nearly invisible when observed through a blue filter. Very high-energy objects, such as **quasars**, are very blue and are seen much more easily through a blue filter than through a red one.

Another remarkable property of spectra is that they can reveal the velocities at which objects are moving, via the **Doppler effect**. Because of the **expansion of the Universe**, almost all galaxies are moving away from us, with the most distant galaxies receding the fastest. Thus, the **redshifts** in galactic spectra can even tell us which galaxies are more distant than others. Perhaps more importantly, understanding the velocities of galaxies can help us to understand how the mass of the Universe is distributed. Almost all motion in the Universe grew from the tug of **gravity**. By tracing where galaxies and stars are being lured, and how fast, we can therefore detect hidden masses in the Universe (see **dark matter**, and Essay 6). On the largest scales we can track the velocities of galaxies relative to one another, and measure how much mass binds them together. We can also study the mass within an individual galaxy by tracking the velocity (i.e. the redshift) across it, thus mapping the rotation of the stars and gas within it. On the very smallest scales, the tiniest changes in velocity have recently been used to make the first tentative identifications of planets orbiting stars other than our own, by detecting the small shifts in a star's velocity produced by the gravitational nudges of one or more planets circling around it.

So, with only two means of observing the Universe, through images and through spectra, we can extract a surprising abundance of information from our observations. Astronomers are continually inventing new means of wielding these two tools, and interpreting their results. While occasional leaps of understanding result from these bursts of cleverness among individual astronomers, far more steady progress is made through the inexorable advance of technology. As we are so limited by the types of observation we can make, any development which increases the quality of an observation, or expands it into a different energy regime, creates an explosion of new ideas and insights. It is principally these changes in technology that drive astronomy forward. At any given moment, astronomers will be spending their time addressing questions which it has just become technologically possible to solve.

Because of this, from the time of Galileo astronomers have always felt privileged, as they have continually experienced the pleasure of watching how advances in technology open up new scientific vistas and make routine what was once difficult. While their instruments represented the state of the art, the astronomers who at the beginning of the 20th century were taking photographs through 1-metre aperture telescopes would hardly recognise the images of today, taken with exquisitely sensitive digital cameras mounted on telescopes with a hundred times the collecting area. With the passing of another century, the equipment that is so amazing to us now will seem amusingly quaint to future generations of astronomers.

In astronomy, the scientific excitement created by new observational possibilities arises from technological improvement in three different quarters:

The continually increasing ability to collect more light from the faintest, farthest objects;

The expansion of observations into new wavelength regimes;

The increasing resolution of observations, revealing the tiniest structures in the most distant galaxies.

As we shall see, each type of advance allows a distinct are of study to be opened up.

Bigger is better

When you first go outside and look up at the night sky, some stars are simply too faint for your eyes to see. Similar limitations apply to astronomical telescopes. For any particular telescope there will be stars and galaxies which are too faint to be detected. Some are too far, like extremely young galaxies lurking at the edge of the Universe. (Because of the time it takes light to travel from the most distant galaxies to the Earth, we are actually observing those galaxies when they were extremely young, as young as 10% of the age of the Universe). Others are just too small, like tenuous dwarf galaxies which are nearly 50,000 times less massive than our own Milky Way.

Luckily, in the same way that fainter stars are revealed as the pupil of your eye dilates and admits more light, astronomers too can increase the size of their 'eyes' by building larger telescopes. The bigger the telescope, the more light can be collected from a distant star or galaxy, and thus the fainter the objects that can be seen. Creating the enormous mirrors used by modern telescopes is a tremendous technological achievement, and in the past decade astronomers and engineers have begun to master the skills needed to build mirrors 6 to 12 metres in diameter. Some, like the 10-metre Keck Telescopes on Hawaii, are made from dozens of mirror segments which work together like a single adjustable mirror. Others are single monolithic pieces of glass, either honeycombed with air to reduce their weight, or incredibly thin and supported by hundreds of computer-controlled actuators which continually push on the back of the mirror, precisely correcting its shape. The coming decade will see the completion of more than ten different telescopes in this size range, opening up the most distant reaches of the Universe to us. Radio telescopes, which can detect both atomic and molecular gas in interstellar and intergalactic space, are also growing larger. They are being networked into large arrays, allowing several telescopes to act a single, much larger one.

Another means of collecting more light is to improve the efficiency of the detectors that astronomers use to collect light, whether for images or for spectra. Modern optical astronomy has been revolutionised by the invention of the charged-coupled device (CCD). These exquisitely sensitive digital cameras can collect almost every photon that reaches them, in remarkable contrast to the photographic film used for much of the 20th century, which was capable of detecting only one or two out of every hundred photons which reached it. Similar revolutions are taking place at other wavelengths. In the field of **infrared astronomy**, efficient digital cameras have been developed which for the first time have allowed wide-field mapping and spectroscopy in this waveband. This extension of digital technology into the infrared is particularly exciting for work on very distant (or 'high-redshift') galaxies. Because of the increasing redshift of light with distance, the part of a galaxy's spectrum that would be in the visible at optical wavelengths if it were nearby is shifted well into the infrared for a galaxy very far away. Thus, the ability to compare the properties of distant, young galaxies with the characteristics of the local Universe requires those infrared capabilities that are just reaching maturity today.

Extragalactic spectroscopy has been one of the largest beneficiaries of the recent growth in telescope size and detector sensitivity.

Spectroscopy is inherently more difficult than imaging because light is spread out over many wavelengths, whereas in imaging the same amount of light is concentrated in a single image. For example, a galaxy spectrum which sorts light into 'bins' 0.1 nm wide is 1000 times more difficult to measure accurately than an image of the galaxy taken through a single filter which is 100 nm wide. Spectroscopy is therefore impossible for the faintest galaxies visible in an optical image. Because of this discrepancy, spectroscopy stands to gain the most from any increase in our ability to detect large numbers of photons efficiently. Furthermore, because spectroscopy is more closely linked to the physical processes which underlie all that we see, the largest leaps in our understanding tend to come from improved spectroscopy.

The current spectroscopic revolution has produced a number of extremely exciting results. For example, with incredibly high-resolution spectra of distant quasars, astronomers have made tremendous progress towards understanding the Lyman-alpha forest (see **intergalactic medium**). Quasars are prodigiously luminous objects appearing as point-like spots of light, thought to be produced when gas falls into a massive black hole. Any hydrogen gas along the line of sight to a quasar absorbs a specific frequency of light from the quasar, producing extremely narrow absorption lines whose wavelength depends on the distance to the hydrogen cloud, and thus on its redshift. These blobs of gas produce a 'forest' of absorption lines, making what would otherwise be a smooth quasar spectrum appear like a field of grass. While astronomers know that most of these absorption lines are caused by hydrogen (a much smaller fraction are caused by heavier elements, such as magnesium and carbon), we have known very little about the origin of the absorbing gas. Were the lines produced by failed galaxies, by the outskirts of normal giant galaxies or by the gaseous vapour that lies in the nearly empty spaces between galaxies?

Using the incredibly accurate, high-resolution spectra that can now be obtained with the largest telescopes, astronomers have begun to unravel the mysteries of the systems producing absorption features in quasar spectra. For the first time, elements other than hydrogen have been detected within a single gas cloud. Because almost all elements heavier than hydrogen were produced in stars, by measuring the relative proportions of these new elements (see **light element abundances**), astronomers can probe the history of star formation within the clouds, looking for clues to their origins. Recent data suggest that the clouds seem to be among the most hydrogen-rich objects in the Universe, showing little sign of contamination with heavier elements ejected by dying stars. The high redshift (and thus young age) of these clouds makes them perfect laboratories for studying the primordial state of gas in the Universe. One particularly important measurement, the abundance of primordial deuterium (a heavier isotope of hydrogen, containing an extra neutron in the nucleus), has recently become possible. The abundance of deuterium has an extremely important bearing on Big Bang **nucleosynthesis**, and has tremendous cosmological implications.

Another vital cosmological test which has only recently become possible is the direct measurement of the **curvature of spacetime**. General relativity shows that the apparent brightness of a star or galaxy falls off more quickly with distance than we would naïvely expect. The rate at which it falls off depends on fundamental cosmological parameters, such as the **density parameter** Ω, a measure of the total density of the Universe, and the **cosmological constant** Λ, a measure of the fraction of the density which is due to vacuum energy. Thus, by measuring how the apparent brightness of an object changes with distance, we can put limits on the values of these important parameters, which control the eventual fate of the Universe. This is just one example of a battery of cosmological tests which go under the name of **classical cosmology**.

The difficulty in this approach is twofold. First, the test requires us to identify a class of objects which have the same intrinsic luminosity at any distance from us; and second, the objects must be detectable at very large distances, where the effects of curvature become strongest. The ideal objects turn out to be **supernovae** – cataclysmic explosions that mark the death of a stars that have pursued particular lines of **stellar evolution**. Supernovae are tremendously luminous, sometimes rivalling the brightness of the entire galaxy that contains them, in spite of being produced by a single star. Because supernovae are controlled by the internal physics of stars, the peak luminosity of supernovae should be much the same at all times in the Universe's history. Therefore, supernovae can be used as *standard candles*: objects whose intrinsic brightness is known for all times and distances. By combining specialised software and the newest, largest telescopes, astronomers have been able to detect dozens of distant supernovae at the peak of their brightness, finding them at redshifts approaching 1. Over the first decade of the new millennium, as the numbers of high-redshift supernovae detected increases, this technique will provide one of the strongest possible measures of the total density of the Universe.

The key to all these remarkable discoveries is the dramatic increase in the number of photons which astronomers can collect and focus in their observations. Another class of experiments represents a different approach to increasing the number of photons. These Herculean projects devote tremendous amounts of time to observing a single patch of sky to unbelievably faint limits, or to mapping enormous areas of the sky. The most dramatic example of the former is the Hubble Deep Field, which is the deepest view of the Universe ever created (see Figure 1). The image was obtained during an unprecedented ten days of constant observation with the **Hubble Space Telescope** (HST). Given that the telescope rarely makes observations lasting more than a few hours, the Hubble Deep Field is a stunning achievement.

At the other extreme are astronomical surveys which devote years of a telescope's lifetime towards a single goal, studying the large-scale distribution of galaxies. The Sloan Digital Sky Survey (see **redshift surveys**), begun in 1997, represents the acme of this approach. Over the course of several years, the survey is planned to map the entire northern sky through many different coloured filters, and then use spectra to measure distances to over a million galaxies and quasars found in the images. The resulting data set will be an immense storehouse of

Figure 1 Three sections of the Hubble Deep Field, showing suspected high-redshift galaxies.

knowledge about the nearby Universe and the galaxies within it, and will be vital in improving our understanding of the **large-scale structure** of the distribution of galaxies and for building theories of **structure formation** (see Essay 2).

Opening up new windows

Our eyes are well matched to the spectrum of the Sun, and are most sensitive in a small range of energies where the Sun emits most of the light that can easily reach the surface of the Earth. While this limitation of biological engineering is quite sensible from an evolutionary perspective, it has the unfortunate side-effect of severely biasing our view of the Universe. Astronomy naturally began as a study of what we could see with our eyes. However, the wavelengths of the light that the Universe creates vary by a factor of ten million from the longest to the shortest, compared with a factor of about two for what we can detect visually. There is clearly much to be learned from the worlds which are hidden from our unaided eyes. Historically, tremendous revolutions in our understanding have followed advances in technology which opened new wavelength ranges to our gaze.

The atmosphere of the Earth is one of the most pernicious obstacles to observing the Universe outside visible wavelengths. While the atmosphere is reasonably transparent to the optical light which our eyes can detect, and to very low-energy radio waves, it acts as a blanket over the Earth at most other wavelengths of light. At infrared wavelengths there are a handful of narrow 'windows' where the atmosphere becomes transparent, but with this exception, observing the Universe at short or long wavelengths (high or low energy) requires astronomers to lift their telescopes above the Earth's atmosphere. While high-altitude balloon experiments are sufficient at some wavelengths, it was the inception of the 'rocket era' of the 1950s that first allowed astronomers to propel their instruments into space, and escape from the obscuring mantle of the atmosphere. Not until the 1960s were the first detections of high-energy X-rays and gamma-rays made of objects other than Sun.

Because of the close link between wavelength and energy, new physical processes are revealed when astronomers can shift their vision to other wavelengths. At the highest energies, new developments are being made in **gamma-ray astronomy**. In particular, astronomers are intensively studying the properties of *gamma-ray bursts*. These short (0.1–100 seconds), intense bursts of gamma-radiation emit almost all of their flux at energies of 50 keV, or 50 times the energy of penetrating

X-rays. For many years it was assumed that gamma-ray bursts probably originated within our own Galaxy, otherwise the energy released in the burst would have to be on an unimaginable scale. While the existence of gamma-ray bursts has been known since the 1970s, it was not until the 1990s that astronomers were able to pin-point the exact position of these bursts and to measure the weakest of them. After the Burst and Transient Source Experiment (BATSE) on the Compton Gamma-Ray Observatory satellite was launched in 1991, it was discovered that the bursts were distributed uniformly over the sky, suggesting that either they were so close that they did not reflect the shape of the flat disk of the Milky Way Galaxy, or that they were actually outside our the Galaxy and so far away that their positions reflected the uniformity of the Universe. If the bursts are extragalactic, then their energies must be so high that they must originate with incredibly violent processes, such as the coalescence of pairs of neutron stars, the superdense end products of dying stars, whose density rivals that of the nucleus of an atom.

Gamma-rays are notoriously hard to focus into an image, so it has been incredibly difficult for astronomers to localise the source of gamma-ray bursts with sufficient accuracy to find their optically visible counterparts. However, since its launch in 1996 the X-ray satellite BeppoSAX has been able to home in on the position of bursts by immediately observing their X-ray afterglow, improving by a factor of 20 the accuracy with which their position can be measured. This new targeting has allowed optical astronomers to make rapid follow-up observations of the immediate region of a gamma-ray burst, looking for an object which is fading along with the burst. Recently, astronomers have been able to take the very first spectrum of a gamma-ray burst source, and have discovered that it is at an extremely high redshift, and thus astoundingly luminous. Although the nature of what causes the burst is still a puzzle, with the continual improvement in the ability to observe the Universe at high energies astronomers are optimistic that it will not be long before the nature of these phenomenal events is understood.

At slightly less extreme energies, studies of the X-ray Universe provide information on the evolution of some of the largest and most massive structures in the Universe – clusters of galaxies. As the planets in our Solar System are gravitationally bound to the Sun, so too are galaxies bound together. Occasionally, thousands of giant galaxies bind tightly together into a cluster. The gravitational pull of this enormous mass of galaxies traps extragalactic gas, and heats it to incredibly high temperatures, so hot that electrons separate from their nuclei and form a hot, X-ray emitting plasma. **X-ray astronomy** is currently blessed with a wealth of X-ray telescopes, with continually improving abilities to focus X-rays into images, and to measure the spectra of X-ray sources. After decades of hard work astronomers can now begin to study the X-ray Universe in the same detail as they can study the easily accessible optical world. While less difficult to capture and study than X-rays and gamma-rays, studies of the near ultraviolet (UV) light have greatly benefited from recent improvements in the sensitivity of CCD cameras to these high-energy photons. New methods in **ultraviolet astronomy** have followed from new technological approaches to the manufacture of the silicon wafers that lie at the heart of all CCD cameras. These developments have allowed routine observations to be made of what had previously been a difficult energy range to study. The UV spectrum is of particular importance in astronomy: the youngest, hottest stars emit most of their light in the UV, and thus UV observations can provide sensitive probes of the rate at which galaxies are forming stars, both now and in the past.

By exploiting a particular property of the UV spectrum, astronomers have been able to use the new sensitivity to UV radiation to find infant galaxies at very large distances, using the intervening absorption of the Lyman-alpha forest. The absorption produced by the intervening hydrogen eats away at the far UV spectra of galaxies, producing a sharp drop at high energies, and when a galaxy is at a large distance the redshift of light moves this drop into the optical region. By using very deep UV and optical images, astronomers have identified very distant galaxies by looking for this drop, which manifests as a near *absence* of light in the UV (see Figure 2). Almost every galaxy which is easily visible through blue and red filters, but is invisible in deep UV images, has been shown to be at a remarkably high redshift. This new approach, made possible by increased UV sensitivity, has revolutionised our ability to study galaxies like our own Milky Way near the moment of their birth.

There are many exciting developments at longer wavelengths as well. Infrared detectors are recapitulating the rise of CCD technology, doubling in size and sensitivity every few years. In the past, infrared detectors could view only very tiny areas of the sky. Now, however, they are beginning to rival optical CCD cameras in size, allowing wide-field studies of the properties of stars and galaxies to be carried out in the infrared. One particularly exciting application of the new detectors is the search for *brown dwarfs* – failed stars which lack sufficient mass to ignite a central thermonuclear engine. While stars continuously produce energy in their cores, brown dwarfs only have the energy they

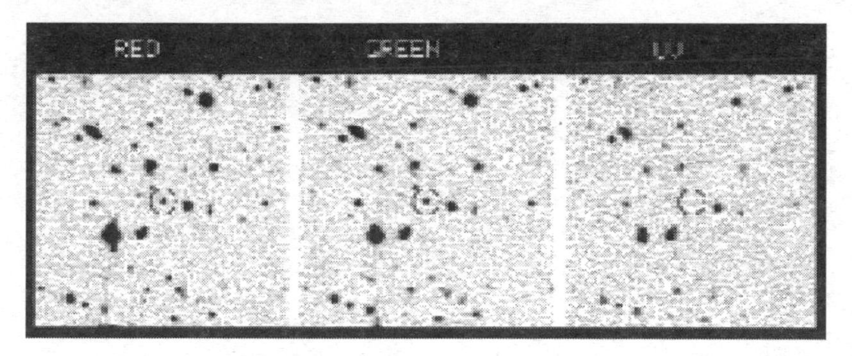

Figure 2 A distant 'ultraviolet dropout' galaxy. The high redshift moves the Lyman-alpha forest from the ultraviolet into the visible, so that the galaxy is much dimmer in the UV than at optical wavelengths.

were born with. Over their lifetime they can only cool down, becoming progressively dimmer and eventually invisible. This makes old brown dwarfs a candidate form of dark matter. Because they are cooler than stars, newly formed brown dwarfs emit most of their light in the infrared. The new generation of infrared cameras has enabled wide-area searches for younger, brighter brown dwarfs to be begun, and the first promising candidates have been identified.

A second burgeoning field of infrared astronomy is spectroscopy. The new large-format detectors are now being coupled with innovative designs of spectrograph, and are giving astronomers the ability to study the infrared spectra of galaxies in the same detail possible in the optical range. These spectrographs will be of particular help in the study of the youngest, high-redshift galaxies. Because of their high redshift, the light that these galaxies emits at visible wavelengths is displaced into the infrared. Thus, by studying the infrared spectra of these young galaxies we can directly compare their 'optical' properties to those of older, closer galaxies, and see how star-formation rates and chemical compositions have changed and evolved through the intervening years. The new generation of infrared satellites, such as the Infrared Space Observatory (ISO, launched in 1995), the Space Infrared Telescope Facility (SIRTF, planned for launch in 2002) and the Far Infrared and Submillimeter Telescope (FIRST, currently scheduled for 2007) will take these studies even further by raising the telescopes above the atmosphere, opening up the regions of the infrared spectrum which are currently denied to earthbound instruments.

At the longest wavelengths of all, in the radio portion of the spectrum, there is also much to be learned. Radio wavelengths are of particular importance for studying the cool gas that makes up much of a galaxy, and from which all stars are born. Although the gas in dense X-ray emitting clusters of galaxies is extremely hot, within galaxies it can be quite cool. At low temperatures, instead of being a turbulent plasma of stripped nuclei and electrons, the gas can form not only atoms, but also complex molecules, many of which occur on the Earth. These molecules have their own radio emission lines, resulting from their characteristic rates of vibration and rotation. These emission lines have been detected for such well-known molecules as water, cyanide and even sulphuric acid. Studying these galactic molecules reveals the conditions in the gas clouds from which stars are born. By extending radio spectroscopy to fainter levels and different wavelengths, astronomers can unravel the conditions in these stellar nurseries in greater detail, and can map how these conditions relate to the properties of the galaxy as a whole.

While much of the progress in wavelengths outside the optical has moved steadily forward, **radio astronomy** has unfortunately suffered some reversals. As any modern city-dweller knows, radio waves have become the carriers of the business of everyday life. Cellular telephones, pagers, cordless phones and satellite television all receive and transmit at radio wavelengths. This background of noisy human chatter blocks out the quiet whispers of the Universe in large chunks of the radio spectrum, prohibiting astronomers from ever again observing these wavelengths from the surface of the Earth. Because of the redshift of light, not only can these polluted wavelengths block the emission of specific molecules, they can block the emission of more distant galaxies, whose emission has shifted into the noisy part of the radio spectrum. It seems that radio astronomers are now faced with a permanent struggle against the continuing encroachment of the global communications industry into the wavelength bands that are their sole means of probing the Universe.

The Universe through a microscope

It is a sad fact of geometry that far-away objects inevitably appear blurred. As another person gradually moves away from us, what appeared at close range as individual pores becomes a smooth cheek, and then a poorly delineated part of a face framed with hair. Our eyes have only a limited ability to distinguish between objects which are separated by only a small angular distance. Features on a person (or an object) blur together when there is little distance between them compared with the distance to the object itself. Like tail lights on a receding car, at some angular separation two objects will merge and appear like one to our eyes. The angle at which this loss of information occurs is a measure of the *resolution* of any sort of light detector. While most telescopes have a far higher resolution than the human eye, over astronomical distances a gorgeously detailed spiral galaxy is a featureless blotch of light to most detectors. The ability to distinguish between different types of galaxy is therefore greatly diminished at large distances.

Worse than the limited image quality imposed by the finite resolution of a telescope is the blurring effect of the Earth's atmosphere. The ocean of air which sheaths our planet is as turbulent as the seas. When rays of light from a star pass through the atmosphere, the moving layers of air shift the point-like image of the star around, acting as constantly changing lenses, and creating the twinkling we are all familiar with. By moving the image of the star about, and continually distorting its image, atmospheric turbulence makes stars appear not as a point, but as a blurry spot.

The most straightforward solution to this problem is to move astronomical telescopes above the atmosphere by launching them into orbit around the Earth as satellites. Figure 3 shows the best-known example, the **Hubble Space Telescope** (HST). Astronomers' excitement about the HST comes almost entirely from its remarkable resolution, as opposed to its size (the mirror of Hubble is only a fifth of the diameter of the largest ground-based telescope). While they are usually larger, the very best resolution reached with 'ground-based' telescopes is at best only a fifth of what can be achieved with the HST, and more typically it is only a tenth. As a result, an image taken of a distant galaxy with the HST reveals the same amount of detail as a ground-based image of a galaxy which is five to ten times closer. This dramatic increase in resolution allows us to observe the distant Universe as if it were our own galactic backyard. Some of the most visually appealing studies of the distant Universe have been in the field of gravitational lensing, where the HST's resolution is ideal for mapping the image distortions produced by dark matter (see Essay 6).

Even for closer galaxies, the HST's resolution has nurtured many exciting new projects. From a cosmological point of view, one of the most important of these is concerned with refining the **extragalactic distance scale** of the Universe. While astronomers have long realised that, because of the expansion of the Universe, the recessional velocities of galaxies, or their redshifts, are measures of the galaxies' distances, the exact proportionality between redshift and distance – the value of the

Figure 3 The Hubble Space Telescope being deployed from the Space Shuttle.

Hubble constant H_0 – is a long-standing question. The quest for H_0 has been greatly furthered by the HST's ability to resolve individual stars in galaxies well beyond our immediate galactic neighbours. By observing Cepheid variable stars, the same technique that Edwin Hubble used to obtain the first measurement of the distance to the Andromeda Galaxy, astronomers can now measure distances to galaxies which are about ten times farther away than this (see Figure 4), and the total number of galaxies amenable to this kind of measurement has increased by a factor of about a thousand. These galaxies can then be used to calculate the intrinsic brightness of supernovae, which can be seen almost out to redshifts of 1. The recessional velocities of these distant galaxies are dominated by the expansion of the Universe, and not by the gravitational pushes and pulls of other galaxies. As such, they provide untainted measurements of the Hubble constant. While the tedious labour of calibrating these steps of the 'distance ladder' is ongoing, the measurcments have already narrowed the uncertainty in H_0 to roughly 20%, whereas it used to be uncertain by at least a factor of two.

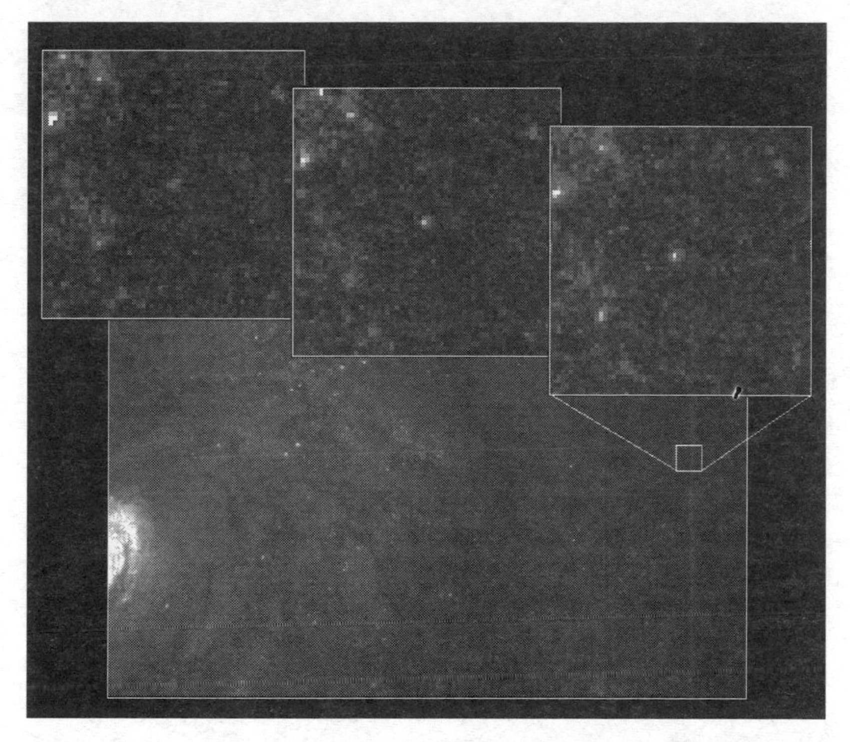

Figure 4 A Cepheid variable in the galaxy M100, imaged by the Hubble Space Telescope's WFPC2 camera. Observation of the star has yielded a distance of 52.5 light years.

Another benefit of high resolution is that it allows astronomers to probe smaller physical scales than ever before. With the tenfold increase in the HST's resolution comes a tenfold decrease in the angular distances that can be resolved. In our own Galaxy, astronomers are now close to mapping the formation of individual solar systems around infant stars. Astronomers have also been able to resolve the very centres of nearby galaxies, on scales approaching the distance of the nearest stars to the Sun. Such studies are beginning to provide strong evidence for the existence of **black holes** at the centres of many galaxies. Spectroscopy across the centre of galaxies has shown gas and stars revolving around the centre at a tremendous rate, so fast that there must be a very dense, very large mass at the centre. Such high densities seem to be compatible only with there being a massive black hole at the centre of the galaxy, whose gravity tightly holds these rapidly orbiting stars. Without the resolution of the HST, velocities so close to the centre could not be measured, and any identification of a central black hole would be much more speculative.

Similar detections of central black holes have been made using radio measurements of the velocity of molecular gas. Unlike ground-based optical telescopes, radio telescopes can achieve extremely high resolution from the ground, because the long radio wavelengths are much less affected by atmospheric blurring. Furthermore, radio telescopes are often linked together in a giant array, which greatly increases resolution. The best resolution that can in theory be reached by a telescope is inversely proportional to its diameter, so that bigger telescopes have better resolution. With much additional software processing, a large array of radio telescopes can be made to function much like a single telescope whose diameter is as large as the largest separation between the elements of the array. For radio telescope arrays, separations measured in kilometres are achievable, so the resolution is very good indeed. The most spectacular type of radio array is achieved by very long baseline interferometry (VLBI), in which radio telescopes are linked in a network spanning the entire globe, giving a resolution limited only by the diameter of the Earth! With VLBI, astronomers have resolutions which are a hundred times better than even that of the HST. As with the jump from ground-based optical telescopes to the HST, the advent of VLBI has made possible studies which could not have been realistically contemplated before, such as mapping the actual expansion of a supernova remnant millions of light-years distant.

At much higher energies, the resolution limitations are quite different. X-rays and gamma-rays are easily observable only from above the Earth's atmosphere, so they have never been subject to the atmospheric blurring which affects optical and infrared images. However, because of the high energy and penetrating nature of high-energy photons, the difficulty lies in building mirrors which can actually focus the light into images. At any wavelength, a mirror is a surface that changes the direction of light without absorbing much of the light's energy. For high-energy photons it is difficult to find materials for a mirror which have the required properties. In the X-ray spectrum, one of the most effective means of focusing is through 'grazing-incidence' mirrors. Incoming X-rays skim the surface of these mirrors, getting a slight deflection in the process which focuses the light to a point a very long distance from the mirror; because the X-rays hit the mirror at such a slight angle, very little of their energy gets transferred to the mirror. For gamma-rays, the technological obstacles are even greater because of their higher energies. To date, the typical resolution of gamma-ray detectors is tens to hundreds of times worse than for optical light.

One promising advance at optical and infrared wavelengths is the development of 'adaptive optics'. The theoretical resolution of an earth-bound telescope is far greater than the practical limit of resolution imposed by the turbulent atmosphere. Therefore, if astronomers can 'undo' the blurring of the atmosphere, they can potentially reach the same resolution as the HST, but from the ground, where telescopes are larger, cheaper to build and more plentiful. The key to 'undoing' the damage the atmosphere has done is to track the effects of blurring on individual stars. Because stars should be perfect points of light, a telescope can track deviations in a star's shape and position, and use sophisticated software and hardware to compensate for and potentially reverse the effects of the atmosphere. The most common forms of adaptive optics are 'tip–tilt' systems, which make frequent minute adjustments to the orientation of the telescope's secondary mirror in order to hold the image of a star perfectly still. More advanced systems actually deform the shape of a telescope's mirrors, to keep both the star's shape and position constant. Currently, the use of adaptive optics is limited to small regions of the sky which happen to have a very bright reference star nearby. However, as the sensitivity of detectors improves, this limitation should recede. Some systems solve this problem by actually creating artificial stars, by shining pinpoint laser beams up into the atmosphere. In the coming decade, adaptive optics will certainly become a prominent feature of ground-based astronomy, and scientific enquiries which were impossible with the HST's small mirror will surely become possible.

The future

History has shown us that in astronomy the future is always bright. Human invention continually improves and triumphs over technological limitations. Over the coming decades, we are assured that telescopes will become bigger, and that detectors will become more sensitive. New types of detector, which can potentially gather imaging and spectroscopic information simultaneously, will gradually pass from experimental devices in engineering laboratories to the workhorses of astronomical research. Astronomers and engineers are also guaranteed to lift telescopes into novel environments. The largest telescopes on Earth today will eventually have counterparts circling the Earth, and possibly even mounted on the Moon. As adaptive optics comes of age, earthbound telescopes will be able to achieve the remarkable results currently only available from space. But while we can be certain that the tools available to astronomers will continue to improve, history has also shown that it is impossible to predict the surprises and wonders that such improvements will reveal.

FURTHER READING

Florence, R., *The Perfect Machine: Building the Palomar Telescope* (Harper-Collins, New York, 1994).

Graham-Smith, F. and Lovell, B., *Pathways to the Universe* (Cambridge University Press, Cambridge, 1988).

Hubble, E., *The Realm of the Nebulae* (Yale University Press, Newhaven, CT, 1936).

Preston, R., *First Light: The Search for the Edge of the Universe* (Random House, New York, 1996).

Tucker, W. and Tucker, K., *The Cosmic Inquirers: Modern Telescopes and Their Makers* (Harvard University Press, Harvard, 1986).

5

THE COSMIC MICROWAVE BACKGROUND

CHARLES H. LINEWEAVER

WHAT IS THE COSMIC MICROWAVE BACKGROUND?

The **cosmic microwave background radiation** is the oldest fossil we have ever found, and it has much to tell us about the origin of the Universe. The cosmic microwave background (CMB) is a bath of photons coming from every direction. These photons are the afterglow of the Big Bang, and the oldest photons we can observe. Their long journey towards us has lasted more than 99.99% of the age of the Universe and began when the Universe was one thousand times smaller than it is today. The CMB was emitted by the hot plasma of the Universe long before there were planets, stars or galaxies. The CMB is an isotropic field of **electromagnetic radiation** – the redshifted relic of the hot Big Bang.

In the early 1960s, two scientists at Bell Laboratories in Holmdel, New Jersey, were trying to understand the detailed behaviour of a very sensitive horn antenna used to communicate with the first generation of Echo communications satellites. The noise level in their antenna was larger than they could account for: there was too much hiss. For several years they tried to resolve this excess noise problem. They asked their colleagues. They removed pigeon guano from their antenna. They even dismantled it and reassembled it. Finally, with the help of a group led by Robert **Dicke** at Princeton, it was concluded that the hiss was coming from outside of New Jersey. The intensity of the signal did not depend on direction: it was isotropic, and thus could not be associated with any object in the sky, near or far. In their article in the *Astrophysical Journal* in 1965 announcing the discovery, Arno **Penzias** and Robert **Wilson** wrote: 'A possible explanation for the observed excess noise temperature is the one given by Dicke, Peebles, Roll and Wilkinson . . . in a companion letter in this issue.' The explanation was – and still is – that in observing this excess noise we are seeing the Big Bang. This hiss was subsequently measured by many groups at many different frequencies, and was confirmed to be isotropic (the same in all directions) and to have an approximately **black-body** spectrum. Penzias and Wilson received the 1978 Nobel Prize for Physics for their serendipitous discovery of 'excess noise', which is now known as the cosmic microwave background radiation.

To understand what 'seeing the Big Bang' means, we need first to understand how the CMB fits into the standard **Big Bang theory** of the Universe, described in previous essays. The standard cosmological models are based on the concept of an evolving Universe. In particular, the observed **expansion of the Universe** has the profound implication that the Universe had a beginning about 15 billion years ago. Independent age determinations support this idea: there do not appear to be any objects with an age greater than about 15 billion years (see **age of the Universe**). In addition, an expanding Universe must have been smaller, denser and hotter in the past.

The Big Bang (see also **singularity**), the name given to the very beginning, happened everywhere about 15 billion years ago. You can see photons from the hot Big Bang in all directions. They come from the *photosphere* of the hot early Universe, which is known as the **last scattering surface**. When you look into a fog, you are looking at a surface of last scattering. It is a surface defined by all the molecules of water that scattered a photon into your eye. On a foggy day you can see for 100 metres, on a really foggy day you can see for only 10 metres. If the fog is so dense that you cannot see your hand, then the surface of last scattering is less than an arm's length away. Similarly, when you look at the surface of the Sun you are seeing photons last scattered by the hot plasma of the Sun's photosphere. The early Universe was as hot as the Sun, and it too had a photosphere beyond which (in time and space) we cannot see. This hot (3000 K) photosphere is the Universe's last scattering surface. As its name implies, this surface is where the CMB photons were scattered for the last time before arriving in our detectors.

Anisotropies and the COBE discovery

Figure 1 shows the historical progression of what happened as the CMB was observed with more and more precision. The isotropic hiss of the CMB discovered by Penzias and Wilson has the spectrum of a flat, featureless black body – there are no anisotropies; the temperature is constant in every direction at $T_0 = 2.73$ K. The top panel of Figure 1 is an actual map of the CMB. No structure is visible: we seem to live in an unperturbed, perfectly smooth Universe. The CMB is smoother than a cue ball. If the Earth were as smooth as the temperature of the CMB,

Figure 1 The top map shows what Penzias and Wilson discovered in 1965: an isotropic background. In the 1970s the dipole due to the Doppler effect of our motion through the Universe became apparent (middle map). In 1992, with the high sensitivity of the COBE satellite's Differential Microwave Radiometer (DMR), we were able to measure and remove the dipole accurately and discover cold and hot spots in the otherwise smooth background. These were the long-sought-for anisotropies in the CMB – temperature fluctuations made by the largest and oldest structures ever detected. Lighter regions indicate a higher than average temperature; darker regions indicate lower temperatures.

then the whole world would be flatter than the Netherlands. If the CMB were *perfectly* smooth, there would be no structure in the present Universe.

On a dark night you can see the Milky Way stretching across the sky. With a telescope you can see that the sky is full of galaxies, clusters of galaxies, great walls of galaxy clusters and giant voids where no galaxies exist. On scales less than about 300 million light years the Universe is clumpy – it is neither isotropic nor homogeneous. A fundamental question in cosmology is how this **large-scale structure** came to be. The most widely accepted answer is that these structures all collapsed gravitationally from initially small overdensities by a process of gravitational **Jeans instability**. In other words, wispy clouds of matter fell in on themselves under the force of their own gravity. A simple answer, but is it correct? CMB observations provide a test of this scenario of **structure formation** in the following way. If structures formed gravitationally from under- and overdensities of matter, then smaller under- and overdensities must have existed in the distant past at the recombination era (see **thermal history of the Universe**), and must have produced slightly hotter and colder spots on the last scattering surface. And with sufficiently sensitive detectors we should be able to detect these *anisotropies*.

The largest anisotropy in the CMB, and the first to be detected, is the *dipole anisotropy*, representing a sinusoidal variation across the sky on a scale of 180°. The CMB is observed to be about three-thousandths of a degree (3 mK) hotter in one direction and about 3 mK colder in the opposite direction. The dipole is easily seen in the middle sky map in Figure 1. It is hotter at the upper right, in the direction of our velocity, and colder at the lower left, where we are coming from. This anisotropy is most simply explained as a **Doppler effect** due to our velocity. It is a speedometer: it tells us how fast we are moving with respect to the rest frame of the CMB (see below). The inferred velocity of our Local Group (the galaxy cluster containing our Milky Way, the Andromeda Galaxy and over two dozen less substantial collections of stars) is 627 km/s towards the constellations of Hydra and Centaurus, in the general direction of a recently discovered concentration of galaxies known as the Great Attractor (see **peculiar motions**).

Measurements of the CMB dipole define the rest frame of the Universe. In physics classes we learn that there is no preferred *rest frame* in the Universe. Special relativity is based on the idea that there is no preferred inertial **frame of reference** (a preferred frame would be one in which the **laws of physics** take on some special form). The

Michelson–Morley experiment of 1881 is sometimes cited as an experiment which showed that the hypothetical **aether** (defining the rest frame of the Universe) does not exist, and thus that there is no preferred reference frame. The detection of the dipole anisotropy in the CMB has often been called the 'new aether drift' because it does define the rest frame of the Universe (or at least the rest frame of the observable Universe). This rest frame is not a preferred frame since the laws of physics seem to be the same in this frame as in any other, but it is fundamental, special and well-defined. Is this fundamental reference frame some initial condition of the Universe, or has it been selected by some kind of process we know nothing about, which happens to select a highly symmetric state for the Universe? The same kinds of question can be asked about the topology of the Universe, or even why the Universe is expanding.

If the CMB has the same temperature in every direction, then how did galaxies and other large-scale structure come to be formed? Before the results obtained by the **Cosmic Background Explorer** (COBE) satellite, the absence of anisotropy in the CMB was taken to be one of the important failings of the Big Bang model. In 1980, Geoffrey **Burbidge** pointed out that 'if no fluctuations can be found, we have no direct evidence at all that galaxies are formed at early epochs through gravitational Jeans instability.'

Observers had been searching for the expected small-amplitude temperature anisotropies ever since the discovery of the CMB. As limits on the fluctuation level decreased, theoretical predictions for the level of CMB anisotropy decreased apace. Knowing the instrument sensitivities and calculating the expected CMB fluctuation level from the observed large-scale structure of the Universe (without invoking **dark matter**) led some cosmologists to conclude that CMB fluctuations should have been discovered 15 years before. As observers found the last scattering surface to be smoother and smoother, stuff called 'non-baryonic dark matter' was invoked to keep the predicted level of temperature variations below observational limits.

The argument went as follows. We count galaxies around us and measure how clumpy they are. Models tell us how fast this clumpiness grows, so we can predict the level of clumpiness at recombination. This in turn gives us a prediction for the induced temperature fluctuations ($\Delta T/T_0$) at recombination. The value of $\Delta T/T_0$ came out to be about 10^{-4} (i.e. about one part in ten thousand). This was too big, and had already been ruled out by observations. So the clumpiness of matter at recombination had to be much smaller – thus it had to have grown faster than the models were telling us. Enter non-baryonic dark matter. If such stuff existed, it would have *decoupled* from (ceased to interact with) the radiation earlier and started to clump together earlier than normal (baryonic) matter. The normal matter would then have been able to have smaller-amplitude fluctuations at recombination and fall into *pre-existing* overdense regions to quickly reach the large level of clumpiness it has today. Thus, for a given measurement of the clumpiness today, invoking dark matter lowers the level of $\Delta T/T_0$ expected at the last scattering surface.

This argument may seem complicated, but the gist of it can be conveyed by a thought experiment. Imagine a large field in which trees have been planted and have now all grown to full maturity. Suppose the field is 100 light years in radius. In looking far away we are looking into the past. The most distant trees must be 100 years younger, and thus smaller, than the trees we see nearby. If we know how fast trees grow, we can predict how big the trees were 100 years ago. But then someone makes an observation of the most distant trees and finds that they are much smaller than our prediction. What is wrong? Perhaps we forgot that when the trees were planted, they had fertiliser around them. This is essentially the story of the search for CMB anisotropies. Local galaxies are the fully grown trees, CMB anisotropies are the seedlings and non-baryonic dark matter is the fertiliser. One of the problems with the idea is that although we have smelt the fertiliser, we have never detected it directly.

As a graduate student, I was part of the COBE team that analysed the data obtained by the Differential Microwave Radiometer (DMR), a type of radio receiver, flown aboard the COBE satellite. The DMR was presciently constructed in the pre-dark-matter epoch with enough sensitivity to probe the lower, dark-matter predictions. The main goal of the instrument was to find temperature variations in the CMB – variations that *had* to be there at some level, as had been argued. COBE was launched in November 1989. After processing six months' worth of data we had the most accurate measurement of the dipole ever made. But when we removed the dipole, there were no anisotropies. It was not until we had a year's data to process that we began to see a signal. It wasn't noise, and it didn't seem to be systematic error. When we modelled and removed the foreground emission from our Galaxy, it was still there. In the Spring of 1992 the COBE DMR team announced the discovery of anisotropies in the CMB. ('Structure in the COBE Differential Microwave Radiometer first-year maps' by Smoot *et al.* (1992) is the technical version of the discovery, but see *Wrinkles in*

Time by Smoot and Davidson for a more accessible Sherlock Holmes version.)

The anisotropies discovered by the DMR are the more prominent dark and light blurry spots above and below the horizontal plane of the Galaxy in the bottom map of Figure 1. They are the oldest and largest structures ever detected: the oldest fossils of the early Universe. If the theory of the **inflationary Universe** is correct, then these structures are quantum fluctuations and are also the *smallest* structures ever measured. If inflation is correct, we are seeing the Universe as it was about 10^{-33} seconds after the Big Bang.

The DMR discovery of CMB anisotropies can be interpreted as strong evidence that galaxies formed through gravitational instability in a dark-matter dominated Big Bang model. This discovery has been hailed as 'proof of the Big Bang' and the 'Holy Grail of cosmology' and elicited comments like: 'If you're religious, it's like looking at the face of God' (George Smoot) and 'It's the greatest discovery of the century, if not of all time' (Stephen Hawking). I knew that we had discovered something fundamental, but its full import did not sink in until one night after a telephone interview for BBC radio. I asked the interviewer for a copy of the interview, and he told me that would be possible if I sent a request to the *religious affairs* department.

Why are the hot and cold spots in the DMR maps so important? The brief answer is that the spots are too big to be causally connected, and so their origin must be closely linked to the origin of the Universe. Two points are causally connected if their past **light cones** intersect – that is, if light has had time to travel between the two since the Big Bang (see **horizons**). The largest causally connected patch on the surface of last scattering subtends an angle of about a degree, about twice the angular size of the full Moon. Patches of the surface of last scattering smaller than this are called *sub-horizon* patches. Larger patches are termed *super-horizon*. Super-horizon-sized hot and cold spots (all the features in the DMR maps) have not had time to reach **thermal equilibrium**, yet they are at the same temperature. They are too large to be explained by the standard Big Bang model without a specific mechanism to produce them (see e.g. **inflationary Universe**).

The distinction between sub- and super-horizon is important because different mechanisms are responsible for producing the structure on different scales. As explained below, the CMB structure on sub-horizon scales can be produced by sound waves, while to explain super-horizon-sized structures we need to invoke inflation and/or special initial conditions.

How did anything get to be larger than the horizon? Inflation answers this question in the following way. If two neighbouring points in space are in causal contact, they can exchange information and reach thermal equilibrium. Inflation takes two neighbouring points that have been in thermal equilibrium before inflation and, in a brief period of ultra-rapid expansion, makes them recede from each other faster than the speed of light. Their temperatures decrease at the same rate and so remain similar, but the two points are so far away today that they appear (if we do not take inflation into consideration) as if they had never been in causal contact. If inflation is correct, the apparent causal disconnection of the spots in the DMR maps is illusory.

Inflation not only provides an explanation for these apparently acausal features, but offers a plausible mechanism for the origin of all structure – one of the most important missing ingredients in the standard Big Bang model. If inflation is correct, CMB anisotropies originate much earlier than the time of last scattering. The structure in the DMR maps may represent a glimpse of quantum fluctuations at the inflationary epoch about 10^{-33} seconds after the Big Bang. These are objects some 10^{16} times smaller than the atomic structure visible with the best microscopes. Such quantum fluctuations act as the seed perturbations, which grow into the large-scale structure we see around us. The COBE results show us the seeds of galaxies, but they do not explain how the seeds got there (inflation does – see Essay 3).

*What produced the spots?

The bulk properties of the Universe can be deduced by comparing observations of hot and cold spots in the CMB with computer models. However, we cannot begin to extract information if we do not know what these spots are. If we do not know what physical processes produced them, we cannot make models of them. First, we need to look at the physical processes that were operating at the last scattering surface and were thus directly responsible for the CMB anisotropies. The dominant physical effects depend on the scale of the anisotropies. On super-horizon scales (the only scales visible in the DMR maps) **gravity** is the dominant effect, while on sub-horizon scales sound waves (acoustic oscillations of the matter and CMB photons) produce the anisotropies. The explanations of the underlying physics were provided first by Rainer Sachs and Art Wolfe, and more recently by Wayne Hu, Naoshi Sugiyama, Joseph **Silk** and several other CMB theorists. The following discussion makes use of the fundamental concept of horizon to distinguish large super-horizon scales from smaller sub-horizon scales.

The temperature of the background radiation can be influenced by any physical effect that disturbs the density or frequency of **electromagnetic radiation**. There are three relevant phenomena:

Gravity induces gravitational **redshifts** and/or blueshifts (see **Sachs–Wolfe effect**);

Variations in *density* produce heating (compression) and cooling (rarefaction) (see **Sakharov oscillations**);

Velocities can change the temperature of photons by inducing a Doppler shift during scattering processes.

These effects all occur to a greater or lesser extent in different models at the last scattering surface. In other words, the net effect can be thought of as

$$\text{Temperature} = \text{Gravity} + \text{Density} + \text{Velocity} \qquad (1)$$

Gravity produces the dominant effect on super-horizon scales. Since the gravity and density fluctuations we are concerned with here are super-horizon-sized, they are too large to have been caused by the infall of matter or any other physical mechanism. Instead, they are 'primordial' in that they were presumably produced by inflation and/or laid down as initial conditions. On these scales the cold and hot spots in the CMB maps are caused by the redshifting and blueshifting of photons as they escape primordial gravitational potential fluctuations. That is, photons at the last scattering surface lose energy when they leave overdense regions and gain energy when they enter underdense regions. Thus cold spots correspond to overdensities (seeds of superclusters of galaxies), and hot spots to underdensities (seeds of giant voids).

Figure 2 illustrates these effects. This diagram exaggerates the thickness Δz_{dec} of the last scattering surface in order to illustrate the physical effects underlying the spots we see in the CMB. As an observer on the left we see microwave photons coming from the last scattering surface. The largest grey circle subtends an angle of θ. Both this circle and the larger white one beneath it are meant to be at super-horizon scales, while all the other smaller circles are at sub-horizon scales; the grey circles are matter overdensities, while the white circles are underdensities. Overdensities produce gravitational potential valleys,

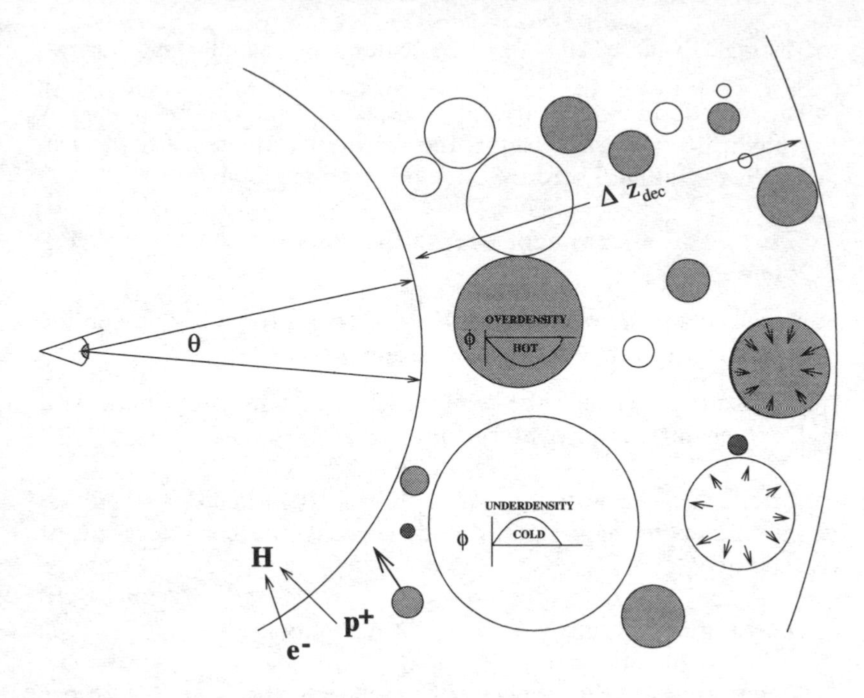

Figure 2 Spots on the last scattering surface at super- and sub-horizon scales.

indicated by the plot of ϕ in the largest grey circle. A potential hill of ϕ is plotted in the largest white circle. The assumed adiabatic initial conditions have hotter photons at the bottoms of potential valleys and cooler photons on the tops of potential hills. These are labelled 'HOT' and 'COLD' in the plots of ϕ. However, while climbing out of the potential valleys, the initially hot photons become gravitationally redshifted and end up cooler than average. Similarly, in falling down the potential hills the initially cooler photons become hotter than average. Thus on super-horizon scales the cool spots in the COBE maps are regions of overdensity (grey circles). Bulk velocities of the matter are indicated by the arrow on the grey spot at the lower left. On sub-horizon scales, matter is falling into potential valleys and falling away from potential hills, producing velocities indicated by the radial arrows in the valley (grey circle) and hill (white circle) on the right. Figure 3 explains how these radial velocities lead to acoustic oscillations. At the lower left of Figure 2, an electron and proton are recombining to form neutral hydrogen, making the Universe transparent. Thereafter photons from the surface of last scattering are free to propagate to the observer.

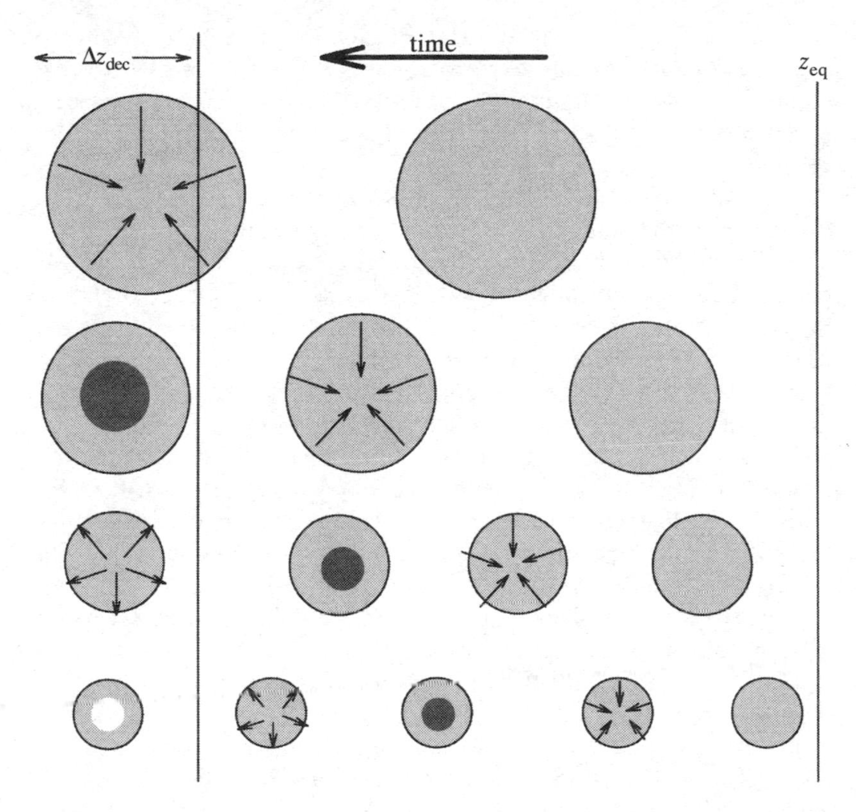

Figure 3 Seeing sound (see the text).

Accompanying the **primordial density fluctuations** are primordial fluctuations in the radiation, but the relation between these two is not obvious. Usually two types of initial condition are considered: *adiabatic* and *isocurvature*. In the more popular adiabatic models the radiation is hotter at the bottom of potential valleys. This means that the gravity and density terms in Equation (1) partially cancel, leading to relatively low-amplitude CMB fluctuations at super-horizon scales. In isocurvature models the radiation is cooler at the bottom of potential valleys, and the gravity and density terms in Equation (1) have the same sign. This produces large-amplitude CMB fluctuations at super-horizon scales which do not seem to fit the data well.

The third term on the right-hand side of Equation (1) is the standard Doppler effect. The velocity can be split into the velocity of the observer and the velocity of the plasma at decoupling. The velocity of the observer produces the large observed dipole, also known as the 'great cosine in the sky' (the middle map of Figure 1). On super-horizon scales we can ignore the velocity of the plasma because no substantial coherent velocities are expected.

On small sub-horizon scales, sound waves – acoustic oscillations of the density and velocity fields – produce the dominant effect. After matter–radiation equality (z_{eq}), the growth of cold dark matter (CDM) potential valleys and hills drives the acoustic oscillations (shown in Figure 3 at four different scales). Since the protons couple to the electrons and the electrons couple to the CMB photons, compression and rarefaction of this 'baryon–photon' fluid creates hot and cold spots that can be seen in Figure 3, which shows how sound waves become visible at the last scattering surface. As in Figure 2, the exaggerated thickness of the last scattering surface is indicated by Δz_{dec}. An observer would be on the left. The grey spots are cold dark matter overdensities (potential valleys) of four different sizes. The radial arrows represent velocities of the baryon–photon fluid. At z_{eq}, when CDM begins to dominate the density of the Universe, the normal matter (coupled to the photons) can begin to collapse into CDM potential valleys smaller than the horizon (notice that the radial infall begins first in the smallest objects). When maximum compression is reached (dark grey) the velocities reverse and we begin to have acoustic oscillations or sound waves. The oscillations stop when the Universe becomes transparent during the interval Δz_{dec}.

When the CMB photons decouple from the matter, the imprint of the oscillations is left in their spatial distribution. Thus we see sound in the snapshot of the Universe visible to us at the last scattering surface. If an overdensity has radial arrows in the interval Δz_{dec}, the CMB photons last scatter off moving plasma and we have a velocity (Doppler effect) contribution to the total power. If an overdensity has a dark or white spot, the CMB photons are compressed or rarefied and we have a density contribution. The top row in Figure 3 corresponds to the largest-scale velocity contribution. It is caught at decoupling with maximum velocity, and contributes power at the angular scale corresponding to the boundary between super- and sub-horizon scales. The second row is the most important. It is caught at maximum compression (a hot spot) when the velocities are minimal. It is the largest angular-scale compression (its angular scale is about 0°.5, and it produces the dominant peak in the power spectrum shown in Figure 6). Due to earlier misconceptions this peak is inappropriately called the first Doppler peak (see

Sakharov oscillations). Underdensities of the same size (not shown here) produce a rarefaction peak (a cold spot). The third row corresponds to a velocity effect similar to that shown in the top row, but here the smaller overdensity has had time to go through an oscillation. The last row is a rarefaction (a white spot) which corresponds to the second Doppler peak. Underdensities of the same size (not shown) would be at maximum compression and contribute equally to this second peak. The 90° phase difference between the velocity and density contributions is the defining characteristic of sound waves.

In a nutshell, we are able to see the sound because the CMB photons are part of the oscillating baryon–photon fluid, and thus trace the oscillations. When you speak, sound waves propagate through the air. When you hear, you decode pressure variations temporally, one after the other. On the last scattering surface, sound can be decoded spatially, high frequencies corresponding to variations on small angular scales. The real last scattering surface is a random superposition of spots, not ordered according to size and not exclusively overdensities as depicted in Figure 3.

To summarise: three effects are responsible for producing CMB anisotropies: gravity, density and velocity. On super-horizon scales, gravity dominates and produces the spots in the DMR maps (visible in the lower map of Figure 1). The photons are gravitationally redshifted as they climb out of valleys or blueshifted as they fall down the hills in the gravitational potential. These primordial hills and valleys were presumably laid down by inflation. The normal Doppler effect produces the dipole. On sub-horizon scales, sound waves – acoustic oscillations of density and velocity – produce anisotropies. Understanding these processes allows us to make models of the CMB power spectrum. These models can then be compared with observational data to determine cosmological parameters.

Determining cosmological parameters from CMB observations

We have described how the CMB fits into the Big Bang model and why it is one of the fundamental pillars of the Big Bang model. In addition, recent theoretical, numerical and observational advances are combining to make the CMB a powerful tool for determining the most important parameters of Big Bang cosmological models. These parameters are measurable quantities relevant to all models, and include:

The Hubble constant H_0 (often given as $h = H_0/100$ km/s/Mpc);

The average density of the Universe, in terms of the cosmological **density parameter** Ω;

The **cosmological constant** Λ.

Determination of these parameters tells us the age, size and ultimate destiny of the Universe. For example, if Ω is less than or equal to 1 the Universe will expand for ever, whereas if Ω is greater than 1 the Universe will recollapse in a hot Big Crunch.

The different contributions to the temperature, as given in Equation (1), depend on these parameters, and on the angular scale studied. So by studying fluctuations on different scales we can attempt to learn about the various models. The CMB **power spectrum** is a way to keep track of the amplitude of temperature fluctuations at different angular scales. For example, Figure 4 translates three simple CMB sky maps into their corresponding power spectra. The first map has just a dipole, and its power spectrum has power only at large angular scales. Smaller spots yield power at smaller scales, as demonstrated by the peaks at large angular frequencies l. Figure 5 is a schematic version of the CMB power spectrum. On large angular scales there is a plateau caused by the Sachs–Wolfe effect. On scales between 0°.1 and 1° there are acoustic oscillations producing the so-called Doppler peaks, and on the smallest scales there is no power because the hot and cold spots are superimposed on others along the line of sight through the finite thickness of the last scattering surface, and therefore tend to cancel out.

There is a new enthusiasm and a sense of urgency among groups of cosmologists making CMB measurements at angular scales between 0°.1 and 1°. Over the next few years their CMB measurements will help to determine cosmological parameters to the unprecedented accuracy of a few per cent, and hence to calculate the age, size and ultimate destiny of the Universe with a similar precision. In such circumstances it is important to estimate and keep track of what we can already say about the cosmological parameters.

To extract information about cosmological parameters from CMB data we need to compare the data with families of models. The angular power spectrum provides a convenient means of doing this. The estimated data and the best-fitting model are plotted in Figure 6. This diagram looks rather complicated and messy because of the substantial observational uncertainties (indicated by the vertical and horizontal error bars through the various data points) and the large number of

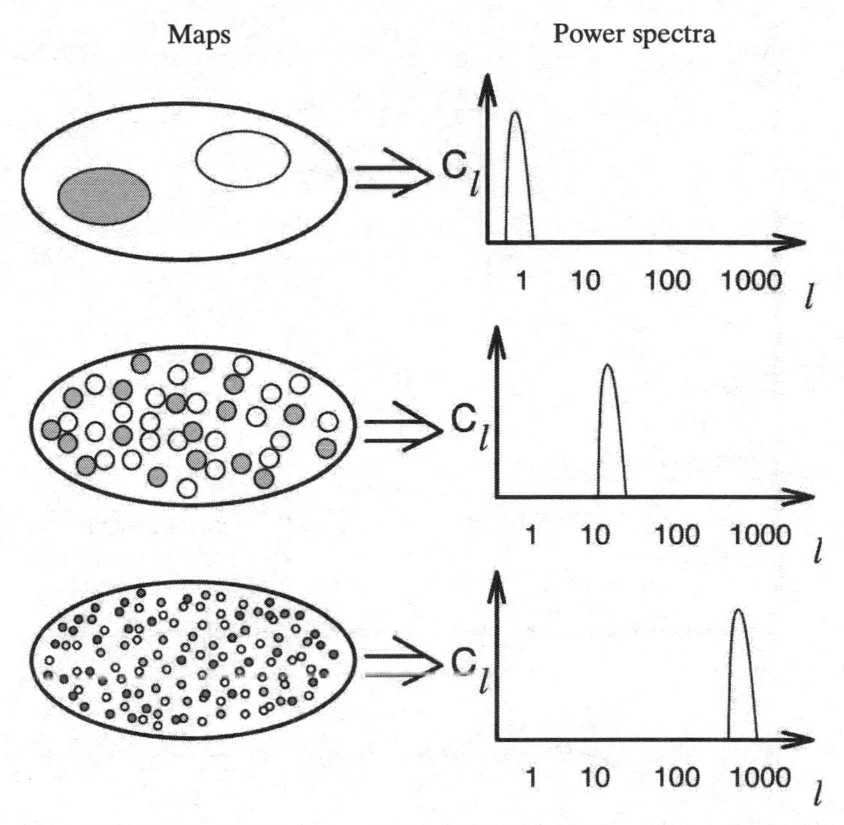

Figure 4 Simple maps and their power spectra. If a full-sky CMB map has only a dipole (top), its power is concentrated on large angular scales ($l \approx 1$). If a map has temperature fluctuations only on an angular scale of about 7° (middle), then all of the power is at $l \approx 10$. If all the hot and cold spots are even smaller (bottom), then the power is at high l.

different sources of data. The important thing is that we can discern a peak such as that represented schematically in Figure 5. Indeed, we can go further than this by using statistical arguments to rule out some models entirely, but it is too early to draw firm conclusions about the values of Ω and h until better data are available.

Before the COBE discovery, there were no data points to plot on Figure 6. The COBE points are at large angular scales on the left, and have been confirmed by the balloon-borne Far Infrared Survey experiment (FIRS) and measurements made from Tenerife. Fluctuations have also been detected at much smaller angular scales (on the right). The measurement of CMB anisotropies is an international effort. The

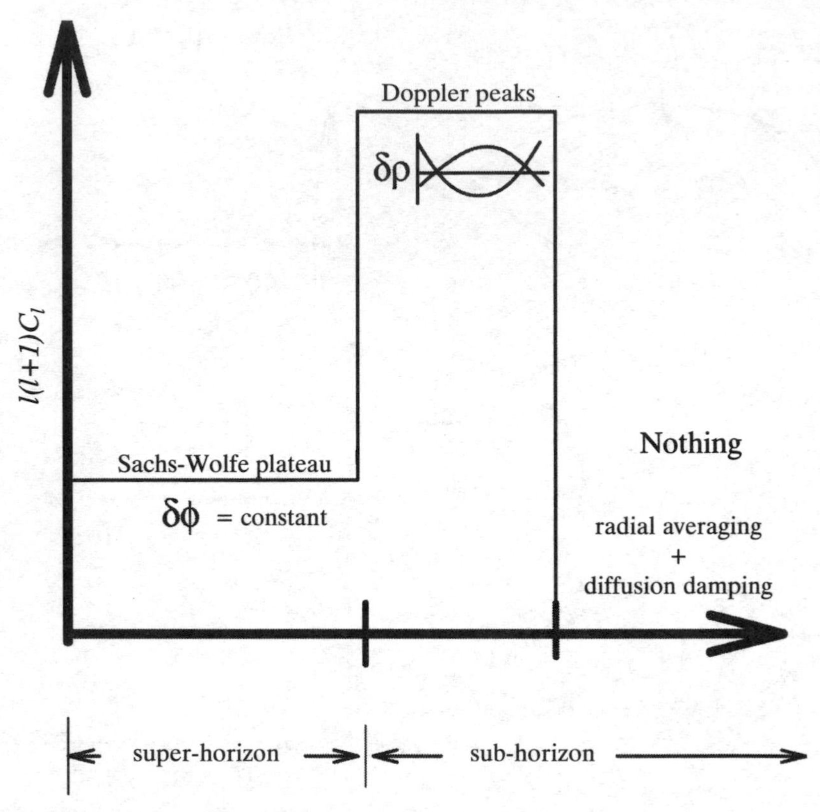

Figure 5 A simplified CMB power spectrum. The CMB power spectrum can be crudely divided into three regions. At large super-horizon scales there is the Sachs–Wolfe plateau caused by primordial gravitational potential fluctuations. The Doppler peaks, on scales slightly smaller than the horizon, are caused by acoustic oscillations (see Figure 3). At smaller scales there is nothing, because small-scale hot and cold spots are averaged along the line of sight through the finite thickness of the last scattering surface. Diffusion damping (photons flowing out of small-scale fluctuations) also suppresses power on these scales.

twelve observational groups that have obtained measurements in Figure 6 are collaborations from many institutions and five countries (USA, England, Spain, Italy and Canada). The frequencies of these observations range between 15 GHz (Tenerife) and 170 GHz (FIRS). Two major types of detector are used: high electron mobility transistors (HEMTs) and bolometers. HEMTs are coherent detectors (like radio receivers, they are sensitive to the phase of the photons) and are used at frequencies up to about 100 GHz. Bolometers are incoherent

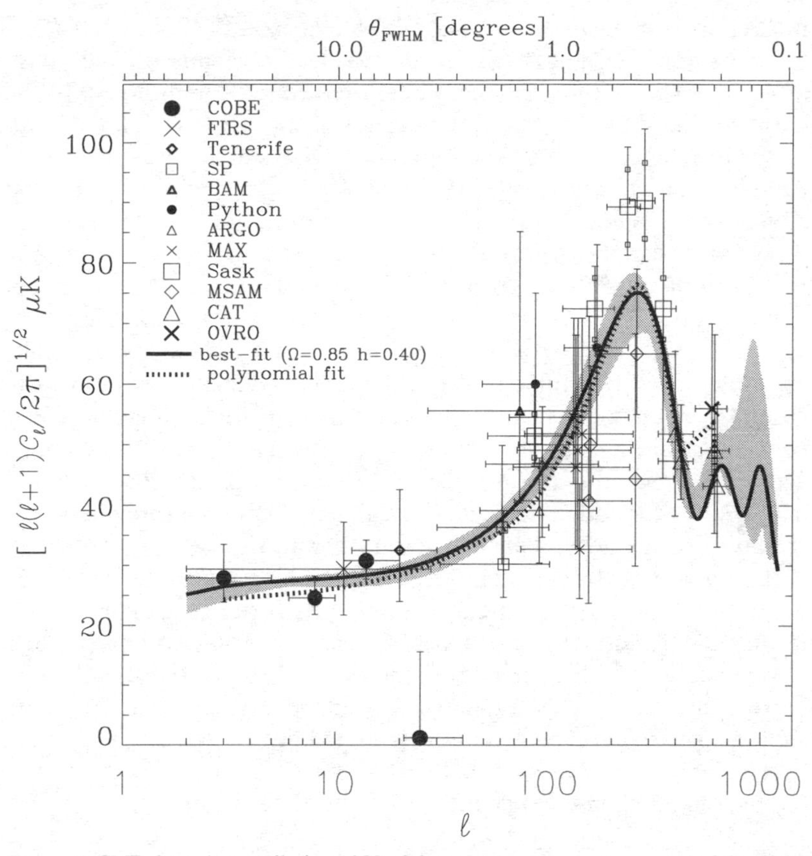

Figure 6 CMB data. A compilation of 32 of the most recent measurements of the CMB angular power spectrum spanning the angular scales 90° to 0°.2 (corresponding to a range of l from about 2 to 600). The angular scale is marked at the top. A model with $h = 0.40$, $\Omega = 0.85$ is superimposed on the data and provides the best fit within a popular family of open and critical-density CDM models bounded by the grey area. The highest peak of the model at $l \approx 260$ is the first Doppler peak (second row of Figure 3) while the smaller peak at $l \approx 700$ is the second Doppler peak (fourth row in Figure 3). Two satellites, MAP and Planck Surveyor, are expected to yield precise spectra for all angular scales down to 0°.3 and 0°.2 respectively.

detectors (very sensitive 'heat buckets', like thermometers) used for high-frequency measurements above about 100 GHz. They can be cooled to around 0.1 K.

The six ground-based observations in Figure 6 were made with HEMTs, while the five balloon-borne observations were made with bolometers. The measurements were made from observing sites all over the Earth: from the high dry Antarctic plateau near the South Pole (SP) to Saskatoon in Canada (Sask), from volcanic islands off the coast of Africa (Tenerife) to the foggy pastures around Cambridge (CAT). Good ground-based sites are at altitudes of about 3 km, balloons drift up to altitudes of about 40 km, while the COBE satellite was at 900 km. Angular scales range from COBE's full-sky coverage to the Cambridge Anisotropy Telescope's (CAT) resolution of about 0°.2. Notice that these two observations probe very different parts of the power spectrum in Figure 6: COBE probes primordial fluctuations at super-horizon scales, while CAT is probing acoustic oscillations between the first and second Doppler peaks at sub-horizon scales. The choice of angular scale or resolution is determined by several factors. At angular scales below about 0°.1 there is presumably no power to measure. At angular scales between about 2° and 90° the ability to distinguish models is small. The vertical extent of the grey band in Figure 6 indicates that the optimal window for making observations aimed at distinguishing between cosmological models is between about 0°.1 and 1°.

Big Bang models, like all models, need to be consistent with observations. A number of variations on the Big Bang theme are contending for the title of most favoured model of structure formation, but each has some problems with matching all the data. If one of these models is correct, then as the CMB data get better the data points in Figure 6 should come to agree more closely with that best-fit model.

It took more than 25 years to detect anisotropies in the CMB. Among the factors which make such measurements difficult are:

The signal is tiny – only 10^{-5} of the average temperature of 3 K. All sources of error have to be reduced below this level. Very-low-noise detectors are needed, and the dipole needs to be removed accurately.

Our Galaxy is a strong source of foreground contamination. Since we cannot yet make measurements from outside our Galaxy, galactic contamination needs to be removed, but it can only be removed to the accuracy with which it has been measured. Galactic emissions, in the form of synchrotron radiation, bremsstrahlung and radiation from dust particles, have frequency dependences that can be used to separate them from the CMB signal. Galactic contamination has a minimum between 50 and 100 GHz. Multi-frequency measurements need to be made in and around this minimum to permit this separation.

Galaxies similar to ours emit similar radiation which contaminates small angular scale measurements (since distant galaxies subtend small angular scales). Some of these extragalactic point sources are particularly problematic since their emission is spectrally flat, like the CMB signal. Supplementary very-high-resolution observations need to be made in order to remove them.

For ground-based and even balloon-borne instruments, the water and oxygen in the atmosphere constitute a strong source of foreground contamination. This contamination limits ground-based observations to low-frequency atmospheric windows. To minimise this problem, high-altitude observing sites with low humidity are sought.

Miscellaneous systematic errors associated with the details of the detectors are notoriously ubiquitous. Examples are errors caused by variations in the Earth's magnetic field, and thermal instability when the detector goes in and out of shadow. Clever design and proper controls can reduce and keep track of such systematic errors.

THE FUTURE

All the excitement about the CMB is because it is measurable. Now that the technology is good enough to detect anisotropies, observational groups are gearing up and much effort is being put into precise measurements of anisotropies at angular scales between about 0°.1 and 1°. The race is on. There is gold in the Doppler hills! Data points are being added to plots like Figure 6 about once a month, and the pace is increasing. A growing community of observers with increasingly sophisticated equipment is beginning to decipher the secrets of the Universe encoded in the CMB. There are more than twenty groups making or planning to make CMB anisotropy measurements. There is a healthy diversity of instrumentation, frequency, observing sites and observing strategies. Current instruments are being upgraded and new detectors are being built. Instruments are being sent up under balloons and, in an effort to reduce systematic errors, even on top of balloons. Instrument packages of 1000 kg which made short balloon flights of a few hours are now being modified and slimmed down to 200 kg to make long-duration balloon flights of a few weeks or even months around Antarctica and across North America. HEMT technology is being extended to achieve lower noise levels at lower temperatures and higher frequencies. Bolometers operating at less than 1 K are pushing for lower frequencies and lower noise. New bolometers designed like spider's webs will reduce cosmic-ray interference.

Groups at Cambridge, Caltech and Chicago are building interferometers which are sensitive to very-small-scale fluctuations. The size and complexity of current interferometers are being increased, and they are being moved to higher and drier sites (such as Tenerife, Spain or the Atacama Desert of northern Chile). This next generation of interferometers should be operational in the early years of the new millennium. In addition to this formidable diversity of ground-based and balloon-borne instrumentation, two new CMB anisotropy satellites are being built: the Microwave Anisotropy Probe (MAP), a NASA satellite, and Planck Surveyor, a European Space Agency satellite. MAP has HEMT detectors with five frequency channels between 22 and 90 GHz, at angular resolutions down to 0°.3. Planck has HEMT and bolometric detectors and nine frequency channels between 31 and 860 GHz, at resolutions down to 0°.2. The Hubble Space Telescope improved angular resolution by a factor of 5 or 10 over ground-based telescopes. The MAP and Planck satellites will improve on COBE by a factor of 20 to 30.

Both MAP and Planck will be launched into orbits six times farther away than the Moon – to a position well away from the thermal and magnetic variations of the Earth and Moon which were the dominant systematic errors in the COBE-DMR data. MAP will have results five years before Planck, but untangling the spatial and frequency dependence of the contaminating foreground signals, as well as the complicated parameter dependences of the models, will certainly be made easier by the higher resolution and broader frequency coverage of Planck. With two new CMB satellites due to be launched in the near future and more than twenty observing groups with upgraded or new instrumentation coming on-line, this is the age of discovery for CMB cosmology.

The DMR results revealed a wholly new class of object. Along with galaxies, quasars, pulsars and black holes, we now have hot and cold spots on the last scattering surface. These objects are in fact over- and underdensities of matter – the seeds of large-scale structure, proto-great-walls and proto-giant-voids. The first high-resolution observations of the first decade of the 21st century will replace the amorphous blotchy spots in the DMR maps with interesting and weird shapes, proto-filaments – individual objects. Their study and detailed characterisation will establish a new branch of astronomy. There will be catalogues with names and numbers. The DMR maps have shown us the seeds, and we are about to embark on what we might call the quantitative embryology of large-scale structures.

A model of the Universe has to be compatible with all reliable cosmological measurements. A coherent picture of structure formation needs to be drawn, not just from CMB anisotropies but from galaxy surveys, bulk velocity determinations, age determinations, measurements of the density of the Universe and the Hubble constant, and from many other cosmological observations. The level of anisotropy measured by the DMR is consistent with the local density field, but the price of this consistency is non-baryonic dark matter – still quite speculative stuff. There is other evidence that some kind of dark matter lurks about: the outlying parts of galaxies and galactic clusters are orbiting too fast to be constrained gravitationally by the visible matter. The question of the existence and nature of dark matter appears increasingly urgent. Research groups are seeking it in the laboratory, in caves, in stars and in the halo of our Galaxy. If it turns out that dark matter does not exist, then some alternative solution, whether it is drastic or anodyne, will have to be found.

The CMB is a newly opened frontier – a new gold-mine of information about the early Universe. Comparing CMB measurements with various cosmological models can already be used to rule out some models, and sharpen our values of the Hubble constant and the density of the Universe. This technique may soon become cosmology's most powerful tool. The angular power spectrum of the CMB will tell us the age, size and ultimate destiny of the Universe as well as details we have not had space to discuss such as re-ionisation, energy injection from decaying particles, rotation of the Universe, gravitational waves and the composition of the Universe.

The biggest prize of all may be something unexpected. We know that our model of the Universe is incomplete at the largest scales, and that it breaks down as we get closer and closer to the Big Bang. It seems very probable that our model is wrong in some unexpectedly fundamental way. It may contain some crucial conceptual blunder (as has happened so many times in the past). Some unexpected quirk in the data may point us in a new direction and revolutionise our view of the Universe on the largest scales. I know of no better way to find this quirk than by analysing increasingly precise measurements of the CMB. Surely this is the Golden Age of cosmology. But there is a caveat:

> The history of cosmology shows us that in every age devout people believe that they have at last discovered the true nature of the Universe.
>
> E.R. Harrison, in *Cosmology: The Science of the Universe*

Further reading

Dicke, R. H., Peebles, P. J. E., Roll, P. G. and Wilkinson, D. T., 'Cosmic black-body radiation', *Astrophysical Journal*, 1965, **142**, 414.

Harrison, E. R., *Cosmology: The Science of the Universe* (Cambridge University Press, Cambridge, 1981).

Lineweaver, C. H. and Barbosa, D., 'What can cosmic microwave background observations already say about cosmological parameters in critical-density and open CDM models?', *Astrophysical Journal*, 1998, **496** (in press).

Penzias, A. A. and Wilson, R. W., 'A measurement of excess antenna temperature at 4080 Mc/s', *Astrophysical Journal*, 1965, **142**, 419.

Smoot, G. F. *et al.*, 'Structure in the COBE Differential Microwave Radiometer first-year maps', *Astrophysical Journal*, 1992, **396**, L1.

Smoot, G. F. and Davidson, K., *Wrinkles in Time* (William Morrow, New York, 1993).

6

The Universe Through Gravity's Lens

Priyamvada Natarajan

Introduction

The **dark matter** problem is one of the most important outstanding questions in cosmology today, because the precise composition and the amount of dark matter determine the ultimate fate of our Universe – whether we continue to expand, begin to contract or start to oscillate. The standard framework of modern cosmology revolves around a small set of defining parameters that need to be determined observationally in order to obtain a complete description of the underlying **cosmological model** of the Universe. These three key cosmological parameters are the Hubble parameter (or **Hubble constant**) H_0, the mass **density parameter** Ω (the total matter content of the Universe, counting both the luminous and dark matter contributions) and the value of the **cosmological constant** Λ. These parameters together define the physical nature and the basic geometry of the Universe we inhabit.

Dark matter is defined as such since it does not emit in any part of the spectrum of **electromagnetic radiation**. It can therefore be probed only indirectly, principally via the gravitational force it exerts on the other masses (**galaxies**, stars) in its vicinity. The mass density inferred by taking into account all the visible matter in the Universe is much less than 1, therefore if $\Omega = 1$, as suggested by models of the **inflationary Universe**, then dark matter is necessarily the dominant component of the Universe and its distribution is expected to have a profound influence on the formation of all the known structures in the Universe.

The first suggestions for the existence of copious amounts of dark matter in galaxies were made in the 1920s. In 1933 Fritz **Zwicky** showed that there was conclusive evidence for dark matter on even larger scales, in galaxy clusters. More than fifty years on, several key issues remain unanswered:

How much of our Universe is truly dark?

What is dark matter composed of?

How is it distributed?

Can we detect the presence of dark matter in indirect ways?

Gravitational lensing has emerged as a powerful means of answering these questions, as it enables mass itself to be detected, as opposed to light emitted. It is an elegant technique, based on very few assumptions, and the only physics required is that of **general relativity**. Lensing can, in principle, tell us about the distribution of mass in galaxies and in clusters of galaxies, and in the near future it might also provide information on still larger-scale structures in the Universe. Although it cannot directly address the question of the nature of dark matter, some lensing experiments can definitely constrain the sizes and the possible distribution of the objects that comprise it, thereby narrowing down the potential candidates.

Several dark matter candidates have been proposed, ranging from 'dark' stars – stellar-mass objects and black holes – to neutrinos, axions and many other exotic species of elementary particle. Stars which have such low masses that they are incapable of igniting the nuclear fuel in their cores, known as *brown dwarfs*, are the favoured candidates for the dark matter component in our Galaxy. In the context of a hot **Big Bang theory**, neutrinos are produced in the early Universe more abundantly than **baryons**, so if they do turn out to possess mass, even though that mass may be very low, they can still contribute significantly to the mass density of the Universe. No cosmologically interesting limits on neutrino masses have yet been obtained, either from high-energy accelerator experiments or from the quest for solar neutrinos. Neutrinos therefore remain a viable dark matter candidate on large scales (see **weakly interacting massive particle**; **massive compact halo object**).

In our own Galaxy, evidence for the presence of dark matter comes from the observed motion of neutral hydrogen gas clouds. These clouds of un-ionised hydrogen gas follow their own orbits around the Galaxy. If they were to move only under the influence of the gravity of the visible mass, then outside the optical limits of the Galaxy their speeds ought to fall off as the square root of their distance from the galactic centre. However, these outlying clouds, detected at radio wavelengths (1400 MHz), are observed to have roughly the same orbital speed all the way out, even beyond the optically visible limit, implying the existence of an extended and invisible dark halo (see **rotation curves**). The orbital motions of the **globular cluster** systems and the small satellite galaxies orbiting our own are also consistent with the presence of an extended dark halo that extends much farther than either the outermost stars or the limits of X-ray emissions from the hot gas that permeates our Galaxy (see **X-ray astronomy**).

Gravitational lensing theory

Gravity is one of the **fundamental interactions**. Because it acts at long range, it is essential to the understanding of almost all astrophysical phenomena. Albert **Einstein**'s theory of general relativity places the gravitational force in a physical context by relating it to the local properties of **spacetime**. The equivalence principle and the **Einstein field equations** form the core of the theory of general relativity. The *equivalence principle* is the statement that all objects of a given mass fall freely with the same acceleration, along trajectories called *geodesics*, regardless of their composition. The **curvature of spacetime** – any departure from flatness – is induced by the local presence of mass. In other words, gravity distorts the structure of spacetime. Einstein's field equations relate the curvature of spacetime called the **metric** to the distribution of mass and the energy content of the Universe. As a consequence, the total matter content is what determines the evolution and fate of the Universe.

The presence of mass concentrations like massive galaxies or clusters of galaxies causes light rays travelling from background sources (typically, distant galaxies or quasars) to be deflected, not unlike the bending effects caused by an optical lens. The amount of deflection produced is directly proportional to the 'strength' of the lens, which in this case is the mass, as well as to the relative orientation of the lens to the object emitting the light. The propagation of light through a lumpy Universe can easily be understood by drawing an analogy with geometrical optics – the study of the propagation of light through media with differing densities.

It is possible to make a series of simplifying assumptions to enable us to understand the lensing phenomenon. First, we assume that light propagates directly from the source to the lens, and from the lens to the observer. Second, the Universe is taken to be as described by a given mathematical prescription for the underlying spacetime, which in this case is what is called the **Robertson–Walker metric.** Finally, the 'thickness' of the region in which the photons passing through are affected is assumed to be very small compared with the total distance they travel. Thus, we assume that the lens can be approximated by a small perturbation to a locally flat spacetime, and also that the perturbation induced by the gravitational potential of the lensing mass along the line of sight is small. (The *gravitational potential* represents the amount of energy that the light has to expend in escaping from a concentration of mass – a *potential well*.) The propagation of light can then be calculated in terms of an effective refractive index n, as in geometrical optics, where the path of a light ray is deflected when it crosses the boundary between two media with different properties. As in geometric optics, the propagation of light through the medium (in this case the potential) is then simply a function of the 'effective refractive index' of the lens.

We find that light slows down in a potential well, just as it slows down when it passes from one medium into a denser one. The presence of the potential well causes a deflection from a straight line in the direction of propagation by some angle, say α. This angle is given by an integral, along the path of the light ray, of the gradient of the refractive index n evaluated perpendicular to the path. Since all deflection angles are assumed to be small, the computation is along the unperturbed ray. This now means that, for any gravitational lens, all that matters is the column density of mass of the lens enclosed within a cylinder along the line of sight. This approximation is often referred to as the *thin lens approximation*. The typical lensing geometry is shown in Figure 1, where the angles θ_s, α and θ_i define the position of the source from which the light is emitted, the deflection as seen by the observer at point O, and the image position at point I. The three corresponding **angular-diameter distances** denoted by D_{ds}, D_d and D_s are also shown, where the subscript d refers to the deflecting lens and s to the source. The solutions of the *lens equation*

$$\theta_i = \theta_s + (D_{ds}/D_s)\alpha$$

help to determine the mass profile of the deflector if the image positions and relative magnifications are known.

It can be seen from Figure 1 that, if, for instance, $\theta_s = 0$ for all θ_i, then all rays from a source on the optic axis focus at the observer, and the appropriate lens has a uniform mass density per unit area. In most cases, multiple images of the source are seen by the observer only when the surface mass density somewhere within the lens exceeds a critical value, say Σ_{crit}. This happens typically within a small central region, whose extent is described by the *Einstein radius* θ_E. The critical value of the mass density per unit area of the lens and the Einstein radius can be

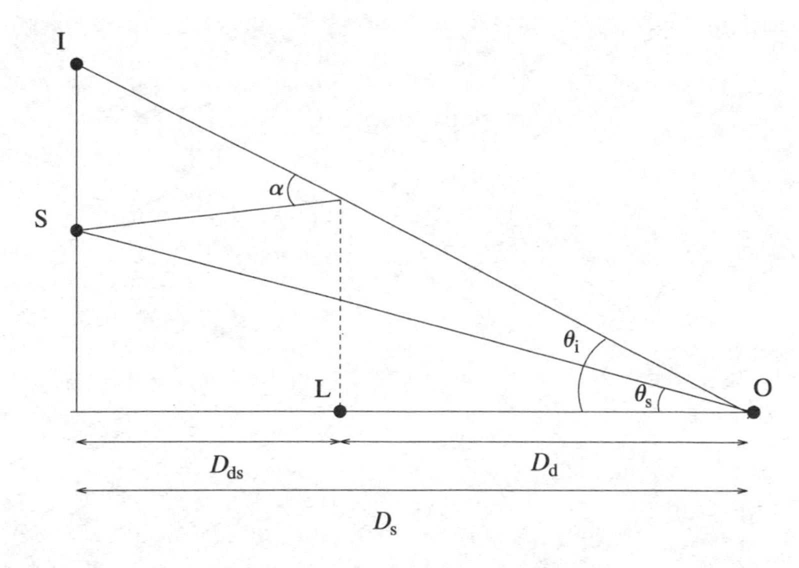

Figure 1 The deflection angles in lensing: a light ray from the source S is expected to travel along the path SO in the absence of any deflecting mass, but the presence of the lens L causes deflection in the path by the angle α, yielding an image at position I.

used to define an effective lensing potential on the plane of the sky. However, in most cases the source lies behind the non-critical region of the lens, in which case no multiple images are produced; instead, the images are magnified and their shapes are distorted. Since the deflection angle is proportional to the slope of the mass distribution of a lens, the scale on which only magnification and weak distortion occur is referred to as the *weak regime*.

The effect of lensing can be thought of physically as causing an expansion of the background sky and the introduction of a magnification factor in the plane of the lens. We are often interested in the magnification of a particular image, given an observed lensing geometry. Lensing conserves the surface brightness of a source along a light ray. The magnification factor of an image is therefore simply the increase in its solid angle on the sky.

For many lens models a source is significantly magnified, often by factors of 2 or larger, if it lies within the Einstein radius θ_E. An *Einstein ring*, such as that shown in Figure 2, can be formed exactly on the Einstein radius. The Einstein radius therefore marks the dividing line between sources that are likely to be multiply imaged, and those which

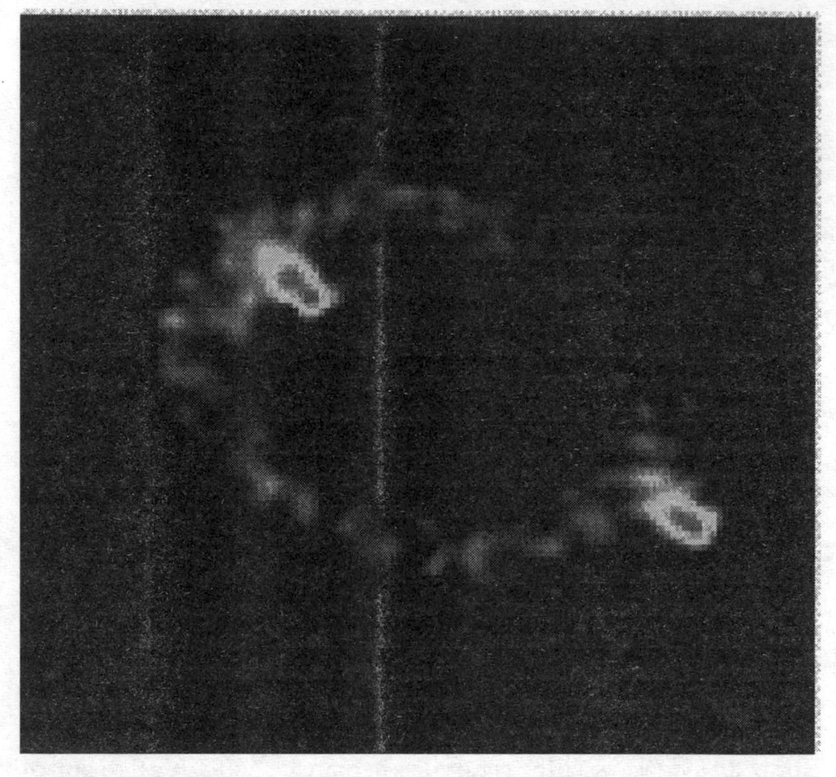

Figure 2 An example of a galaxy, MG 1131+0456, that is lensed into a complete circle due to gravitational bending of light caused by the presence of a cluster of galaxies along the line of sight, producing an Einstein ring.

are singly imaged. Figure 3 shows how the precise location of the source with respect to the lens determines the type of lensing and the observed lensing effects.

For instance, faint circular sources that fall within the strong regime are often seen as highly elongated, magnified 'arcs', whereas small deformations of the shape into ellipses are produced in the weak regime. Therefore, looking through a lens, from the observed distortions produced in background sources (given that the distribution of their intrinsic shapes is known in a statistical sense), a map of the intervening lens can be reconstructed. This lens-inversion mapping provides a detailed mass profile of the total mass in a galaxy or a cluster of galaxies. The comparison of this mass distribution, obtained by solving the lens equation for a given configuration, with that of the observed

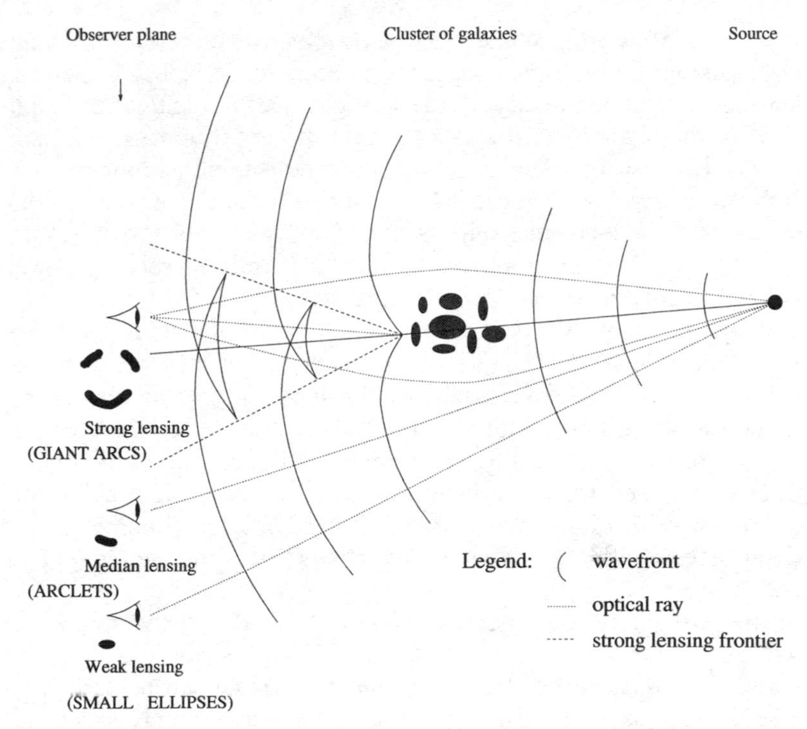

Figure 3 Different types of lensing, illustrated for a cluster of galaxies as the deflecting mass. The region marked by the short-dashed lines is where strong lensing can occur, producing either multiple images or highly distorted and magnified arcs. The observable effects are considerably weaker for light rays passing farther away from the central dense part of the cluster.

light distribution enables constraints to be put on the amount of dark matter that is present in these systems. At present, lensing-based galaxy mass models obtained in this fashion seem to indicate that up to 80% of the mass in a galaxy is probably dark.

Dark matter in galaxies

When a dark mass, like a brown dwarf or a MACHO, passes in front of a background star, the light from the star is gravitationally lensed. This lensing is insufficient to create multiple images, and what is seen is simply a brightening of the background star – a phenomenon known as *microlensing*. Since MACHOs are composed of baryons, the detection of microlensing events can help to determine how much dark matter is in the form of baryons. While the scales involved in microlensing are not large enough for multiple images to be observed, as expected in strong lensing events, the intensity of the starlight can be significantly amplified, showing up as a sharp peak in the light curve of the background star.

This was first suggested as a potentially detectable phenomenon by Bohdan Paczyński at Princeton University in 1986. The image splitting caused by these solar-mass objects in our Galaxy is not observable, since the expected Einstein radius is measured in milli-arcseconds – well below the current resolution of optical telescopes. Paczyński argued that, by continuously monitoring the light curves of stars in the Large Magellanic Cloud (LMC), a satellite galaxy to our own, we would be able to observe increases in brightness that took place whenever a source in the LMC transmitted through the Einstein radius of a MACHO in our Galaxy (see Figure 4 for an example of the amplified light curve when a microlensing event is in progress). Since, inside the Einstein radius, magnification can occur by factors of 2 or larger, microlensing is easily detected as a sudden rise in the light intensity, independent of the observed frequency.

The probability a star being lensed by MACHOs distributed in the outskirts of our Galaxy can be estimated by modelling the lenses as point masses. The quantity needed to compute the number of expected events is referred to as the *optical depth to lensing*, which is simply the chance that a given star in the LMC lies within the Einstein radius of a lens at a given time. The optical depth is calculated along the line of sight, and it depends on the total assumed number density of MACHO lenses.

There are currently several observational research groups searching for microlensing signatures in LMC stars and stars in the galactic bulge by continuously monitoring the light curves of millions of stars. Looking towards the centre of our Galaxy, we seek to detect MACHOs in the disk, and looking in the direction of the LMC we seek MACHOs distributed in the galactic halo. Several large international collaborations, known as MACHO, EROS, DUO and OGLE, are currently engaged in this venture. When the MACHO group analysed the data from their first-year run, consisting of almost 10 million light curves, they detected one event with a significant amplitude in the magnification, and two with modest magnifications. They estimated the total mass of MACHOs inside a radius of 50 kiloparsecs to be around 8×10^{10} solar masses. This result was found to be reliable and fairly independent of the assumed details for the underlying halo model. However, it is clear that the fractional contribution to the halo mass from these MACHOs is small. For instance, within the mass range of 3×10^{-4} to 6×10^{-2} solar masses,

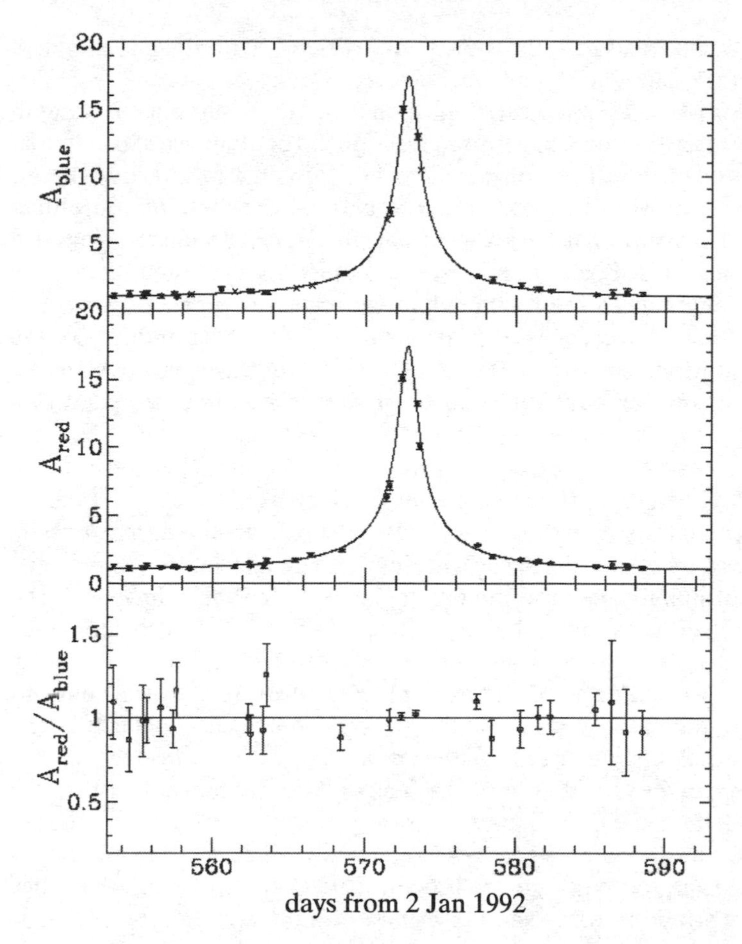

Figure 4 A typical light curve of a microlensing event. The upper and middle curves show the intensity of the microlensed foreground star, in arbitrary units, at blue and red wavelengths. The intensity is independent of frequency (within the limits of observational accuracy), as shown by the lower plot, in which one set of data is divided by the other. Note the sharp rise in amplitude during lensing, and the rapid fall-off after a few days. This light curve is of one of the events seen by the research team known as MACHO, investigating microlensing events towards the centre of the Galaxy.

MACHOs account for significantly less than 50% of the halo. At the end of their second year of accumulating data, now with six to eight events, they estimated a halo fraction of 30% to 90% in the mass range 0.1 to 0.4 solar masses. The picture that emerges of our Galaxy in the light of the results from these microlensing searches is that, perhaps, a significant fraction of the dark matter content of our halo is baryonic, and is distributed in stellar-mass objects.

Lensing by a galaxy, with a typical mass of 10^{12} solar masses, instead of by a star of 1 solar mass, produces splittings of an arc second or so between the multiple images. The first lensing galaxy, designated 0957+561A, was discovered in 1979, and as of early 1998 more than 30 such gravitational lenses were known. Since the lens magnifies a faint background galaxy or quasar, it acts as a gravitational telescope and enables us to see farther than we can ever probe using either ground-based telescopes or instruments in space. For multiple image configurations, since the different light-ray paths that correspond to the different images have different lengths, relative time delays can be measured if the source is variable.

A successfully 'inverted' lens model can be used to measure the Hubble constant H_0, the precise value of which has implications for both the age and the size of the Universe. H_0 can be determined from lensing, in theory, by measuring two quantities: the angular separation between two multiple images, and the time delay between those images. If the source itself is variable, then the difference in the light travel time for the two images comes from two separate effects: the first is the delay caused by the differences in the path length traversed by the two light rays from the source, known as the *geometric time-delay*, and the second is a general relativistic effect – the *gravitational time-delay* – that causes a change in the rate at which clocks tick as they are transported through a gravitational field. And since the two light rays travel through different portions of the potential well created by the deflecting lens, the clocks carrying the source's signal will no longer be synchronised when they emerge from the potential. Once these time delays, the image separations and their relative magnifications are measured, the distance to the lens and the source can be deduced from the lens equation, which then allows an independent estimate of H_0 to be made.

Quasars are ideal subjects for lensing since they are very luminous, lie at cosmological distances and hence have a high lensing probability. The first multiply imaged quasar, QSO 0957+561A,B, was discovered in 1979 by Walsh, Carswell and Weymann. The lensing of this distant quasar at a **redshift** of $z = 1.41$ is caused by a bright elliptical cluster galaxy at $z = 0.34$. This system has been continuously monitored for several years, since it was thought to be an ideal candidate for estimating H_0 from the measured time-delay. Detailed modelling has provided estimates of the properties of the lensing galaxy (such as its mass and density profile) which are in good agreement with the values obtained from independent dynamical studies. For the 0975+561 system, there has been some disagreement between different groups that have attempted to measure the time-delay from the offsets of the light curves of the two images, leading to two estimates of the Hubble constant that differ by 20%. At present there are several systematic surveys under way aimed at detecting both large and small multiple-imaging lenses in the optical and radio wavebands. Therefore, while lensing is at present unable to provide a precise measurement of the Hubble constant on the basis of the candidate multiple image systems detected so far, perhaps the ideal 'golden lens' is waiting to be discovered.

Massive foreground galaxies can also lens fainter background galaxies, and this effect can be used to examine several interesting issues. The frequency of galaxy–galaxy lensing provides a glimpse into the redshift distribution of galaxies, and the distribution of mass at high redshifts, and gives us an idea of typical mass distributions in galaxies. Galaxy–galaxy lensing is expected to produce mainly weak effects, such as an apparent increase in the statistically small likelihood of a ring of faint background galaxies occurring around bright foreground galaxies. A tentative detection of such a signal has been reported, and the results seem to indicate that isolated galaxies have very large dark halos extending out to around a hundred kiloparsecs from their centres. Dynamical estimates of the mass distribution of isolated, non-cluster galaxies obtained by mapping the motion of satellite galaxies in orbit around them also seem to indicate that, while luminous matter dominates in the inner regions of galaxies, in the outer regions dark matter can constitute up to 90% of the total mass.

Dark matter in clusters of galaxies, and beyond

Clusters of galaxies are the most recently assembled and largest structures in the Universe. Clusters are more complex systems and harder to understand than stars, for instance, since their formation necessarily depends on the initial cosmic conditions. A typical rich cluster (see **large-scale structure**) contains roughly a thousand galaxies, plus gravitationally bound, hot, X-ray emitting gas; and there is strong evidence for the presence of significant amounts of dark matter (comprising about 90% of the total mass of the cluster).

The currently accepted theories for **structure formation** in a Universe dominated by cold dark matter postulate that dark haloes essentially seed the formation of visible galaxies. Cosmic structures are also expected to build up hierarchically, small objects forming first and then aggregating together, driven primarily by gravity, to form larger units. In the standard picture, each galaxy forms within a dark halo as a result of the gas collapsing, cooling and fragmenting to form stars. It is believed that when galaxies, along with their dark haloes, hurtle together to form a cluster, the individual haloes merge into a large, cluster-scale dark halo.

Lensing of background galaxies by clusters can be divided into strong lensing, in which giant arcs are observed, and weak lensing, in which images of background galaxies are weakly distorted, producing 'arclets' (see Figure 3). For a general lens model the number of images obtained from a compact source is odd: one image is obtained if the source is far away, but as the distance decreases it crosses curves known as *caustics*. Every time a caustic is crossed, the number of images increases by two. Giant arcs are observed because the magnification of a source is greatest when it lies on a caustic. Giant arcs may be used to investigate the mass distribution in clusters, in much the same way that the lens model inversion method can reveal the mass distribution in galaxies. There are now several successfully modelled lensing clusters, where the mass maps obtained agree well with those determined from the clusters' X-ray emission and by applying the **virial theorem** to the motions of cluster galaxies.

Strong and weak lensing

For weak lensing by an extended lens, and in the thin-lens approximation, ray-tracing methods borrowed from geometric optics may be used to map objects from the source plane into the image plane in the process of solving the lensing equation. Several properties of lensing can be used to refine this mapping:

The conservation of surface brightness, as in conventional optics;

The achromatic nature of lensing (i.e. lensing effects are independent of the frequency of the light emitted by the source);

The fact that the deflection angle does not vary linearly with the impact parameter.

Lensing produces two distinct physical effects: the convergence or magnification (κ) is the focusing term that represents simple magnification produced by matter enclosed within the beam; $\kappa > 1$ corresponds to strong lensing, which gives rise to multiple images and arcs. The second effect is the *shear* (γ), which is the anisotropic distortion of images that lie outside the beam produced by the gradient of the potential; $\kappa \cong 0$ and $\gamma > 0$ corresponds to weak lensing, which gives rise to distorted images (arclets) of the faint background sources. The total amplification is a sum of the contributions from both these effects.

Strong lensing is observed in the multiply imaged region where the surface mass density, Σ, exceeds Σ_{crit}. The number of multiple images is determined by the precise configuration, the redshift distribution of the sources (which is in general unknown) and an underlying cosmological model. Giant arcs have been observed around some 30 clusters, primarily by the exquisite imaging capabilities of the **Hubble Space Telescope** (HST). Giant arcs, which are typically images of spiral galaxies at high redshift, are defined as having an axis ratio (the ratio of the long axis to the short axis) in excess of 10. The curvature of the arc is a measure of the compactness of the mass distribution of the lensing cluster, since the radius of the arc corresponds roughly to the Einstein radius. The rotation curves along the length of arcs have been mapped for the Abell (dense) clusters Abell 2390 and CL 0024 and found to be flat, indicative of the presence of a dark halo. In principle, if the true luminosity of the lensed galaxy is known, this technique can be used to extend the **extragalactic distance scale** to objects with very high redshift.

Detailed modelling of cluster cores requires the following ingredients: arc positions, the number of merging images and whether this number is odd or even, arc widths, shapes and curvature to constrain the location of critical lines on the image plane. Given one or more measured redshifts of the arcs, the mass enclosed within the arc can then be accurately estimated, enabling the lens model to be refined. Many cluster cores have been successfully studied from their strong lensing features: Abell 370, Abell 2218 (see Figure 5), AC 114 and MS 2137-223 to name a few. The HST's imaging power uniquely helps in the identification of multiple images, so these models can be used to assess the smoothness of the dark matter distribution. The results obtained with these models demonstrate that the total mass distribution in a cluster closely follows the luminous mass. The overall ratio of the total mass to the total light measured in the visual band in solar units (i.e. in terms of the Sun's mass and luminosity) ranges from 100 to 300, in good agreement with the values of 150 to 250 obtained by independent methods.

In weak lensing by clusters, single images are obtained, but they are sheared as well as magnified. The deformation in shape produced by the lens can be related directly to the contributions from the deflecting mass if the shape of the source is known, but unfortunately this is rarely the case. We therefore have to proceed by statistical methods, assuming

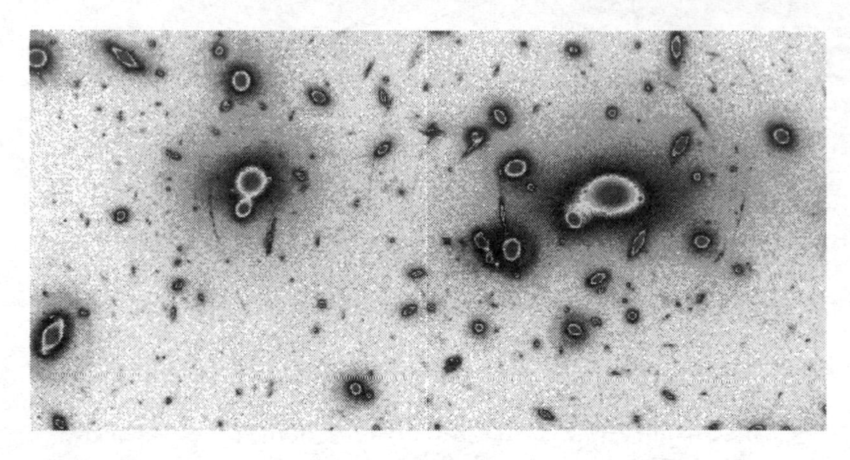

Figure 5 Cluster lensing by Abell 2218: a Hubble Space Telescope image of the lensing cluster at a redshift of $z = 0.175$. Note the large number of highly distorted arcs in the central regions around the two bright galaxies, and the weakly sheared images at the outskirts. Abell 2218 has at present the best constrained mass model since a large number of the arcs now have measured spectroscopic redshifts.

that there is a distribution of shapes. An elegant 'inversion procedure' can be used to obtain a map of the mass density in the plane of the lens from the statistics of these sheared shapes. This map is only relative, since a uniform sheet of dark matter will produce no detectable shear. The mass density obtained by this method is therefore only a lower limit to the true mass: if a uniform sheet of material were added, the observed results would not change.

Several variants and refinements of this basic scheme have been developed and successfully applied. The total amount of matter that is suggested by these measurements is such that the mass-to-light ratio typically lies in the range 200–800 solar units. These values are consistent with estimates obtained on comparable scales from X-ray observations. Since the mass-to-light ratio measured for the luminous parts of galaxies ranges from 1 to 10 solar units, indicating that large amounts of dark matter must be present in clusters, as first proposed by Fritz Zwicky. While most inferred total mass distributions roughly follow the distributions of luminous matter, some clusters seem to have a more centrally concentrated mass distribution than is traced by the galaxies, while others have mass distributions that are much smoother the than the light distribution. Aside from providing mass estimates for individual clusters independently of any assumptions made about their dynamical state, the ultimate goal is to determine the relative numbers of clusters of different masses, since that is a strong test of the underlying cosmological model.

Some recent research has focused on combining the information obtained for a cluster in the strong and weak lensing regimes to build composite mass models. One question that has been tackled is that, if all individual galaxies have massive and extended dark haloes, then what is the fate of these haloes when the galaxies hurtle together to form a cluster? What fraction of the dark matter gets stripped and redistributed? By applying lensing techniques to a very deep, wide-field HST image of the cluster AC114, it is found that on average a bright cluster galaxy has only two-thirds the mass of a comparable non-cluster counterpart, indicative of mass-stripping having occurred. The halo size is also much more compact than that of an isolated galaxy. The conclusion at present is that only 10% to 15% of the total mass of a cluster is associated with the member galaxies, and the rest is probably distributed smoothly throughout the cluster.

Since gravitational lensing is sensitive to the total mass enclosed within a cylinder along the line of sight, we can potentially reconstruct the **power spectrum** of mass fluctuations that over time have been amplified by gravity, leading to the formation of massive **large-scale structures**. In the standard scenario, very massive objects like superclusters and filaments are expected to form, and they can be probed by the weak lensing signal they induce in background galaxies. In this case it is not the surface mass density that is reconstructed, as with clusters, but rather the power spectrum of density fluctuations. The distortions that are measured can be related to the fluctuations of the gravitational potential along the line of sight. At present, there have been no unambiguous detections of shear on scales larger than clusters, but the prospects are encouraging.

Conclusions and future prospects

Great strides have been made in probing dark matter using gravitational lensing to map the mass distributions of galaxies and clusters of galaxies. Theoretical progress in the future is expected primarily in the field of improved mass-map reconstruction techniques and their applications to probe the mass distribution in galaxies, clusters and other large-scale structures. Extending existing methods to detect coherent weak shear induced by still larger-scale structures like filaments and superclusters is the next step. In order to make any further observational headway in the detection of weak shear induced by the intervening large-scale structure, we need wide-field images that probe down to much fainter magnitudes. The new generation of instruments – including the Hubble Advanced Camera for Exploration, due to be installed on the HST in 1999, and the large-collecting-area mosaic CCD detectors currently under construction – are ideally suited for detecting shear to high precision. Lensing has provided a wealth of astrophysical applications. The most significant have been:

Limits have been placed on the baryonic dark matter content of our Galaxy;

The properties of individual lenses can be used to refine the values of cosmological parameters – the Hubble constant H_0, the cosmological constant Λ and the density parameter Ω;

Lensing has provided an independent way of measuring the masses of galaxies and clusters of galaxies that is independent of any assumptions made about the dynamical state of the system;

It simulates a giant gravitational telescope that offers a view of the distant Universe that would otherwise remain inaccessible.

It has provided essential clues to the evolution of galaxies by enabling the mass profiles (inferred from lensing) in dense environments like cluster cores to be compared with those of isolated, non-cluster galaxies.

Further reading

Blandford, R. and Narayan, R., 'Cosmological applications of gravitational lensing', *Annual Reviews of Astronomy and Astrophysics*, 1992, **30**, 311.

Fort, B. and Mellier, Y., 'Arc(let)s in clusters of galaxies', *Astronomy and Astrophysics Review*, 1994, **5**, 239.

Kneib, J.-P. and Ellis, R. S., 'Einstein applied', *Astronomy Now*, May 1996, p. 435.

Walsh, D., Carswell, R. F. and Weymann, R. J., '0957+561A,B: Twin quasi-stellar objects or gravitational lens?', *Nature*, 1979, **279**, 381.

II
Key Themes and Major Figures

A

Absorption line A narrow feature at a particular wavelength in the spectrum of **electromagnetic radiation** emitted by an object; it indicates that much less energy is being received at that wavelength than at others. Absorption lines are usually produced by a particular atomic process, such as when an electron jumps from one energy state to another, higher state. Since, according to **quantum physics**, these states occur only at discrete levels, the electron needs to absorb a photon of a specific wavelength in order to perform the jump.

The spectrum of radiation passing through some form of matter therefore suffers a sharp deficit as photons of particular wavelengths are preferentially removed. Jumps between different energy levels correspond to absorption at different wavelengths, so a series of lines can be produced. The resulting pattern is a precise diagnostic of the atomic constituents of matter (see **spectroscopy**). The spectrum of optical radiation from the Sun, for example, contains thousands of absorption lines (*Fraunhofer lines*), the analysis of which yields the composition of the solar atmosphere.

In cosmological applications not such a huge number of different lines can be observed, but studies of absorption lines are still very important. In particular, a series of lines associated with the transitions of electrons between different states of the hydrogen atom – the ultraviolet *Lyman series* – can be used to probe the properties of matter along the line of sight to distant **quasars**. The brightest Lyman line, known as *Lyman alpha*, occurs at a wavelength of 121.6 nm. Quasars usually display a very strong **emission line** at this wavelength, caused by very hot material, but, because quasars are often at high **redshifts**, this feature is shifted into the optical part of the spectrum as seen by the observer. More importantly, however, clouds of cool neutral hydrogen gas will absorb radiation at the Lyman alpha frequency. If these clouds lie on the line of sight from the observer to the quasar they will be at lower redshifts, so the absorption lines in an observed spectrum appear to be at shorter wavelengths than the emission line. The spectrum of a quasar therefore shows a sharp peak corresponding to the emission feature, and a series of absorption lines at shorter wavelengths, each one corresponding to a single intervening cloud of absorbing material. A single spectrum may contain scores of such lines, usually called the *Lyman alpha forest*. Statistical studies of these lines indicate that most of the systems responsible for producing them are very small, and may be rather wispy in structure, but there are also larger systems – called *damped systems* –

which may contain about as much mass as normal **galaxies**. There is also some evidence that absorbing material may be distributed in sheet-like structures reminiscent of the **large-scale structure** of the Universe observed in the distribution of galaxies. Because the Lyman alpha forests are seen in the spectra of objects at large redshifts, and therefore at such early times, they may help to resolve the problem of cosmological **structure formation**.

SEE ALSO: **intergalactic medium**.

ABUNDANCE see **light element abundances**

ACTIVE GALAXY, ACTIVE GALACTIC NUCLEUS (AGN) There are different kinds of active galaxy, but they are all characterised by the prodigious amounts of energy they emit, often in many different parts of the spectrum of **electromagnetic radiation**, from radio to X-ray wavelengths. This highly energetic behaviour sets them apart from the so-called normal **galaxies**, whose energy output is largely accounted for by normal stellar radiation. Moreover, much of the energy broadcast by active galaxies is associated with a relatively small central region of the galaxy, called the *nucleus*. The term active galactic nucleus (AGN) is therefore often used to describe these regions. Sometimes the central nucleus is accompanied by a *jet* of material being ejected at high velocity into the surrounding **intergalactic medium**. Active galaxies include Seyfert galaxies, radio galaxies, BL Lac objects and **quasars**.

Seyfert galaxies are usually spiral galaxies with no radio emission and no evidence of jets. They do, however, emit radiation over a continuous range of frequencies, from the infrared to X-rays, and have strong and variable **emission lines**.

Most *radio galaxies*, on the other hand, are elliptical galaxies. These objects are extremely dramatic in their appearance, frequently having two *lobes* of radio-emitting material extending far from opposite sides of a central compact nucleus. There is also sometimes the appcarance of a *jet* of material, extending from the core into the radio lobes. It appears that material is ejected from the nucleus along the jet, eventually being slowed down by its interaction with the intergalactic medium, which is what gives rise to the radio lobes. The central parts of radio galaxies seem to have properties similar to those of Seyfert galaxies.

BL Lac objects have no emission lines, but emit strongly in all wavebands from radio to X-ray frequencies. Their main characteristic is their extremely strong and rapid variability. (They were first identified as variable stars – the name is an abbreviation of BL Lacertae, a variable-star designation.) A possible explanation for these objects is that we are seeing a jet of material travelling head-on at close to the velocity of light. This would account for the rapid variability, because **special relativity** suggests that the observed timescale should be shortened in this situation. If the radiation from the jet is beamed towards the observer, then it would also be expected to swamp the emission lines we would otherwise expect to see in the spectra of BL Lac objects.

The various kinds of active galaxy were discovered at different times by different people, and were originally thought to be entirely different phenomena. Now, however, there is a unified model in which these objects are all interpreted as having basically similar structures but different orientations to the observer's line of sight. The engine that powers the activity in each case is thought to be a **black hole** of up to about 100 million solar masses. This seems very large, but it is just a small fraction of the mass of the host galaxy, which may be a thousand times larger. Material surrounding the black hole is attracted towards it and undergoes a process of accretion, gradually spiralling in and being swallowed. As it spirals in, it forms a so-called *accretion disk* around the black hole. This disk can be very hot, producing the X-ray radiation frequently observed coming from AGNs, but its presence prevents radiation from being transmitted through it. Radiation tends therefore to be beamed out of the poles of the nucleus, and does not appear from the equatorial regions, which are obscured by the disk. When the beamed radiation interacts with material inside the host galaxy or in the surrounding medium, it forms jets or radio lobes. By considering how the thickness of the disk, the size of the 'host' galaxy, the amount of gas and dust surrounding the nucleus, and the orientation at which the whole system is viewed can all vary, we can account, at least qualitatively, for the variety of active galaxies observed.

It is not known what fraction of normal galaxies undergoes activity at some stage in their careers. Although active galaxies are relatively uncommon in our neighbourhood, this may simply be because the active phase lasts for a very short time compared with the total lifetime of a galaxy. For example, if activity lasts only for one-thousandth of the total lifetime, we would expect only one galaxy in a thousand to be active at any particular time. It is perfectly possible, therefore, that the kind of extreme activity displayed by these galaxies is merely a phase through which all galaxies pass. If so, it would suggest that all normal galaxies also possess a massive black hole at their centre, which is not powering an accretion disk because there is insufficient gas in the surrounding regions.

A somewhat milder form of activity is displayed by *starburst galaxies* which, as their name suggests, are galaxies undergoing a vigorous period of star formation. Such activity is not thought to involve an AGN, but is probably triggered by a tidal interaction between two galaxies moving closely past each other.

FURTHER READING: Robson, I., *Active Galactic Nuclei* (Wiley-Praxis, Chichester, 1996).

AETHER (OR **ETHER**) Light (and **electromagnetic radiation** in general) behaves like a wave. This was realised in the 17th century, following work by Christiaan Huygens (1629–95) and others, but the first complete description was provided in the shape of James Clerk **Maxwell**'s theory of electromagnetism (see **fundamental interactions**) in the 19th century. Maxwell showed that electromagnetic radiation was described mathematically in terms of the so-called *wave* *equation*, which also described a wide range of other physical phenomena, such as sound propagation and ocean swells.

Wave phenomena generally consist of periodic fluctuations in a material medium that travel at a well-defined speed. For example, acoustic (sound) waves consist of variations in pressure that in air travel with a velocity of about 300 m/s. Since all other waves travel through some kind of medium, it was supposed by most scientists that light also must also travel through something: the idea of a wave travelling through empty space seems nonsensical, as empty space contains nothing that can fluctuate. The hypothetical 'something' that was supposed to support the transmission of light waves was dubbed the *aether.*

In the 1880s, the physicists Albert Michelson (1852–1931) and Edward Morley (1838–1923) set about the task of measuring the velocity of the Earth through this ubiquitous medium, using a very simple idea which can be illustrated as follows. Imagine that a source of light and a detector are mounted in a fast-moving rocket a distance d away. Suppose that the detector is at the front of the rocket, and the source is at the back. If we send a light signal to the detector when the rocket is stationary with respect to the aether, then the time taken for light to travel from source to detector is just d/c, where c is the speed of light. Now suppose that the rocket travels with a velocity v through the aether. If a light signal is now sent from the back of the rocket to the front, it will take longer than time d/c to reach the detector, because the front of the rocket will have moved with respect to the aether during this time. The effective speed of light now appears to be slower than it was when the rocket was at rest: it takes longer for a signal to travel from the back to the front of the rocket.

The *Michelson–Morley experiment*, performed in 1887, used not rockets but a system of mirrors to measure the time taken for light to travel the same distance in two different directions on the Earth. Because the Earth moves around the Sun, it must also be moving through the aether, so it can play the role of the rocket in the above illustration. To the surprise of physicists of the time, Michelson and Morley found no difference at all in the light travel times for a beam sent in the direction of the Earth's motion and a beam sent at right angles to it. This shows that the velocity of light does not depend on the velocity of the apparatus used to measure it. The absence of the expected 'aether drift' was explained in 1905 by Albert **Einstein**'s theory of **special relativity**. Among other things, this theory forbids the existence of any preferred **frame of reference**. Theories involving the aether have such a preferred frame – the frame in which one is at rest relative to the aether – so they are incompatible with the principle of relativity. Modern relativistic theories do not require any medium to support the oscillations of the electromagnetic field: these waves propagate in a vacuum.

AGE OF THE UNIVERSE The time that has elapsed since the Big Bang **singularity**, usually given the symbol t_0. There are two ways to work out the value of t_0. One is a theoretical argument based on properties of **cosmological models**, and the other is predominantly observational. If we have a consistent model of the Universe, then the two approaches should give results which agree with each other.

The first argument depends on the value of the **Hubble constant** H_0, and hence on the construction of a reliable **extragalactic distance scale**. This has not yet been satisfactorily achieved, but the uncertainty in H_0 is now down to manageable proportions, probably taking a value between 60 and 70 kilometres per second per megaparsec (the usual units). Since kilometres and megaparsecs are both measures of distance, the Hubble constant has units of inverse time. The reciprocal of the Hubble constant therefore defines a characteristic time called, not surprisingly, the *Hubble time*, usually denoted by the symbol t_H. For values of the Hubble constant in the range 60–70 km/s/Mpc, t_H is between about 14 and 17 billion years.

If the **expansion of the Universe** proceeded at a constant rate, the Hubble time would be precisely equal to the age of the Universe, t_0. This would only be true, however, in a completely empty universe which contained no matter to cause a gravitational deceleration. In the more realistic **Friedmann models** the expansion is decelerated by an amount which depends on the value of the **deceleration parameter** q, which in turn depends on the **density parameter** Ω and the **cosmological constant** Λ.

If $\Lambda = 0$, then the expansion is always decelerated ($q > 0$) and the actual age is always less than the Hubble time ($t_0 > t_H$), as shown in the (see Figure 1). The effect of deceleration is, however, not particularly large. In a **flat universe**, with $\Omega = 1$ and $q = 0.5$, t_0 is just two-thirds of t_H so that, for the range of values of H_0 given above, the age of the Universe should be between about 9 and 11 billion years.

An independent method for estimating the age of the Universe is to try to date some of the objects it contains. Obviously, since the Big Bang represents the origin of all matter as well as of spacetime, there should be nothing *in* the Universe that is older *than* the Universe. Dating astronomical objects is, however, not easy. We can estimate the ages of terrestrial rocks by using the radioactive decay of long-lived isotopes, such as uranium-235, which have half-lives measured in billions of years. The method is well-understood and similar to the archaeological use of radiocarbon dating, the only difference being that a vastly larger timescale is needed for a cosmological application requiring the use of elements

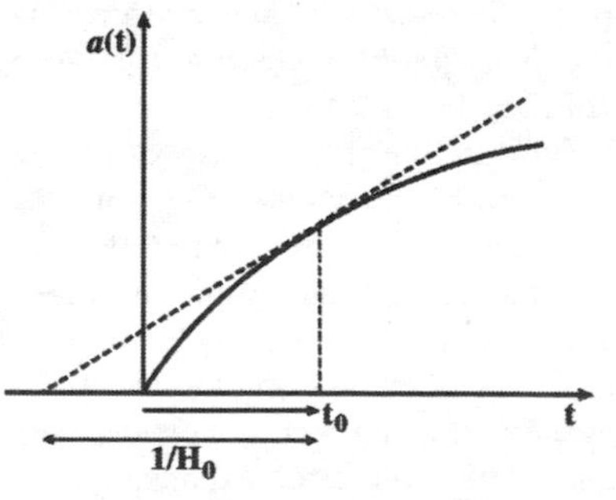

Age of the Universe (1) The effect of deceleration on the age of the Universe, t_0, is to decrease the actual age relative to the Hubble time, $1/H_0$. $a(t)$ is the cosmic scale factor.

with half-lives much longer than that of carbon-14. The limitation of such approaches, however, is that they can only be used to date material within the Solar System. Lunar and meteoritic rocks are older than terrestrial material, but as they were formed comparatively recently during the history of the Universe they are not particularly useful in this context.

The most useful method of measuring the age of the Universe is less direct and exploits arguments based on the theory of **stellar evolution**. The best guide to the value of t_0 comes from studies of **globular clusters**. The stars in these clusters are thought to have all formed at the same time, and the fact that they generally of low mass suggests that they are quite old. Because they all formed at the same time, a collection of these stars can be used to calculate how long they have been evolving. This puts a lower limit on the age of the Universe, because there must have been some time between the Big Bang and the formation of these clusters. Such studies suggest that globular clusters are around 14 billion years old, though this estimate is somewhat controversial (see **globular clusters** for more details).

We can see that this poses immediate problems for the flat universe favoured by many theorists and predicted by models of the **inflationary Universe**. (Note, however, that inflation does not greatly affect the age of the Universe because the period of accelerated expansion lasts for only a tiny fraction of a second.) Globular cluster stars are simply too old to fit into the short lifetime of such a universe. This argument has lent some support to the view that we in fact live in an **open universe**, with $\Omega < 1$. On the other hand, we should not forget the possible existence of a **cosmological constant**. Cosmological models incorporating such a term may enter a phase where the expansion of the Universe is no longer decelerated and may be accelerating now. It is then possible to have a flat universe in which $t_0 > t_H$, which is impossible without a cosmological constant Λ. It would be premature, however, to rule out particular models on the basis of these arguments because there are still substantial uncertainties both in the construction of the extragalactic distance scale leading to H_0, and in the accuracy of the determination of the ages of globular clusters (see Figure 2).

The fact that there is even rough agreement between the ages of the oldest stars and the inverse of the Hubble constant lends some support to the **Big Bang theory**, rather than old rival the **steady state theory**. In

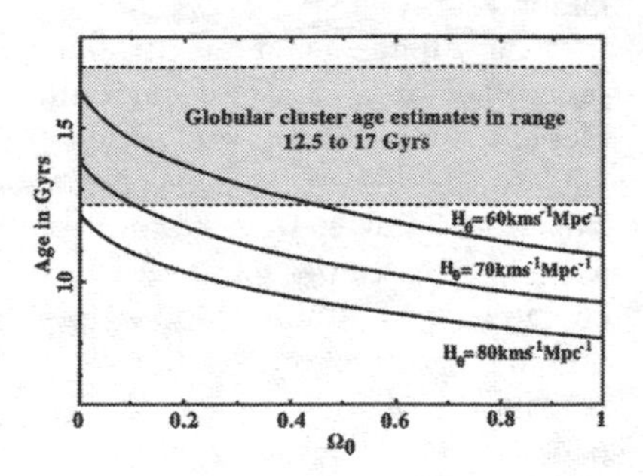

Age of the Universe (2) Comparison between ages, in billions of years (Gyrs), estimated from globular clusters and ages calculated from models. The three curves show the predicted value of t_0 for three values of H_0 and for various values of Ω_0. Low values of both H_0 and Ω_0 are favoured by current estimates.

the latter cosmology, the Universe is eternal and its age cannot therefore be defined; it is, however, an expanding model, so within it we can define the Hubble constant and hence calculate the Hubble time. Since the steady state model requires the continuous creation of matter for all eternity, the stars we could see would present a spread of ages, with the oldest being very much older than the Hubble time. Any agreement at all between the ages of the oldest stars and the inverse of the Hubble constant is simply an unexplained coincidence in this model.

Alpher, Ralph Asher (1921–) US scientist. In the late 1940s, with Hans **Bethe** and George **Gamow**, he developed the 'alpha, beta, gamma' model of **nucleosynthesis**, which correctly yielded the **light element abundances**; and with Gamow and Robert **Herman** he predicted that the event now called the Big Bang would leave a residual radiation with a temperature of about 5 K.

Angular-diameter distance Suppose we have a rod of known length, say one metre. If we see this rod at an unknown distance from us, how can we calculate how far away it is? The answer is found by simple trigonometry: we work out the angle it subtends, and from this angle the distance can be calculated straightforwardly. This is the basic principle of surveying.

Now suppose that we know that certain galaxies are of a particular size. Can we use the same argument to calculate their distance from us? The answer is not so straightforward, for two main reasons. First there is the **curvature of spacetime**. This may mean that familiar methods of distance estimation used in surveying, such as triangulation, do not give the results anticipated for flat, Euclidean space. Secondly, there is the finite velocity of light. If we observe an object at a sufficiently large distance for cosmological effects to be relevant, then we are also observing it as it was in the past. In particular, because of the **expansion of the Universe** the object would have been nearer to the observer when its light was emitted than it is when the light is received.

When we add these effects together, we find strange phenomena occurring. For example, we would imagine that galaxies of the same size observed at higher and higher **redshifts** would subtend smaller and smaller angles. But this is not necessarily so. Galaxies observed at high redshift, which are therefore extremely distant now, had to be almost on top of the us when the light we now observe was emitted. The angle subtended by such objects may increase at high redshifts. This is the basis of one of the classical cosmological tests (see **classical cosmology**).

Correcting for these complications to obtain the **proper distance** is not straightforward unless we assume a particular **cosmological model**. Astronomers therefore usually define the *angular-diameter distance* of an object which subtends a given angle to be the distance at which the object would lie in a non-expanding, Euclidean universe if it subtended the same angle as is observed. This distance will not in general be equal to the **proper distance**, and will also differ from the **luminosity distance**, but it is a useful quantity in many applications.

FURTHER READING: Berry, M. V., *Principles of Cosmology and Gravitation* (Adam Hilger, Bristol, 1989); Narlikar, J. V., *Introduction to Cosmology*, 2nd edition (Cambridge University Press, Cambridge, 1993).

ANISOTROPY According to the **cosmological principle**, the Universe is roughly homogeneous (it has the same properties in every place) and isotropic (it looks the same in every direction). These mathematical features are built into the standard **cosmological models** used to describe the bulk properties of the cosmos in the standard **Big Bang theory**.

But our Universe is not exactly homogeneous and isotropic. A glance at the night sky shows that the sky does not look the same in every direction. Any observed departure from isotropy is covered by the generic term *anisotropy*. The plane of the Milky Way, clearly visible, represents a large-scale anisotropy of the stars in our galaxy. If **galaxies** rather than stars are plotted on the celestial sphere, they also appear anisotropically distributed, but they do not follow the pattern of the Milky Way. Relatively nearby galaxies tend to lie in a band on the sky roughly at right angles to the Milky Way, in a direction called the *supergalactic plane*. However, as we look at more and more distant sources, their distribution on the sky becomes smoother and smoother, tending to the idealised case of pure isotropy. Counts of radio galaxies (see **active galaxies**), for example, are the same to within a few per cent in different directions on the sky. The extragalactic **X-ray background** is isotropic to a similar level of accuracy. The temperature of the **cosmic microwave background radiation**, which comes from an even greater distance than the sources of the X-ray background, is isotropic to within one part in a hundred thousand.

While the small levels of observed anisotropy are good evidence in favour of the **cosmological principle**, the statistical properties of these deviations from pure isotropy are important for theories of structure formation. In particular, the small variations in the temperature of the **cosmic microwave background radiation** discovered by the **Cosmic Background Explorer** satellite (the famous **ripples** discussed at length in Essay 6) provide very important clues which might lead to a complete theory of cosmological **structure formation**.

SEE ALSO: **inhomogeneity**.

ANTHROPIC PRINCIPLE The assertion that there is a connection between the existence of **life in the Universe** and the fundamental physics that governs the large-scale cosmological behaviour. The first to use this expression was Brandon Carter, who suggested adding the word 'anthropic' to the usual **cosmological principle** to stress the fact that our Universe is 'special', at least to the extent that it has permitted intelligent life to evolve within it.

There are many otherwise viable **cosmological models** that are not compatible with the observation that human observers exist. For example, we know that heavy elements like carbon and oxygen are vital to the complex chemistry required for terrestrial life to have developed. We also know that it takes around 10 billion years of **stellar evolution** for generations of stars to synthesise significant quantities of these elements from the primordial gas of hydrogen and helium that existed in the early stages of a Big Bang model. We know, therefore, that we could not inhabit a Universe younger than about 10 billion years. This argument, originally put forward by Robert **Dicke**, places some restrictions on the **age of the Universe** in standard Big Bang models. Since the size of the Universe is related to its age, if it is expanding then this line of reasoning sheds some light on the question of why the Universe is as big as it is. It has to be big, because it has to be old if there has been time for us to evolve within it.

This form of reasoning is usually called the *'weak' anthropic principle* (WAP), and is essentially a modification of the *Copernican principle* that we do not inhabit a special place in the Universe. According to the WAP, we should remember that we can inhabit only those parts of **spacetime** compatible with human life. As an obvious example, we could not possibly exist near the centre of a massive **black hole**. By the argument given above, we obviously could not exist at a much earlier epoch than we do. This kind of argument is relatively uncontroversial, and can lead to useful insights.

One example of a useful insight gleaned in this way relates to the **Dirac cosmology**. Paul **Dirac** was perplexed by a number of apparent coincidences between large dimensionless ratios of physical constants. He found no way to explain these coincidences using standard theories, so he decided that they had to be a consequence of a deep underlying principle. He therefore constructed an entire theoretical edifice of time-varying fundamental constants on the so-called *large number hypothesis*, However, the simple argument by Dicke outlined above dispenses with the need to explain these coincidences in this way. For example, the ratio between the present size of the cosmological **horizon** and the radius of an electron is roughly the same as the ratio between the strengths of the gravitational and electromagnetic forces binding protons and electrons. (Both ratios are huge: of order 10^{40}.) This does indeed seem like a coincidence, but remember that the size of the horizon depends on the time: it gets bigger as time goes on. And the lifetime of a star is determined by the interplay between electromagnetic and gravitational effects. It turns out that both these ratios reduce to the same value precisely because they both depend on the lifetime of **stellar evolution**: the former through our existence as observers, and the latter through the fundamental physics describing the structure of a star.

Some cosmologists, however, have sought to extend the anthropic principle into deeper waters. While the weak version applies to physical properties of our Universe such as its age, density and temperature, the *'strong' anthropic principle* (SAP) is an argument about the **laws of physics** according to which these properties evolve. It appears that these fundamental laws are very finely tuned to permit complex chemistry, which, in turn, permits the development of biological processes and ultimately human life. If the laws of electromagnetism and nuclear physics were only slightly different, chemistry and biology would be impossible. On the face of it, the fact that the laws of nature do appear to be tuned in this way seems to be a coincidence, in that there is nothing in our present understanding of fundamental physics that requires the laws to be conducive to life. This is therefore something we should seek to explain.

In some versions of the SAP, the reasoning is essentially teleological (i.e. an argument from design): the laws of physics are as they are because they *must* be like that for life to develop. This is tantamount to requiring that the existence of life is itself a law of nature, and the more familiar laws of physics are subordinate to it. This kind of reasoning may appeal to those with a religious frame of mind, but its status among scientists is rightly controversial, as it suggests that the Universe was designed specifically in order to accommodate human life.

An alternative and perhaps more scientific construction of the SAP involves the idea that our Universe may consist of an ensemble of *mini-universes*, each one having different laws of physics to the others. Obviously, we can have evolved in only one of the mini-universes compatible with the development of organic chemistry and biology, so we should not be surprised to be in one where the underlying laws of physics appear to have special properties. This provides some kind of explanation for the apparently surprising properties of the laws of nature mentioned above. This latter form of the SAP is not an argument from design, since the laws of physics could vary haphazardly from mini-universe to mini-universe, and in some respects it is logically similar to the WAP. Reasoning of this kind applies in some recent versions of the **inflationary Universe** theory.

FURTHER READING: Barrow, J. D. and Tipler, F. J., *The Anthropic Cosmological Principle* (Cambridge University Press, Cambridge, 1986); Gribbin, J. and Rees, M. J., *The Stuff of the Universe* (Penguin, London, 1995).

ANTIMATTER The fundamental building-blocks of matter are the **elementary particles**, of which there are many different varieties possessing different kinds of physical properties. These properties, such as electric charge and spin, are each described by a number, usually called a quantum number. Each kind of particle possesses a unique combination of these quantum numbers and, when different particles interact with one another during processes described by any of the **fundamental interactions**, the sum of all the quantum numbers is conserved.

To take electric charge as an example, a neutron can decay into a proton and an electron. (Another particle – a form of neutrino – is also produced which conserves the total spin, but we can ignore it for this discussion.) The proton has a positive charge, the electron has an equal but negative charge, and the neutron has no charge. So the net charge going in (zero) is equal to the net charge coming out (zero), as is the case for all other quantum numbers.

Electrons are familiar to us from high-school physics, but the **laws of physics** describing the fundamental interactions are equally valid for a mirror-image particle wherein all the quantum numbers describing the electron change sign. Such a particle is a form of antimatter, called an anti-electron (or positron), and is known to exist in nature. All other particles possess antiparticle counterparts, even if they have no electric charge like the electron, because there are always other quantum numbers that can be reversed. The one property that is identical for particles and antiparticles, however, is their mass: electrons and positrons have the same mass.

If a particle and an antiparticle of the same species (e.g. an electron and a positron) collide, they will annihilate each other, producing pure radiation in the form of gamma rays. It is also possible to induce the reverse effect, creating a particle–antiparticle pair from radiation alone since, according to **special relativity**, mass and energy are equivalent. This latter effect is particularly relevant for cosmology, because pair creation is expected to be very efficient in various stages of the **thermal history of the Universe** in the **Big Bang model**.

The complete symmetry between particles and antiparticles in the laws of physics raises a perplexing question: why is the real Universe dominated by ordinary matter and not by antimatter? If there were equal mixtures of both, then the Universe would be entirely filled with radiation, and all the matter and antimatter would have annihilated. The existence of even small amounts of antimatter in the intergalactic medium is ruled out because the radiation it would produce by interacting with ordinary matter is not seen. The observed asymmetry between matter and antimatter was a challenge to early supporters of the Big Bang model, and eventually led to the theory of **baryogenesis**.

B

Baade, (Wilhelm Heinrich) Walter (1893–1960) German-born astronomer, who worked mainly in the USA. He made fundamental contributions to the study of **stellar evolution** and the **evolution of galaxies**, and was responsible in the 1950s for correcting an error in Edwin **Hubble**'s determination of the **extragalactic distance scale.**

Baby universe The usual approach to the construction of **cosmological models** starts from the assumption that spacetime has an essentially simple structure. For example, **closed universe** models are generally thought to have a structure similar to that of a sphere. (In mathematical language, a sphere is a compact space which has a topological structure that is simply connected.) In theories of **quantum gravity**, however, spacetime is not expected to have such benign properties. It is thought that the smooth and well-behaved structure may break down on very small scales, and instead there is a kind of spacetime 'foam'. Rather than being topologically simple, our Universe may therefore consist of a complex collection of intersecting bubbles linked by tubes called **wormholes**. These bubbles may be undergoing a continuing process of nucleation, expansion and recollapse in which each behaves like a low-budget version of an entire universe. They are, however, very small indeed: no greater than the **Planck length** in size, and generally lasting for about the **Planck time**. These tiny bubbles are often called *baby universes*.

If the ideas associated with the **inflationary Universe** models are correct, then our observable **Universe** may have begun as one of these tiny bubbles, which then underwent a period of dramatic expansion, ending up thousands of millions of light years across. The entire Universe may therefore be an infinite and eternal set of mini-universes connected to each other in a very complicated way. Although each individual bubble behaves according to the **Big Bang theory**, the overall picture closely resembles the **steady state theory**, except that the continuous creation of matter does not occur on an atom-by-atom basis, but involves creating whole separate universes.

This theory is speculative, but has led some researchers to discuss the possibility of trying to create a baby universe experimentally. Such a project has not yet been granted funding.

FURTHER READING: Hawking, S. W., *Black Holes and Baby Universes and Other Essays* (Bantam, New York, 1993).

Background radiation see **cosmic microwave background radiation**,

infrared background, **X-ray background**.

Baryogenesis The **laws of physics** describing **fundamental interactions** between **elementary particles** possess certain properties of symmetry. For example, Maxwell's equations, which describe electromagnetic interactions, are symmetric when it comes to electrical charge. If we could change all the plus signs into minus signs and vice versa, then Maxwell's equations would still be correct. To put it another way, the choice of assigning negative charge to electrons and positive charges to protons is arbitrary – it could have been done the other way around, and nothing would be different in the theory. It therefore seems to make sense that our Universe does not have a net electrical charge: there should be just as much positive charge as negative charge, so the net charge is expected to be zero.

The laws of physics seem also to fail to distinguish between matter and **antimatter**. But we know that ordinary matter is much more common than antimatter. In particular, we know that the number of baryons exceeds the number of antibaryons. Baryons actually possess an extra kind of 'charge', called their *baryon number*, B. The Universe therefore carries a net baryon number. Like the net electric charge, B should be a conserved quantity. So if B is not zero now, there seems to be no avoiding the conclusion that it cannot have been zero at any time in the past. The problem of generating this asymmetry – the problem of *baryogenesis* (sometimes called *baryosynthesis*) – perplexed scientists working on the **Big Bang theory** for some considerable time.

Andrei **Sakharov** in 1967 was the first to work out under what conditions there could actually be a net baryon asymmetry and to show that, in fact, baryon number need not be a conserved quantity. He was able to produce an explanation in which the **laws of physics** are indeed baryon-symmetric. At early times the Universe had no net baryon number, but as it cooled a gradual preference for baryons over antibaryons emerged. His work was astonishingly prescient, because it was performed long before any **grand unified theories** were constructed. He was able to suggest a mechanism which could produce a situation in which, for every billion antibaryons in the early Universe, there were a billion and one baryons. When a baryon and an antibaryon collide, they annihilate in a puff of **electromagnetic radiation**. In Sakharov's model, most of the baryons would encounter antibaryons and be annihilated. We would eventually be left with a Universe containing billions of photons for every baryon that managed to survive. This is actually the case in our Universe: the **cosmic microwave background radiation** contains billions of photons for every baryon.

An answer to the question of whether Sakharov's conditions could be met in realistic physical situations had to wait until further developments in grand unified theories. In particular, because he required the Universe to be cooling and not in **thermal equilibrium**, developments in the theory of cosmological **phase transitions** were required for the idea to be fully developed into a complete theory.

FURTHER READING: Barrow, J. D. and Silk, J., *The Left Hand of Creation: The Origin and Evolution of the Expanding Universe* (Basic Books, New York, 1983).

BARYON see **elementary particles**.

BARYON CATASTROPHE According to some models of cosmological **structure formation**, the Universe is dominated by an unseen component of **dark matter** which is in the form of **weakly interacting massive particles** (WIMPs). Although it is invisible by virtue of being unable to produce **electromagnetic radiation**, this material can in principle be detected by its gravitational effect on visible matter. But calculating the amount of dark matter in this way is a difficult business, particularly if the object in question has many different components, such as stars and gas as well as the putative WIMPs.

One kind of astronomical object that permits a detailed inventory to be made of its component matter is a massive cluster of **galaxies** such as the Coma Cluster (see **large-scale structure**). The Coma Cluster is a prominent concentration of many hundreds of galaxies. These galaxies are moving around in a hot plasma whose presence is detectable by **X-ray astronomy** methods. The luminous matter in the galaxies and the more tenuous plasma in the intracluster medium are both made of baryons, like all visible matter. As would be expected in a Universe dominated by WIMPs, baryonic cluster matter is only a small part of the total mass, most of which does not radiate. The total mass of the cluster can be estimated using dynamical arguments based on the **virial theorem**. This is used to infer the total mass of the cluster from the large **peculiar motions** of the component galaxies. It does not matter if the galaxies are not responsible for the mass in order for this to be done. All that is necessary is that they act like test particles, moving in response to the **gravity** generated by whatever mass is there.

When such a detailed audit of the mass of the Coma Cluster was carried out, the conclusion was that the baryonic components contributed about 25% of the total mass of the cluster, a result that was dubbed the *baryon catastrophe* by scientists responsible for analysing the data. So what is catastrophic about this result? The answer relates to the theory of primordial **nucleosynthesis**, one of the main pillars upon which the **Big Bang theory** is constructed. The predictions of calculations of the **light element abundances** produced in the early stages of the primordial fireball agree with observations only if the fractional contribution of baryons, Ω_b, to the critical density (see **density parameter**) is only 10% or so. According to some models of structure formation, the total density parameter, Ω, is equal to 1 (so that we live in a **flat universe**), which means that 90% of the mass of the Universe is in the form of WIMPs. What is more, in these theories there seems to be no way of concentrating baryons relative to the non-baryonic matter in an object the size of the cluster. So the fraction of baryons in the Coma Cluster should be no more than 10%

or so. The observed value (25%) therefore appears to rule out a Universe with $\Omega = 1$. While this conclusion may be catastrophic for those die-hard adherents of a flat universe, many others simply take the view that we must be living in an **open universe** with $\Omega = 0.2$ or so. The fraction of baryons in the Coma Cluster would then be reconcilable with the 10% of the critical density it needs to be in order to fit with nucleosynthesis calculations.

Similar studies have been carried out on other clusters which cast some doubt on the original interpretation of data from Coma. These studies show that the baryon fraction seems to vary significantly from cluster to cluster, which it should not do if it represents the global fraction of baryonic matter in the Universe at large.

FURTHER READING: White, S. D. M. *et al.*, 'The baryon content of galaxy clusters: A challenge to cosmological orthodoxy', *Nature*, 1993, **366**, 429.

BARYOSYNTHESIS see **baryogenesis**.

BETHE, HANS ALBRECHT (1906–) German-born US physicist, professor of physics at Cornell University for 40 years. With Ralph **Alpher** and George **Gamow** he developed the 'alpha, beta, gamma' model of nucleosynthesis, which correctly yielded the **light element abundances**.

BIG BANG THEORY The standard theoretical framework within which most cosmologists interpret observations and construct new theoretical ideas. A more precise term is *hot Big Bang*, to distinguish the current theory from an older version (now discarded) which had a cold initial phase. The existence of the **cosmic microwave background radiation** is extremely strong evidence that the Universe must have been hot in the past (see Essay 5). It is also not entirely correct to call it a 'theory', and many prefer to use the word 'model'. The difference between theory and model is subtle, but a useful definition is that a theory is usually expected to be completely self-contained (it can have no adjustable parameters, and all mathematical quantities are defined *a priori*), whereas a model is not complete in the same way. Owing to uncertain aspects of the Big Bang model, it is quite difficult to make cast-iron predictions from it, and it is consequently not easy to falsify (falsifiability being regarded in many quarters as an essential quality of a scientific theory). Advocates of the **steady state theory** have made this criticism on many occasions. Ironically, the appellation 'Big Bang' was initially intended to be derogatory and was coined by Fred **Hoyle**, one of the model's most prominent detractors, in a BBC radio programme.

In the Big Bang model, the **Universe** originated from an initial state of high temperature and density (the *primordial fireball*) and has been expanding ever since. The dynamics of the Big Bang are described by **cosmological models**, which are obtained by solving the **Einstein equations** in the theory of **general relativity**. The particular models that form the basis of the standard Big Bang theory are the **Friedmann models**, which describe a universe which is both homogeneous and isotropic (see **cosmological principle**). These models all predict the existence of a **singularity** at the very beginning, at which the temperature and density are infinite. Since this event is the feature that most encapsulates the nature of the theory, many people use 'Big Bang' to refer to the very beginning, rather than to the subsequent evolution of the Universe. Most cosmologists interpret the singularity as meaning that the Einstein equations break down at the **Planck time** under the extreme physical conditions of the very early Universe, and that the very beginning must be addressed using a theory of **quantum cosmology**. This incompleteness is the reason why the word 'model' is probably more appropriate. The problem of initial conditions in cosmology is the reason why the **density parameter** and **Hubble constant** are both still unknown quantities, and explains why we cannot answer basic questions such as whether the Universe will expand for ever or will eventually recollapse.

Within the basic framework of a cosmological model, the **laws of physics** known from laboratory experiments or assumed on the basis of theoretical ideas can be used to infer the physical conditions at different stages of the **expansion of the Universe**. In this way the **thermal history of the Universe** is mapped out. The further back into the past we extrapolate, the hotter the Universe gets and the more exotic is the physical theory required to describe it. With our present knowledge of the physics of **elementary particles** and **fundamental interactions**, we can turn the clock back from the present **age of the Universe** (some 15 billion years or so) and predict with reasonable confidence what was happening to within about a microsecond of the Big Bang. Using more speculative physical theory not tested in the laboratory, including **grand unified theories**, cosmologists have tried to push the model to within 10^{-35} seconds of the very beginning, leading to developments of the standard model into the **inflationary Universe**.

Despite gaps in our knowledge of physics at the very highest energies, the theory is widely accepted because it accounts for the expansion of the Universe, as described by **Hubble's law**, for the existence of the cosmic microwave background radiation and for the **light element abundances** predicted by the theory of **nucleosynthesis**. Probably the most important of these is the cosmic microwave background radiation, which provided the most direct evidence for the assumption that there was a hot dense phase in the evolution of the Universe. Most cosmologists abandoned the rival **steady state theory** in the 1960s, mainly as a result of the discovery of this radiation, and also because the evidence for the **evolution of galaxies** was emerging at the time.

There is another important gap in the Big Bang theory, apart from the problem of initial conditions and the breakdown of known laws of physics at the initial singularity. The Big Bang model describes the properties of the Universe only in an average sense, because it is built into the theory that the Universe is the same in every place and looks the same in every direction (i.e. the cosmological principle holds).

But we know that, while this may true in some broad-brush sense, the Universe is not exactly smooth. Instead, it contains a profusion of **large-scale structure**. One of the most active areas in continuing work on the Big Bang theory is to develop a theory of **structure formation**, which will explain how so much complexity developed from such apparently smooth and simple beginnings.

SEE ALSO: Essay 2.

FURTHER READING: Silk, J., *The Big Bang*, revised and updated edition (W. H. Freeman, New York, 1989).

BIG CRUNCH see **closed universe**.

* **BLACK BODY** If a body emits **electromagnetic radiation** in a state of **thermal equilibrium**, then that radiation is described as *black-body radiation* and the object is said to be a *black body*. Of course, this does not mean that the body is actually black – the Sun, for example, radiates light approximately as a black body. The word 'black' here simply indicates that radiation can be perfectly absorbed and re-radiated by the object. The spectrum of light radiated by such an idealised black body is described by a universal spectrum called the *Planck spectrum*, which is described below. The precise form of the spectrum depends on the absolute temperature T of the radiation.

The **cosmic microwave background radiation** has an observed spectrum which is completely indistinguishable from the theoretical black-body shape described by the Planck spectrum. Indeed, the spectrum of this relic radiation is a much closer to that of a black body than any that has yet been recorded in a laboratory (see the Figure overleaf). This serves to prove beyond reasonable doubt that this radiation has its origins in a stage of the **thermal history of the Universe** during which thermal equilibrium was maintained. This, in turn, means that matter must have been sufficiently hot for scattering processes to have maintained thermal contact between the matter and the radiation in the Universe. In other words, it shows that the Big Bang must have been hot. The spectrum and isotropy of the cosmic microwave background radiation are two essential pieces of evidence supporting the standard version of the **Big Bang theory**, based on the **Friedmann models**.

In **thermodynamics**, we can associate the temperature T with a fundamental quantity of energy $E = kT$, where k is a fundamental physical constant known as the *Boltzmann constant*. In thermal equilibrium, the principle of *equipartition of energy* applies: each equivalent way in which energy can be stored in a system is allocated the same amount, which turns out to be one-half of this fundamental amount. This principle can be applied to all sorts of systems, so calculating the properties of matter in thermal equilibrium is quite straightforward. This fundamental property of objects in thermal equilibrium has been known since the 19th century, but it was not fully applied to the properties of thermal black-body radiation until the early years of the 20th century. The reason for this delay was that applying the thermodynamic rules to the problem of black-body radiation was found to give nonsensical results. In particular, if the principle of equipartition is applied to the different possible wavelengths of electromagnetic radiation, we find that a black body should radiate an infinite amount of energy at infinitely short wavelengths. This disaster became known as the *ultraviolet catastrophe*.

The avoidance of the ultraviolet catastrophe was one of the first great achievements of **quantum physics**. Max **Planck** stumbled upon the idea of representing the thermal behaviour of the black body in terms of oscillators, each vibrating at a well-defined frequency. He could then apply thermodynamical arguments to this set of oscillators and get a sensible (non-divergent) answer for the shape of the spectrum emitted. Only later, however, did Albert **Einstein** put forward the idea that is now thought to be the real reason for the black-body spectrum: that the radiation field itself is composed of small packets called *quanta*, each with discrete energy levels. The quanta of electromagnetic radiation are now universally known as *photons* (see **elementary particles**). The energy E of each photon is related to its frequency ν by the fundamental relation $E = h\nu$, where h is the *Planck constant*. Note that this constant has the same dimensions (energy) as kT: the ratio $h\nu/kT$ is therefore a dimensionless number. With the aid of the quantum theory of light, we can obtain the form of the Planck spectrum:

$$B(T) = (2h\nu^3/c^2)[\exp(h\nu/kT) - 1]^{-1}$$

where c is the speed of light. As one

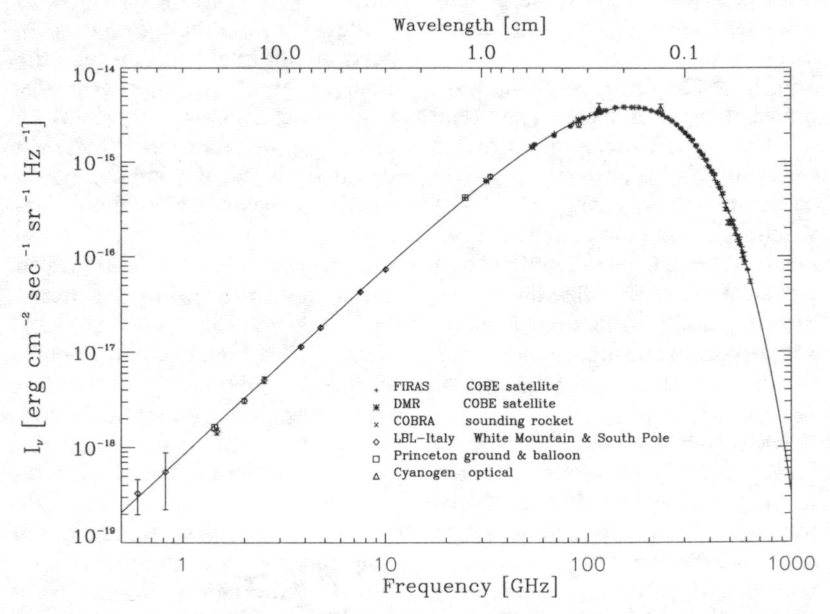

Black body A compilation of experimental measurements of the spectrum of the cosmic microwave background radiation reveals an accurate black-body spectrum.

might have expected, this curve has a maximum at $h\nu \approx kT$, so that hotter bodies radiate most of their energy at higher frequencies. Note, however, that the exponential function guarantees that the function $B(T)$ does not diverge at high values of $h\nu/kT$. At low frequencies ($h\nu/kT >> 1$) this function reproduces the *Rayleigh–Jeans* law, which states that $B(T) = 2\nu^2kT/c^2$, and at high frequencies one finds that $B(T)$ falls off exponentially according to Wien's law:

$$B(T) = (2h\nu^3/c^2)\exp(-h\nu/kT)$$

This latter behaviour is responsible for avoiding the ultraviolet catastrophe.

Everything gives off photons in this way, and the higher the temperature of an object, the more energetic the emitted photons are. Hot furnaces with black walls glow red; hotter furnaces glow brighter and whiter. Furnaces emit a black-body spectrum determined by their temperature, of around 600 K. The Sun is ten times hotter, and emits a nearly black-body spectrum at 6000 K. The **last scattering surface** emitted like a black body at about 3000 K, and now, when the Universe is a thousand times larger, the Universe is pervaded by photons that have maintained their black-body spectrum but have undergone a **redshift** to about 3 K (i.e. 3000K/1000) because of the **expansion of the Universe**.

Measurements of the spectrum of the cosmic microwave background (CMB) radiation indicate that it matches the Planck spectrum perfectly, to an accuracy better than one part in ten thousand, with a temperature T of around 2.73 degrees Kelvin. The accuracy to which this temperature is known allows us to determine very accurately exactly how many photons there are in the Universe and therefore work out the details of the thermal history of the Universe.

Although no deviations from a black-body spectrum have yet been detected, we do not expect the CMB to have an exact black-body spectrum. Features in the spectrum such as **emission lines** and **absorption lines**, an infrared excess (see **infrared astronomy**) or any of the distortions mentioned below would tell us about the details of the sources responsible for them.

The process by which a system reaches thermal equilibrium is called *thermalisation*. Thermalisation of both the photon energy and photon number are required to erase a distortion and make the spectrum black-body. But thermalisation requires scattering between the matter and radiation. As the Universe cools and rarefies, thermalisation (or equipartition of energy) becomes more difficult. Up to 1 year after the Big Bang, both photon energy and photon number were thermalised because scattering was very efficient. This is unfortunate, because it means that all remnants of events in the early Universe have been lost. Between 1 year and 100 years after the Big Bang, photon energy but not photon number were thermalised. Later than 100 years after the Big Bang neither photon energy nor photon number could be thermalised. Any distortions in the spectrum during this epoch should be observable.

There are many possible effects which can cause the CMB spectrum to deviate from a true black-body spectrum. In general, any photon–matter interaction which has a frequency dependence will distort the spectrum. When hot electrons in the **intergalactic medium** heat up the CMB photons, they introduce a distortion called the **Sunyaev–Zel'dovich effect**, sometime also known as the *y-distortion*. Electrons scattering off ions produce radiation known as *bremsstrahlung*, 'braking radiation', which can distort the CMB spectrum. Dust emission produces an infrared excess in the spectrum. Recombination itself produces a Lyman-alpha emission line. The temperature anisotropies discussed in Essay 5 also produce distortions when the spectrum is obtained over different regions of the sky, thus adding together spots of different temperature, but this effect is at a low level. The lack of any distortions places strong limits on any non-thermal processes that might have injected energy into the radiation field at early times. For example, it has been speculated that a so-called μ-*distortion* might have been produced by non-thermal processes, perhaps associated with **Silk damping** or via the decay of some form of elementary particle. Under these circumstances the exponential term in the Planck law becomes modified to $(h\nu + \mu)/kT$. But the limits from spectral measurements on the size of m are exceedingly strong: it has to have an effect of less than 0.01% on the shape of the spectrum, thus effectively ruling out such models.

Despite the many physically possible sources of distortions, none has so far been found in the averaged CMB spectrum. As more precise measurements of the CMB spectrum come to be made, spectral distortions are expected to be found at some level. New instruments which are currently being developed, such as the DIMES satellite, may detect distortions at low frequencies. Their results would tell us much about the thermal history of the Universe. Measurements of the Sunyaev–Zel'dovich (SZ) effect in the directions of galaxy clusters are giving us information about the hot gases in clusters, cluster size and an independent determination of the Hubble constant. SZ measurements in the directions of distant clusters also independently confirm that the CMB is not local in origin.

SEE ALSO: Essay by Lineweaver.

FURTHER READING: Chown, M., *The Afterglow of Creation: From the Fireball to the Discovery of Cosmic Ripples* (Arrow, London, 1993); Rybicki, G. and Lightman, A. P., *Radiative Processes in Astrophysics* (John Wiley, New York, 1979).

BLACK HOLE A region of spacetime where the action of **gravity** is so strong that light cannot escape. The idea that such a phenomenon might exist can be traced back to 1783 and the English clergyman John Michell (1724–93), but black holes are most commonly associated with Albert **Einstein**'s theory of **general relativity**. Indeed, one of the first exact solutions of Einstein's equations describes such an object. This, the *Schwarzschild solution*, was obtained in 1916 by Karl **Schwarzschild**, who died soon after on the eastern front in the First World War. The solution corresponds to a spherically symmetric distribution of matter, and it was originally intended that this could form the basis of a mathematical model for a star. It was soon realised, however, that for an object of any mass M there is a critical radius (R_S, the *Schwarzschild radius*) such that if all the mass is squashed inside R_S then no light can escape. In terms of the mass M, speed of light c and the gravitational constant G, the critical radius is given by $R_S = 2GM/c^2$. For the mass of the Earth, the critical radius is only 1 cm, whereas for the Sun it is about 3 km. So for the Sun to be formed into a black hole would require the solar material to be compressed to a phenomenal density.

Since the pioneering work by Schwarzschild, much research on black holes has been carried out, and other kinds of mathematical solution have been obtained. For example, the *Kerr solution* describes a rotating black hole, and the *Reissner–Nordström solution* corresponds to a black hole with an electric charge. Various theorems have also been demonstrated relating to the so-called *no-hair conjecture*, which states that black holes show very little outward sign of what is inside them.

Although there is as yet no watertight evidence for the existence of black holes, they are thought to exist in many kinds of astronomical object. It is possible that very small black holes, with masses ranging from millions of tonnes to less than a gram, might have been formed very early on in the Big Bang. Such objects are usually called *primordial black holes*. Black holes of stellar mass may be formed as an end-point of **stellar evolution**, after massive stars explode into **supernovae**. More extreme *supermassive black holes* might have formed from the collapse of bodies of, say, 100,000 solar masses (such bodies, sometimes termed *very massive objects* or *superstars*, may have existed before galaxies were formed). Studies of the dynamics of stars near the centre of galaxies indicate the presence of very strong mass concentrations that are usually identified with black holes with masses around 100 million solar masses. The intense gravitational field that surrounds a black hole of about 100 million solar masses is thought to be the engine that drives **active galaxies**.

As well as having potentially observable consequences, black holes also pose deep fundamental questions about the applicability of **general relativity**. In this theory, the light is prevented from escaping from a black hole by the extreme **curvature of spacetime**. It is as if the space around the hole were wrapped up into a ball, so that light can travel around the surface of the ball but cannot escape. Technically, the term 'black hole' actually refers to the *event horizon* (see **horizon**) that forms around the object, ensuring that no communication is possible between the regions of spacetime inside and outside the hole. But what happens inside the horizon? According to the famous singularity theorems proposed by Roger **Penrose** and others, the inevitable result is a **singularity**, where the density of material and the curvature of spacetime become infinite. The existence of this singularity suggests to many that some fundamental physics describing the gravitational effect of matter at extreme density is absent from our understanding. It is possible that a theory of **quantum gravity** might enable physicists to calculate what happens deep inside a black hole without having all mathematical quantities become infinite. Penrose's work on the mathematical properties of black hole singularities led to further work by himself and Stephen **Hawking** which showed that a singularity is also inevitable at the creation event that starts off the Big Bang.

After working on the singularity theorems, Hawking turned his attention back to black holes and, in probably his most famous discovery, he found that black holes are not truly black: they radiate what is now known as **Hawking radiation**. The temperature of this radiation is inversely proportional to the mass of the hole, so that small holes appear hotter. The consequence is dramatic for the small primordial black holes, which are expected to have evaporated entirely into radiation. But the effect on large holes is small. A black hole with a mass of a billion tonnes or more would take longer than the age of the Universe to evaporate.

The equations of general relativity that describe collapse of an object into a black hole are symmetric in **time**. Just as the formation of a black hole is a solution of these equations, so is the time-reverse solution, which describes the spontaneous appearance of matter from nothing. Such a hypothetical source of matter is usually called a *white hole*. No object with properties corresponding to this solution has ever been observed.

FURTHER READING: Thorne, K. S., *Black Holes and Time Warps* (Norton, New York, 1994).

BOHR, NIELS HENRIK DAVID (1885–1962) Danish theoretical physicist. He was responsible for the first model of the structure of the hydrogen atom to include ideas that later became part of **quantum theory**.

BOLTZMANN, LUDWIG EDWARD (1844–1906) Austrian physicist. He established the theory of statistical mechanics, and used the kinetic theory to relate the classical (macroscopic) theory of **thermodynamics** to the microscopic description of matter in terms of atoms.

BONDI, SIR HERMANN (1919–) British cosmologist and mathematician, born in Austria. In 1948 Bondi, Thomas **Gold** and Fred **Hoyle** developed the **steady state theory**. He also studied **general relativity**, working on **Mach's principle** and showing, in 1962, that general relativity predicts the existence of **gravitational waves**.

BOSON see **elementary particles**.

BRANS–DICKE THEORY An alternative to the theory of **general relativity** which is motivated by ideas relating to **Mach's principle**. At the core of this theory is the idea that the *inertial mass* of an object, which governs how it responds to the application of an external force, is not an intrinsic property of the object itself but is generated by the gravitational effect of all the other matter in the Universe. Although Albert **Einstein** was impressed with Mach's principle, and it largely motivated his own early work on the theory of **gravity**, general relativity does not itself address the idea in a satisfactory way.

The theory was proposed in 1961 by Carl Henry Brans (1935–) and

Robert **Dicke**. In order to extend Einstein's theory to take account explicitly of the origin of mass in the gravitational effect of all other particles in the Universe, they introduced a new field in the **Einstein equations**. This field, a **scalar field** ϕ, is generated by all matter, and its function in the theory is to couple the masses of particles to the large-scale distribution of material. This coupling is controlled by a parameter which is not predicted by the theory, but which must be obtained by reference to experiment and observation. The addition to general relativity of this scalar field, which is based entirely on the mathematics of **tensor** equations, has led to it being classified as a scalar–tensor theory. Like Einstein's theory, however, the Brans–Dicke theory can be used to construct viable **cosmological models**.

One of the most important consequences of the Brans–Dicke theory is that the strength of gravity, as described by the gravitational constant G, must depend on time, and the rate at which it changes is controlled by a parameter ω. We cannot use, for example, the **age of the Universe** or **nucleosynthesis** to estimate the magnitude of ω. But we can use more local observations, such as historical data on lunar eclipses, the properties of fossils (changes in the Earth's rotation being reflected in changes in the rate of sedimentation and hence fossilisation), **stellar evolution** (particularly the evolution of the Sun), the deflection of light by celestial bodies, and the advance of the perihelion of Mercury. These observations, which put limits on possible departures from pure Einsteinian gravity, do not rule out the Brans–Dicke theory but they do suggest that $\omega > 500$ (the larger the value of ω, the closer Brans–Dicke theory gets to general relativity).

In recent years interest in the Brans–Dicke theory as an alternative to general relativity has greatly diminished, but cosmologists have investigated the behaviour of certain models of the **inflationary Universe** which involve a **scalar field** with essentially the same properties as the Brans–Dicke scalar field has. Such models are usually called *extended inflation* models, and in them the scalar field is known as the *dilaton field*.

FURTHER READING: Brans, C. and Dicke, R. H., 'Mach's principle and a relativistic theory of gravitation', *Physical Review Letters*, 1961, **124**, 125; Narlikar, J. V., *Introduction to Cosmology*, 2nd edition (Cambridge University Press, Cambridge, 1993), Chapter 8.

BURBIDGE, GEOFFREY RONALD (1925–) and **(ELEANOR) MARGARET** (1919–) English astrophysicists, married in 1948, who spent nearly all their working life in the USA. Their principal achievement, in collaboration with Fred **Hoyle** and William **Fowler** (the 'B^2FH theory'), was to establish the process of **nucleosynthesis** in stars in the various stages of **stellar evolution**. Their other most important work was an early in-depth study of **quasars**; they have also studied galaxies and the **dark matter** problem.

C

Classical cosmology A branch of observational **cosmology** that consists of a battery of techniques called the *classical cosmological tests*. Most of these tests boil down to attempts at using some geometrical property of space, and some kind of 'standard' source of radiation, to probe directly the **curvature of spacetime**, and hence the **deceleration parameter** q and the **density parameter** Ω. The idea is basically to measure the angular size and/or light intensity of very distant sources, and look for differences in these measured quantities from what we would expect to see in a Euclidean, non-expanding Universe. The size of the effect produced by differences in geometry and deceleration rate is usually small unless the objects are observed at high **redshift**, so it is necessary to examine very distant objects. A sound knowledge of the properties of the sources being observed is also necessary. In general, classical cosmologists attempt to identify a set of 'standard candles' (i.e. objects with known luminosity) or 'standard rods' (i.e. objects of known physical size) at different redshifts, and use their properties to gauge the geometry.

This field is fraught with difficulties, and there are potential sources of error or bias in the results. The problem with most of the tests is that, if the **Big Bang theory** is correct, objects at high redshift are younger than those nearby. We should therefore expect to detect evolutionary changes in the properties of galaxies, and any attempt to define a standard rod or candle to probe the geometry will be very prone to the effects of such changes (see **evolution of galaxies**). Indeed, since the early days of cosmology the emphasis has now largely switched to the use of these tests to study the evolution of distant objects rather than the cosmological properties of the Universe as a whole. The simplest classical cosmological test, the straightforward counting of galaxies as a function of their apparent brightness on the sky, is now used almost exclusively as a test for models of galaxy formation and evolution, rather than probing the cosmological parameters (see **source counts**, **Hubble Space Telescope**).

One simple cosmological test is based on the relationship between **luminosity distance** and redshift z for a standard source of light. The difference between the Euclidean expectation and the result for a **Friedmann model** with deceleration parameter q is simply a factor of $(q-1)z$, provided z is not too large. The effect is therefore small (see the Figure), unless the sources being observed are at an appreciable redshift. Using galaxies is problematic because of the effects of evolution. In a recent revival of the

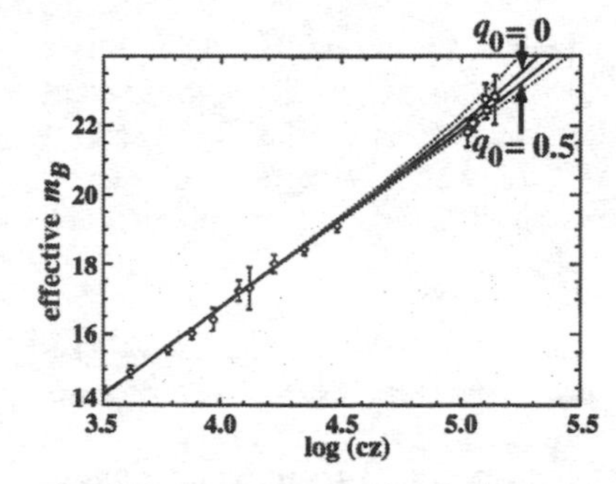

Classical cosmology A classical cosmological test involving the use of distant supernovae as distance indicators. The curves show the expected dependence of apparent magnitude on redshift z for different values of the deceleration parameter q_0. Recent measurements do not pick out a favoured value, but it is anticipated that it will be possible to measure q_0 in a few years' time.

test, the objects observed are **supernovae** in very distant galaxies, which are probably much less affected by evolution than the galaxies themselves (because the supernovae studied are of the same type and mass, their luminosities will be similar, so they can be used as 'standard candles'). This method appears very promising, but is limited by the number of supernovae that can be observed. Preliminary estimates using this approach suggest a positive value of q.

Another interesting method makes use of the **angular-diameter distance** for sources whose apparent size can be measured. If the size of a rod is known, its distance from the observer can be calculated in Euclidean space (see **curvature of spacetime**) by simple triangulation: the farther away it is, the smaller the angle it subtends. What is interesting is that the observed angular diameter for a standard rod may actually *increase* with increasing redshift z. This is counter-intuitive, but is easily explained by the fact that light from a distant source, having farther to travel, must have been emitted before light from a nearby source. The distant source must therefore have been nearer to us when it emitted its light than it is now, when the light is being detected. Moreover, any matter between the source and the observer has the effect of bending these light rays, thus increasing the angular size still further. We can use the properties of standard rods and the different redshifts to infer the geometry and deceleration rate. Again, we have the problem of finding a suitable standard source; this time a standard rod, rather than a standard candle. Radio sources were used in early applications of this idea but, somewhat embarrassingly, the relation between angular diameter and redshift appears to be completely Euclidean for extended sources. This is usually attributed to the evolution in physical size of objects with cosmic time: perhaps they were systematically smaller in the past than they are now. An interesting attempt to apply this test to compact radio sources yields a behaviour which is of the expected type and consistent with the supernova results. We still cannot be certain, however, that we are not seeing some kind of systematic evolution of size with time.

SEE ALSO **gravitational lensing**, **radio astronomy**.

FURTHER READING: Metcalfe, N. *et al.*, 'Galaxy formation at high redshifts', *Nature*, 1996, **383**, 236; Perlmutter, S. *et al.*, 'Discovery of a supernova explosion at half of the age of the Universe',

Nature, 1998, **391**, 51; Kellerman, K. I., 'The cosmological deceleration parameter estimated from the angular size/redshift relation for compact radio sources', *Nature*, 1993, **361**, 134.

Closed universe Any **cosmological model** in which the **curvature of spacetime** is positive. In such a universe the normal rules of Euclidean geometry do not necessarily hold. For example, the sum of the interior angles of a triangle is greater than 180°, and parallel lines can actually intersect. Among the family of **Friedmann models**, the particular cases describing closed universes are those in which the **density parameter** $\Omega > 1$ and the deceleration parameter $q > 0.5$. These models are finite in physical size. They also recollapse in the future: the deceleration generated by **gravity** eventually causes the **expansion of the Universe** to cease, and go into reverse. Eventually these models produce a second **singularity**, sometimes called the *Big Crunch*, in which the density of all matter again becomes infinite.

It was often thought that there could be an infinite series of big bangs followed by big crunches, so that a closed model could, in some sense, be eternal. It is now thought, however, that an infinite *oscillating universe* of this type is not possible because each cycle becomes progressively more disordered than the previous one as a consequence of the second law of **thermodynamics**. Eventually the oscillations would peter out, rather like a bouncing ball which gradually comes to rest as its energy dissipates.

Most theoretical treatments of **quantum cosmology** suggest that the Universe should be closed, but this is difficult to reconcile with present determinations of the density parameter, which suggest the strong possibility that we live in an **open universe**.

Clustering of galaxies see **correlation function**, **large-scale structure**.

Coordinate system see **frame of reference**.

Copernicus, Nicolaus (1473–1543) Polish astronomer. He is famous for proposing a Sun-centred model of the Solar System, thus advocating the idea that we do not occupy a special place in the **Universe**. This is a forerunner of the modern **cosmological principle**.

* **Correlation function** The testing of theories of **structure formation** using observations of the **large-scale structure** of the distribution of galaxies requires a statistical approach. Theoretical studies of the problem of structure formation generally consist of performing numerical ***N*-body simulations** on powerful computers. Such simulations show how **galaxies** would form and cluster according to some well-defined assumptions about the form of **primordial density fluctuations**, the nature of any **dark matter** and the parameters of an underlying **cosmological model**, such as the **density parameter** and **Hubble constant**. The simulated Universe is then compared with observations, and this requires a statistical approach: the idea is to derive a number (a 'statistic') which encapsulates the nature of the spatial distribution in

some objective way. If the model matches the observations, the statistic should have the same numerical value for both model and reality. It should always be borne in mind, however, that no single statistic can measure every possible aspect of a complicated thing like the distribution of galaxies in space. So a model may well pass some statistical tests, but fail on others which might be more sensitive to particular aspects of the spatial pattern. Statistics therefore can be used to reject models, but not to prove that a particular model is correct.

One of the simplest (and most commonly used) statistical methods appropriate for the analysis of galaxy clustering observations is the *correlation function* or, more accurately, the *two-point correlation function*. This measures the statistical tendency for galaxies to occur in pairs rather than individually. The correlation function, usually denoted by $\xi(r)$, measures the number of pairs of galaxies found at a separation r compared with how many such pairs would be found if galaxies were distributed at random throughout space. More formally, the probability of finding two galaxies in small volumes dV_1 and dV_2 separated by a distance r is defined to be

$$dP = n^2(1 + \xi(r))\, dV_1\, dV_2$$

where n is the average density of galaxies per unit volume. A positive value of $\xi(r)$ thus indicates that there are more pairs of galaxies with a separation r than would occur at random; galaxies are then said to be *clustered* on the scale r. A negative value indicates that galaxies tend to avoid each other; they are then said to be *anticlustered*. A completely random distribution, usually called a *Poisson distribution*, has $\xi(r) = 0$ for all values of r.

Estimates of the correlation function of galaxies indicate that $\xi(r)$ is a power-law function of r:

$$\xi(r) \approx (r/r_0)^{-1.8}$$

where the constant r_0 is usually called the *correlation length*. The value of r_0 depends slightly on the type of galaxy chosen, but is around 5 Mpc for bright galaxies. This behaviour indicates that these galaxies are highly clustered on scales of up to several tens of millions of light years in a roughly **fractal** pattern. On larger scales, however, $\xi(r)$ becomes negative, indicating the presence of large voids (see **large-scale structure**). The correlation function $\xi(r)$ is mathematically related to the **power spectrum** $P(k)$ by a Fourier transformation; the function $P(k)$ is also used as a descriptor of clustering on large scales.

FURTHER READING: Peebles, P. J. E., *The Large-Scale Structure of the Universe* (Princeton University Press, Princeton, 1980).

COSMIC BACKGROUND EXPLORER (COBE) A satellite launched by NASA in 1989 to investigate the **cosmic microwave background radiation** (CMB). About two years later, COBE made headlines around the world when its scientific team, led by George Smoot, discovered the famous **ripples** in the CMB (discussed at length in Essay 5). This discovery alone would have justified the relatively modest cost of the COBE mission, but in fact the satellite carried three experiments altogether, all of which have led to important discoveries. As well as the CMB and the **infrared background**, both of direct relevance to cosmology, COBE has improved our understanding of such diverse phenomena as the internal structure of the Milky Way and the diffuse zodiacal light produced in the Solar System.

The experiment that detected the ripples consisted of a set of differential microwave radiometers (the DMR experiment) working at three different frequencies. Using a beam-switching technique, this experiment was able to detect slight variations in the temperature of the CMB in different directions on the sky. The radiometers were attached to horn antennae which opened at an angle of about 10°. The relatively large width of these beams meant that COBE was able to detect **anisotropy** only on a relatively large angular scale. Using the three different frequencies allowed the team to distinguish between angular variations produced by the anisotropy of the Milky Way and those produced by the CMB itself. As discussed in Essay 5, many new experiments are being planned to probe the structure of the CMB on a finer resolution than this.

The other two experiments on COBE were the Far-Infrared Absolute Spectrometer (FIRAS) and the Diffuse Infrared Background Experiment (DIRBE), both of which were intended to study the infrared background. Among other important discoveries, these experiments have confirmed the 'pure' **black-body** spectrum of the CMB, and established its temperature to an accuracy of better than one part in a hundred thousand.

In accordance with its planned mission lifetime of four years, the satellite was switched off in 1993, but many data still remain to be analysed and new results continue to be announced. Indeed, as this book was nearing completion the COBE team announced at the January 1998 meeting of the American Astronomical Society new results pertaining to the infrared background based on DIRBE data.

SEE ALSO: **infrared astronomy**.

FURTHER READING: Smoot, G. F. and Davidson, K., *Wrinkles in Time* (William Morrow, New York, 1993).

COSMIC MICROWAVE BACKGROUND RADIATION (CMB) Discovered accidentally in 1965 by two radio engineers, Arno **Penzias** and Robert **Wilson** (who were later awarded the 1978 Nobel Prize for Physics for their work; see **radio astronomy**), the cosmic microwave background is probably the most important piece of observational evidence in favour of the **Big Bang theory**.

The high degree of uniformity of the microwave background radiation shows that it is not associated with sources within our Galaxy (if it were, it would not be distributed evenly on the sky). It was therefore immediately recognised as having an extragalactic origin. Moreover, the characteristic **black-body** spectrum of this radiation demonstrates beyond all reasonable doubt that it was produced in conditions of **thermal equilibrium** in the very early stages of the Big Bang. This radiation has gradually been cooling as an effect of the **expansion of the Universe**, as each constituent photon suffers a **redshift**. If we could wind back the clock to an earlier stage of the **thermal history of the Universe**, we would reach a stage where this radiation would have been sufficiently hot (i.e. the wavelengths of photons would have been sufficiently short) to completely ionise all the matter in the Universe. This would have happened about 300,000 years after the Big Bang **singularity**, a stage which corresponds to a redshift factor of about a thousand. Under conditions of complete ionisation, matter (especially the free electrons) and radiation undergo rapid collisions which maintain the thermal equilibrium. When the degree of ionisation fell as a result of the recombination of electrons and protons into atoms, photon scattering was no longer efficient and the radiation background was no longer tied to the matter in the Universe. This process is known as *decoupling*. What we see today as the CMB, with a temperature of around 2.73 K, is the radiation that underwent its last scattering off electrons at the epoch of recombination. When it was finally released from scattering processes, this radiation would have been in the optical or ultraviolet part of the spectrum of **electromagnetic radiation**, but since then it has been progressively redshifted by the expansion of the Universe and is now seen at infrared and microwave wavelengths.

How do we know that the CMB comes from a well-defined cosmic epoch? Could it not be an accumulation of radiation from sources along all the lines of sight? The black-body spectrum of the CMB is strong evidence in favour of a surface and against such an accumulation. Consider a fog of iron needles as an example. We know from its isotropy that the CMB cannot have originated very nearby. Suppose it came from somewhere beyond distant galaxies, but closer than the **last scattering surface**. If our hypothetical diffuse medium of iron needles were at a constant temperature but spread out in distance, the needles farther away would be more redshifted and thus their black-body temperature would appear lower. The observed spectrum of the CMB would then be a superposition of redshifted black-body spectra of different temperatures, and not the very nearly black-body spectrum actually observed (see the Figure under **black body**).

The CMB has a black-body spectrum because it was in thermal equilibrium with the hot plasma of electrons and protons before *recombination* (see **thermal history of the Universe**). High densities and temperatures ensured that these particles were colliding and exchanging energy rapidly. Such a state, in which the kinetic energy has been equally distributed between the matter and the radiation, is thermal equilibrium, and is what is required for the emission of a black-body spectrum. Thus the energy distribution of photons, no matter what it was originally, would have had time to relax into a black-body spectrum characterised by a single temperature. At recombination this temperature was about 3000 K, as determined by two factors: the ionisation energy of hydrogen and the photon-to-proton ratio.

Recombination occurred when the CMB temperature fell to the point where there were no longer enough high-energy photons to keep hydrogen ionised. Although the ionisation energy of hydrogen is 13.6 eV (corresponding to a temperature of around 10^5 K), recombination occurred at around 3000 K. The fact that there were around a billion photons for every proton or electron allowed the highest-energy photons of the Planck spectrum (see **black body**) to keep the comparatively small number of hydrogen atoms ionised until this much lower temperature was reached.

The CMB is of paramount importance in the Big Bang theory because it tells cosmologists about the thermal conditions present in the early stages of the Universe. Also, because of its near-perfect isotropy, it provides some evidence in favour of the **cosmological principle**. With the discovery in 1992 by the **Cosmic Background Explorer** satellite of small variations (see **ripples**) in the temperature of the microwave background on the sky, it has also become possible to use maps of the temperature distribution to test theories of cosmological **structure formation**.

SEE ALSO: Essay 6.

FURTHER READING: Chown, M., *The Afterglow of Creation: From the Fireball to the Discovery of Cosmic Ripples* (Arrow, London, 1993); Smoot, G. and Davidson, K., *Wrinkles in Time* (William Morrow, New York, 1993).

COSMOGONY see **cosmology**, **structure formation**.

COSMOLOGICAL CONSTANT (Λ) A constant originally introduced into the equations of **general relativity** by Albert **Einstein** himself in order to produce a **cosmological model** which was static in time. This was before the discovery by Edwin **Hubble** of the **expansion of the Universe**. Later events caused Einstein to regret the addition of this term, but its legacy still lingers.

What Einstein did was to modify the left hand side of the field equations (see **Einstein equations**) by changing the terms that involve the **curvature of spacetime**. This was tantamount to modifying the law of **gravity** slightly so that on sufficiently large scales he could balance the universal attraction of gravity with a repulsive force of his own devising. The effect of the cosmological constant Λ on Newton's law for the gravitational force between two masses is to add a new term to the usual law (which depends on the inverse square of the separation of masses); the new term is directly proportional to the separation, instead of depending on the inverse square. On large scales, therefore, the Λ-term dominates. If it is positive, Λ can be understood as a cosmic repulsion; if it is negative, it acts as a cosmic tension leading to an extra attraction over and above the usual gravitational force. This modification was not entirely arbitrary, however, because it is completely consistent with the fundamental reasoning that led Einstein to general relativity in the first place.

One might imagine that the cosmological constant would have vanished from the scene with the discovery of the expansion of the Universe, but that certainly did not happen. With developments in the **Big Bang theory**, cosmologists began to ponder the consequences of **fundamental interactions** in the very early Universe. These considerations concerned the right-hand-side of Einstein's equations, the part of the theory of general relativity that deals with the properties of matter. It was realised that the cosmological constant term could just as easily be put on the left-hand side of the field equations. The cosmological constant is a vacuum energy density: an energy not directly associated with matter or radiation, but with 'empty' space.

Each time matter changes state (i.e. each time it undergoes a **phase transition**), some amount of vacuum energy is expected to remain. Physicists working on the theory of **elementary particles** tried to calculate the net amount of vacuum energy produced by all the phase transitions the Universe underwent as it cooled. The answer is catastrophically large: about 10^{120} times larger than the density of all the matter in the Universe. Such a result is at odds with observations, to put it mildly.

Some cosmologists believe that a cosmological constant term is necessary in order to reconcile **the age of the Universe** with estimates of the **Hubble constant** and the **density parameter**. But the size of the term required corresponds to a vacuum energy density of the same order as the density of matter, about 10^{120} times smaller than the predictions of particle physics. Many are deeply uncomfortable about the size of this discrepancy, and suggest that it means that the cosmological constant has to be exactly zero.

The cosmological constant also plays a role in the **inflationary Universe**. The mathematical solution that describes a universe undergoing inflation, first found by Willem **de Sitter** in 1917, involves the cosmological constant term (actually that is all it involves – the de Sitter universe is empty apart from the vacuum density). The **scalar field** responsible for driving the inflationary expansion behaves in such a way that the vacuum energy dominates: the solution in this case is identical to the de Sitter model.

FURTHER READING: Einstein, A., 'Cosmological considerations on the general theory of relativity', 1917, reprinted in *The Principle of Relativity*, edited by H. A. Lorentz *et al.*, (Dover, New York, 1950); Weinberg, S. 'The cosmological constant problem', *Reviews of Modern Physics*, 1989, **68**, 1; Goldsmith, D., *Einstein's Greatest Blunder? The Cosmological Constant and Other Fudge Factors in the Physics of the Universe* (Harvard University Press, Cambridge, MA, 1995).

COSMOLOGICAL MODEL A mathematical solution of the field equations (see **Einstein equation**) that make up Einstein's theory of **general relativity**. Cosmological models are required to be self-contained: because they represent the entire Universe, they are not allowed to be influenced by factors external to them.

Many cosmological models are very simple because they are based on the **cosmological principle** that the Universe must be homogeneous and isotropic. This assumption drastically reduces the number of equations to be solved, and leads directly to the *Friedmann solutions* upon which standard cosmological models are based and which have relatively straightforward mathematical properties. The **Friedmann models** are a family of solutions of the Einstein equations, described by their **density parameter**

Ω, and can represent **open universes**, **closed universes** or **flat universes** depending on the value of Ω. The flat Friedmann model, which corresponds to the special case of $\Omega = 1$, is called the *Einstein–de Sitter* solution. The rather bizarre case with $\Omega = 0$ (i.e. no matter at all) is called the *Milne model*. With no matter, there is no gravity in the Milne model, so it is often described as a kinematic model. (See also **curvature of spacetime**, **expansion of the Universe**).

But the family of Friedmann models is not the only possible set of relativistic cosmological models. If we introduce a **cosmological constant** we can obtain a wider range of behaviour. The Einstein universe, for example, is chronologically the first cosmological model and it differs greatly from the Friedmann models in the presence of a cosmological constant. The *Einstein model* is static and finite, and has positive spatial curvature. There is also the *de Sitter* model, which contains no matter at all, but which is flat and expanding exponentially quickly because of the cosmological constant (see also **inflationary Universe**). The *Lemaître universe*, like the Einstein universe, contains both matter and a cosmological constant term. This model has a positive spatial curvature, but expands for ever without recollapsing. The most interesting aspect of such a universe is that it undergoes a *coasting phase* in which the expansion grinds to a halt. For a short time, therefore, this model behaves like the Einstein model but differs in the long run.

All the preceding models are homogeneous and isotropic, and consequently have a spatial geometry described by the **Robertson–Walker metric**. It is much more difficult to construct models which are inhomogeneous and/or anisotropic, but some exact solutions have been obtained for cases of some particular symmetry. The *Bianchi models* are model universes which are homogeneous but not isotropic. There are actually ten distinct Bianchi types, classified according to the particular kinds of symmetry they possess. A special case of the Bianchi type IX models, the *Kasner solution*, is the basis of the *mixmaster universe*, an idea popular in the 1960s, in which the Universe begins in a highly anisotropic state but gradually evolves into one of near isotropy (see **horizon problem**).

Globally inhomogeneous models are harder to construct, but examples of exact inhomogeneous solutions include the *Tolman–Bondi* solutions, which describe spherically symmetric distributions of matter embedded in a smooth background. Other solutions have been found for various special situations such as universes permeated by **magnetic fields** and those containing **topological defects** of various kinds. Usually, however, inhomogeneity cannot be handled by using exact mathematical methods, and approximation techniques have to be resorted to. These methods are used to study the growth of density fluctuations in the Friedmann models during the process of **structure formation**.

The most bizarre cosmological model of all is the rotating *Gödel universe*. This is not a realistic model of the Universe, but is interesting because it acts as a warning of some of the strange things that can be produced by general relativity. In particular, the Gödel universe allows observers to travel backwards in **time.**

FURTHER READING: MacCallum, M. A. H., 'Anisotropic and inhomogeneous cosmologies', in *The Renaissance of General Relativity and Cosmology*, edited by G.F.R. Ellis *et al.* (Cambridge University Press, Cambridge, 1993), p. 213.

Cosmological principle The assertion that, on sufficiently large scales (beyond those traced by the **large-scale structure** of the distribution of **galaxies**), the **Universe** is both *homogeneous* and *isotropic*. Homogeneity is the property of being identical everywhere in space, while isotropy is the property of appearing the same in every direction. The Universe is clearly not exactly homogeneous, so cosmologists now define homogeneity in an average sense: the Universe is taken to be identical in different places when we look at sufficiently large pieces of it. A good analogy is that of a patterned carpet which is made of repeating units of some basic design. On the scale of the individual design the structure is clearly inhomogeneous, but on scales that are larger than each unit it is homogeneous.

There is quite good observational evidence that the Universe does have these properties, though the evidence is not completely watertight. One piece of evidence is the observed near-isotropy of the **cosmic microwave background radiation**. Isotropy, however, does not necessarily imply homogeneity without the additional assumption that the observer is not in a special place: the so-called *Copernican principle*. We would observe isotropy in any spherically symmetric distribution of matter, but only if we were in the middle. A circular carpet with a pattern consisting of a series of concentric rings would look isotropic only to an observer who is standing at the centre of the pattern. Observed isotropy, together with the Copernican principle, therefore implies the cosmological principle. More direct evidence comes from recent **redshift surveys** of galaxies which have produced three-dimensional maps of the distribution of galaxies in space consistent with global homogeneity (see **large-scale structure**).

The cosmological principle was introduced by Einstein and adopted by subsequent relativistic cosmologists without any observational justification whatsoever. Their motivation was entirely to enable them to obtain solutions to the complicated mathematical equations of **general relativity**. Indeed, it was not known until the 1920s that the 'spiral nebulae' were external to our own Galaxy. A term frequently used to describe the entire Universe in those days was *metagalaxy*, indicating that the Milky Way was thought to be essentially the entire cosmos.

The Galaxy is certainly not isotropic on the sky, as anyone who has looked at the night sky will know. Although the name 'principle' sounds grand, principles are generally introduced into physics when there are no data to go on, and cosmology was no exception to this rule. Why the Universe is actually as homogeneous as it is was not addressed by the standard **cosmological models**: it was just assumed to be so at the outset.

More recently, the origin of large-scale smoothness has been called the **horizon problem** and it is addressed, for example, in models of the **inflationary Universe**.

The most important consequence of the cosmological principle is that the **spacetime** geometry must be as described by the **Robertson–Walker metric**, which drastically restricts the set of cosmological models compatible with general relativity. It also requires that the **expansion of the Universe** be described by **Hubble's law**. These assumptions are what lay behind the construction of the Friedmann models and their incorporation into the **Big Bang theory**.

In this context it is perhaps helpful to explain why the existence of quite large structures does not violate the cosmological principle. What counts, as far as cosmological models are concerned, is that the actual **metric** of the Universe should not deviate too far from the idealised mathematical form called the Robertson–Walker metric. A concentration in mass need not generate a large disruption of the metric unless it is of comparable size to the cosmological **horizon**, which is extremely large: thousands of millions of light years. The existence of structure on a scale of a few hundred million light years does not necessarily put the standard models in jeopardy.

Hermann **Bondi**, Fred **Hoyle** and Thomas **Gold**, among others, introduced a further refinement of this idea: the so-called *perfect cosmological principle*, in which homogeneity was extended to time as well as space. This produced the idea of an unchanging and eternal Universe described by the **steady state theory**. Subsequent evidence for the **evolution of galaxies** and, later, the discovery of the cosmic microwave background radiation, led to the rejection of the steady state model in the 1960s by most cosmologists.

SEE ALSO: **anthropic principle**, **hierarchical cosmology**.

COSMOLOGY The branch of astronomy concerned with the origin, evolution and physical properties of the **Universe**. The name is derived from the Greek word *cosmos* (meaning 'Universe'), which also gives its name to specialities within cosmology itself. For example, *cosmogony* is the name given to the study of the formation of individual objects within the Universe (see **structure formation**); this name is also used in older texts to refer to the formation of the Solar System. *Cosmogeny* or *cosmogenesis* describes the origins of the entire Universe; within the **Big Bang theory**, modern treatments of cosmogeny involve **quantum cosmology**. *Cosmography* is the construction of maps of the distribution of stars, **galaxies** and other matter in the Universe.

Broadly speaking, *observational cosmology* is concerned with those observable properties that give information about the Universe as a whole, such as its chemical composition, density and rate of expansion (see **expansion of the Universe**), and also the **curvature of spacetime**. Among the techniques for studying the latter properties are the tests of **classical cosmology**. The task of determining the expansion rate of the Universe boils down to the problem of calibrating the **extragalactic distance**

scale. The cosmography of the spatial distribution of galaxies (particularly the **large-scale structure** of this distribution) is also part of observational cosmology. Improvements in detector technology have made it possible to pursue observational cosmology to larger scales and to a greater accuracy than was imaginable in the 1960s and 1970s.

Physical cosmology is concerned with understanding these properties by applying known **laws of physics** to present-day data in order to reconstruct a picture of what the Universe must have been like in the past. One aspect of this subject is the reconstruction of the **thermal history of the Universe**, using the measured properties of the **cosmic microwave background radiation** and the physical laws describing the **elementary particles** and **fundamental interactions**.

Theoretical cosmology is concerned with making **cosmological models** that aim to provide a mathematical description of the observed properties of the Universe based on this physical understanding. This part of cosmology is generally based on the application of **general relativity**, **Einstein**'s theory of gravitation, because the large-scale behaviour of the Universe is thought to be determined by the fundamental interaction that has the strongest effect on large scales, and that interaction is **gravity**. General relativity enables us to create mathematical models that describe the relationship between the properties of **spacetime** and the material content of the Universe. The **Einstein equations**, however, are consistent with many different cosmological models so additional assumptions, such as the **cosmological principle**, are used to restrict the range of models considered. The 'standard' cosmological theory that has emerged is called the **Big Bang theory**, and is based on a particular family of solutions of the Einstein equations called the **Friedmann models**. This theory is supported by a great deal of observational evidence and is consequently accepted by the vast majority of cosmologists. However, direct knowledge of the physical behaviour of matter is limited to energies and temperatures that can be achieved in terrestrial laboratories. Theorists also like to speculate about the cosmological consequences of various (otherwise untestable) theories of matter at extreme energies: examples include **grand unified theories**, **supersymmetry**, **string theories** and **quantum gravity**. Modern cosmology is highly mathematical because it applies known (or speculative) physical laws to extreme physical situations of very high energies in the early Universe, or to very large scales of distance where the effects of the curvature of spacetime can be significant.

Cosmology also has philosophical, or even theological, aspects in that it seeks to understand why the Universe should have the properties it has. For an account of the development of cosmology from its beginnings in mythology and religious thought to its emergence as a branch of modern physical science, see Essay 1.

CRITICAL DENSITY see **density parameter**.

CURVATURE OF SPACETIME In **general relativity**, the effects of

gravity on massive bodies and radiation are represented in terms of distortions in the structure of **spacetime** in a rather different manner to the representation of forces in the theory of the other **fundamental interactions**. In particular, there is a possibility in Einstein's theory that the familiar laws of Euclidean geometry are no longer applicable in situations where gravitational fields are very strong. In **general relativity** there is an intimate connection between the geometrical properties of spacetime (as represented by the **metric**) and the distribution of matter. To understand this qualitatively, it is helpful to ignore the time dimension and consider a model in which there are only two spatial dimensions. We can picture such a two-dimensional space as a flat rubber sheet. If a massive object is placed on the rubber sheet, the sheet will distort into a curved shape. The effect of mass is to cause curvature. This is, in essence, what happens in general relativity.

The clearest way of tracing the curvature of spacetime is to consider the behaviour not of material bodies but of light paths, which follow curves called *null geodesics* in spacetime. These curves correspond to the shortest distance between two points. In flat, *Euclidean space*, the shortest distance between two points is a straight line, but the same is not necessarily true in curved space. Anyone who has taken a flight across the Atlantic knows that the shortest distance between London and New York does not correspond to a straight line on the map in the central pages of the in-flight magazine. The shortest distance between two points on the surface of a sphere is a *great circle*.

What complicates our understanding of this curvature in general applications of the theory of general relativity is that it affects both space and time, so that the bending of light by a gravitational field is also accompanied by serious difficulties in the choice of appropriate **time** coordinate. In the simplest **cosmological models** (the **Friedmann models**), however, the assumption of the **cosmological principle** singles out a unique time coordinate. The spacetime is then described by a relatively simple mathematical function called the **Robertson–Walker metric**. In this function we can clearly separate the gravitational effects on space and time: in the cosmological setting, it is really only the spatial part of the metric that is curved, so we can talk about the curvature of space alone from now on.

Even spatial curvature is difficult to visualise, because we are dealing with three spatial dimensions. We can all visualise a three-dimensional space without curvature: it is the ordinary Euclidean space with which we are familiar. But it is easier to see the effects of curvature by reducing the dimensionality to two and returning to the analogy of the rubber sheet. Here a flat space is a two-dimensional sheet, in which all of Euclidean geometry holds. For example, the internal angles of a triangle drawn in the surface add up to 180°. Now, imagine rolling up this sheet so that it becomes the surface of a sphere. This is a two-dimensional space with *positive* curvature. It is also finite. The sum of the angles of a triangle drawn in such a space is greater than 180°. The alternative case of *negative* curvature is even harder to visualise: it looks like a saddle, and triangles on it have internal angles adding up to less than 180°. It can also be infinite, whereas the sphere cannot. Now you have to visualise three-dimensional spheres and saddles. Whether or not such curved three-dimensional spaces are easy to imagine visually, they form an essential part of the mathematical language of cosmology.

If a Friedmann model universe has a density greater than a particular critical value (see **density parameter**), then space has *positive* curvature. This means that the space is curved in on itself like the sphere mentioned above. Such a universe has finite spatial size at any time, and also has a finite lifetime: it is a **closed universe**. A universe with a density less than the critical density has *negative* spatial curvature, is spatially infinite and expands for ever: it is an **open universe**. A critical density Friedmann universe is spatially flat (Euclidean) and infinite in both space and time (see **flat universe**).

Apart from the incorporation of curvature in cosmological models, there is another application of the idea which is relevant to cosmology. Suppose we have a flat cosmological model, represented (again) by a two-dimensional rubber sheet. This describes the gross structure of the Universe, but it does not describe any local fluctuations in density due to **galaxies**, **large-scale structure** or even **black holes**. The presence of such objects would not change the overall curvature of the sheet, but it would create local bumps within it. Light travelling past such a bump would be deflected. To visualise this, imagine placing a heavy cannonball on the rubber sheet and then trying to roll a marble past it. The marble would not travel in a straight line: it would deflected towards the cannonball. The deflection of light by massive objects is called **gravitational lensing**; cosmological applications of this idea are described in Essay 6.

D

DARK MATTER Material whose existence is inferred from astrophysical arguments, but which does not produce enough radiation to be observed directly. It is sometimes called the *missing mass*, which seems inappropriate as it is not missing at all. Several independent arguments suggest that our Universe must be dominated by dark matter whose form is not yet understood.

First, there are arguments based on **classical cosmology**. The classical cosmological tests use observations of very distant objects to measure the curvature of space, or the rate at which the **expansion of the Universe** is decelerating. In the simplest of these tests, the ages of astronomical objects (particularly stars in **globular clusters**) are compared with the age predicted by cosmological theory. If the expansion of the Universe were not decelerating, the **age of the Universe** would simply be the inverse of the **Hubble constant** H_0 (usually called the *Hubble time*). But the standard **Friedmann models** are always decelerating, so the actual age of the Universe is always less than the Hubble time by an amount which depends on the rate of deceleration, which in turn depends on the **density parameter** Ω. However, the predicted age depends much more sensitively on H_0, which is still quite uncertain, than on Ω; and in any case the ages of old stars are not known with any great confidence, so this test is not a great help when it comes to determining Ω.

In other classical tests, the properties of very distant sources are used to estimate directly the rate of deceleration or the spatial geometry of the Universe. Some of these techniques were pioneered by Edwin **Hubble** and developed into an art form by Allan **Sandage**. They fell into some disrepute in the 1960s and 1970s when it was realised that, not only was the Universe at large expanding, but objects within it were evolving rapidly. Since we need to probe very large distances to measure the geometrical effects of spatial curvature, we are inevitably looking at astronomical objects as they were when their light started out on its journey to us. This could be more than 80% of the age of the Universe ago, and there is no guarantee that the objects then possessed the properties that would now make it possible for us to use them as standard sources for this kind of observation. Indeed, the classical cosmological tests are now used largely to study the **evolution of galaxies**, rather than to test fundamental aspects of cosmology. There is, however, one important exception: the use of **supernova** explosions as standard light sources. No good estimates of Ω have yet been obtained from these supernova studies, but they do enable us to narrow down considerably the range within which the actual value of the **cosmological constant** must lie.

Next are arguments based on the theory of **nucleosynthesis**. The agreement between observed **light element abundances** and the predictions of these abundances in the early Universe is one of the major pillars of evidence supporting the **Big Bang theory**. But this agreement holds only if the density of matter is very low – no more than a few per cent of the critical density. This has been known for many years, and at first sight it seems inconsistent with the results obtained from classical cosmological tests. Only relatively recently, however, has it been realised that this constraint applies only to matter that can participate in nuclear reactions – to the specific class of **elementary particles** called *baryons* (protons and neutrons). With developments in particle physics came the suggestion that other kinds of particle might have been produced in the seething cauldron of the early Universe. At least some of these particles might have survived until now, and may make up at least some of the dark matter. If nucleosynthesis arguments and astrophysical arguments are both correct, then at least some of the Universe must be made from some form of exotic non-baryonic particle. 'Ordinary' matter, of which we are made, may even be but a small contaminating influence compared with the vast bulk of cosmic material whose nature is yet to be determined.

It also seems that the amount of baryonic matter required to make the predictions of nucleosynthesis fit the observations of light element abundances is larger than the amount of mass contained in stars. This suggests that at least some of the baryons in the Universe must be dark. Possible baryonic dark objects might be stars of very low mass, Jupiter-like objects or small **black holes**. Such objects go under the collective name of **massive compact halo objects**, or MACHOs. If they exist, they would reside in the haloes of **galaxies** and, though they would not produce enough light to be seen directly, they might be observable through their microlensing effects (see **gravitational lensing**, and Essay 6).

The third class of arguments is based on astrophysical considerations. The difference between these arguments and the intrinsically cosmological measurements discussed above is that they are based on looking at individual objects rather than the properties of the space between them. In effect, we are trying to determine the density of the Universe by weighing its constituent parts one by one. For example, we can attempt to use the internal dynamics of galaxies to work out their masses by assuming that the rotation of a galactic disk is maintained by gravity in much the same way as the motion of the Earth around the Sun. It is possible to calculate the mass of the Sun from the velocity of the Earth in its orbit, and by simply going up in scale the same idea can be extended to measure the amount of mass in galaxies (see **rotation curves**).

The principle of using velocities to infer the amount of gravitating mass can also be extended to clusters of galaxies and **large-scale structure** in the Universe, by using the **virial**

theorem or by analysing large-scale **peculiar motions**. These investigations overwhelmingly point to the existence of much more matter in galaxies than we can see in the form of stars. Moreover, rich clusters of galaxies – huge agglomerations of galaxies more than a million light years across – also contain more matter than is associated with the galaxies in them. Just how much more is unclear, but there is very strong evidence that there is enough matter in the rich cluster systems to suggest that Ω is certainly as big as 0.1, and possibly even larger than 0.3. The usual way of quantifying dark matter is to calculate the *mass-to-light ratio* (M/L): the ratio of the total mass (inferred from dynamics) to the total luminosity (obtained by adding up all the starlight). It is convenient to give the result in terms of the mass and luminosity of the Sun (which therefore has M/L equal to unity). For rich clusters (M/L) can be 200 or higher, so there is roughly ten times as much dark matter in clusters as there is in individual galaxies, for which (M/L) is typically about 20. These dynamical arguments have been tested and confirmed against independent observations of the gravitational lensing produced by clusters, and by measurements of the properties of the very hot, X-ray emitting gas that pervades them. Intriguingly, the amount of baryonic matter in clusters as a fraction of their total mass seems much larger than the value allowed by nucleosynthesis for the Universe as a whole if the Universe has the critical density. This so-called **baryon catastrophe** means either that the overall density of matter is much lower than the critical value, or that some unknown process has concentrated baryonic matter in clusters.

The fourth and final class of arguments is based around the problem of **structure formation**: how the considerable lumpiness and irregularity of the Universe can have developed if the Universe is required by the **cosmological principle** to be largely smooth. In the Big Bang models this is explained by *gravitational instability*. Since gravity attracts all matter, a region of the Universe which has a density slightly higher than average will accrete material from its surroundings and become still denser. The denser it gets, the more it will accrete. Eventually the region will collapse to form a gravitationally bound structure such as a galaxy. The rate at which the density increases inside these proto-structures depends on the overall density of matter, Ω. The detection by the **Cosmic Background Explorer** satellite of features in the **cosmic microwave background radiation** tells us how big the irregularities were when the microwave background was produced, about 300,000 years after the Big Bang. And we can try to measure the irregularities by measuring the clustering properties of galaxies, and also of other systems.

In principle, then, we can determine Ω; but in practice the calculation is incredibly complicated and prone to all kinds of uncertainty and bias. Models can and have been constructed which seem to fit all the available data with Ω very close to 1. Similarly consistent models can be built on the assumption that Ω is much less than 1. This may not sound encouraging, but this kind of study probably ultimately holds the key to a successful determination of Ω. If more detailed measurements of the features in the cosmic microwave background can be made, then the properties of these features will tell us immediately what the density of matter must be. And, as a bonus, it will also determine the Hubble constant, bypassing the tedious business of constructing the **extragalactic distance scale**. We can only hope that the satellites planned to do this, the Microwave Anisotropy Probe (NASA) and Planck Surveyor (ESA), will fly successfully in early years of the next millennium (see Essays 2 and 5).

Since the first attempt to gauge the amount of dark matter by using these various approaches, the estimated amount has been creeping steadily upwards. This is partly because of the accumulating astronomical evidence, as described above, but also because of the realisation that, in theory, elementary particles other than baryons might be produced at the very high energies that prevailed in the Big Bang (see **weakly interacting massive particles**). Further support for a high density of dark matter comes from the theory of cosmological inflation (see **inflationary Universe**; see also **flatness problem**).

FURTHER READING: Coles, P. and Ellis, G.F.R., *Is the Universe Open or Closed?* (Cambridge University Press, Cambridge, 1997).

DECELERATION PARAMETER (q) The parameter that quantifies the rate at which the **expansion of the Universe** is being slowed down by the matter within it. The deceleration parameter, along with the **Hubble constant** H_0 and the **density parameter** Ω, describes some of the mathematical properties of the **Friedmann models**. Like the density parameter, to which it is closely related, it is a pure (dimensionless) number. In the standard Friedmann models, in which the **cosmological constant** term is set to zero, the value of q is simply equal to half the value of Ω. An empty universe would have $q = 0$ (no deceleration), while a **flat universe** would have $q = 0.5$. In the context of these models the deceleration parameter is redundant, as it provides no more information about the behaviour of the model than does Ω itself.

The situation is different, however, once a cosmological constant term (Λ) is included in the equations of **general relativity**. Such a term can produce an acceleration in the **expansion of the Universe** if the cosmic repulsion it produces is strong enough to overcome the self-attractive gravitational forces between the ordinary matter. To put it another way, acceleration can be produced if the vacuum energy density associated with a cosmological constant exceeds the energy density of normal matter. In such cases the equivalence of Ω and q is broken: though Ω is then always positive, we can still make models in which q is negative.

Despite it being similar in meaning to Ω, the deceleration parameter is still a very useful number, principally because some of the tests in **classical cosmology** depend on it directly. In the early days of modern cosmology, many observational astronomers hoped that by using these tests they could learn directly about q. Indeed,

Allan Sandage once wrote a paper entitled 'Cosmology: The search for two numbers', the two numbers in question being q and H_0. Unfortunately, cosmologists still do not know exactly what these two numbers are because of the confusing effect of the **evolution of galaxies** on the interpretation of observational data. Contemporary astronomers, however, usually concentrate on measuring directly the amount of **dark matter**, leading to estimates of the density parameter rather than the deceleration parameter.

DECOUPLING see **cosmic microwave background radiation**, **thermal history of the Universe**.

DENSITY PARAMETER (Ω) In the standard **Big Bang theory**, the evolution of the Universe is described by one of the family of **Friedmann models** and, in particular, by the Friedmann equation which all these models obey. The Friedmann equation can be thought of as expressing the law of conservation of energy for the Universe as a whole. There are basically two forms of energy that are significant on cosmological scales: the kinetic energy of the **expansion of the Universe**, and the potential energy arising from the attractive force that all matter exerts on all other matter. In cosmological terms, the kinetic energy depends crucially on the expansion rate or, in other words, on the **Hubble constant** (H_0). The potential energy depends on how much matter there is on average in the Universe per unit volume. The phrase 'on average' is important here because the Friedmann models describe a universe in which the **cosmological principle** is exactly true. However, our Universe is not exactly homogeneous and isotropic, so we have to imagine taking averages over regions that are sufficiently large for fluctuations in density from region to region to be small. Unfortunately, this average density is not known at all accurately: it is even less certain than the value of H_0. If we knew the mean density of matter and the value of H_0, we could calculate the total energy of the Universe. This would have to be constant in time, in accordance with the law of conservation of energy (or, equivalently, to be consistent with the Friedmann equation).

Setting aside the technical difficulties that arise whenever **general relativity** is involved, it is possible to discuss the evolution of the Universe in broad terms by using analogies familiar from high-school physics. For instance, consider the problem of launching a vehicle from the Earth into space. Here the mass responsible for the gravitational potential energy of the vehicle is the Earth. The kinetic energy of the vehicle is determined by the power of the rocket we use. If we give the vehicle only a modest boost, so that it does not move very quickly at launch, then the kinetic energy is small and may be insufficient for the rocket to escape the attraction of the Earth. The vehicle would then go up some way and then come down again. In terms of energy, what happens is that the rocket uses up its kinetic energy, given expensively at its launch, to pay the price in potential energy for its increased height. If we use a more powerful rocket, the vehicle would go higher before crashing down to the ground. Eventually we find a rocket big enough to supply the vehicle with enough energy for it to buy its way completely out of the gravitational clutch of the Earth. The critical launch velocity here is usually called the *escape velocity*. Above the escape velocity, the rocket keeps going for ever; below it, the rocket comes crashing down.

In the cosmological context the picture is similar, but the critical quantity is not the velocity of the rocket (which is analogous to the Hubble constant, and is therefore known, at least in principle), but the density of matter. It is therefore most useful to think about a critical density of matter, rather than a critical velocity. The critical density required to make the Universe recollapse equals $3H_0^2/8\pi G$, where G is the Newtonian gravitational constant. For values of the Hubble constant H_0 between 50 and 100 km/s/Mpc, the critical density is extremely small: it corresponds to just a few atoms of hydrogen per cubic metre, or around 10^{-29} grams per cubic centimetre. Although the stars and planets contain matter whose density is of the same order as that of water (1 gram per cubic centimetre), the average density is extremely small because most of space is very much emptier than this.

If the real density of matter exceeds the critical density, the Universe will eventually recollapse: its gravitational energy will be sufficient to slow down, stop and then reverse the expansion. If the density is lower than this critical value, the Universe will carry on expanding for ever. And now, at last, we can introduce the quantity Ω: it is simply the ratio of the actual density of matter in the Universe to the critical value that marks the dividing line between eternal expansion and ultimate recollapse. $\Omega = 1$ therefore marks that dividing line: $\Omega < 1$ defines an ever-expanding universe, while $\Omega > 1$ defines a universe that eventually recollapses at some time in the future in a *Big Crunch* (see the Figure). Whatever the precise value of Ω, however, the effect of matter is always to slow down the expansion of the Universe, so that these models always predict a cosmic deceleration.

But the long-term viability of the cosmological expansion is not the issue whose resolution depends on Ω. These arguments based on simple ideas of energy resulting from Newtonian physics are not the whole story. In general relativity the total energy density of material determines a quantity for which there is no Newtonian analogue at all: the **curvature of spacetime**. A space of negative global curvature results in models with $\Omega < 1$. Such models are called **open universe** models. Positive curvature results in **closed universe** models with $\Omega > 1$. In between, poised

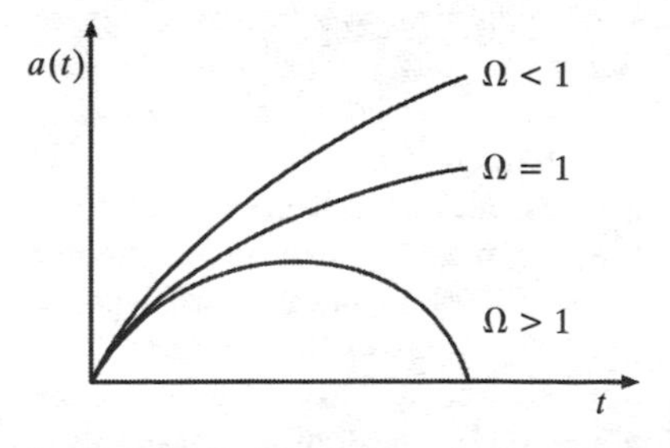

Density parameter Schematic illustration of how the cosmic scale factor $a(t)$ depends on time for different values of Ω in the standard Friedmann models.

between eternal expansion and eventual recollapse, are **flat universe** models with Ω exactly equal to 1. A flat universe has the familiar *Euclidean geometry* – that is, one in which Euclid's theorems apply.

So, the quantity Ω determines both the geometry of space on cosmological scales and the eventual fate of the Universe. But it is important to stress that the value of Ω is not *predicted* by the standard Big Bang model. It may seem a fairly useless kind of theory that is incapable of answering the basic questions that revolve around Ω, but that is an unfair criticism. In fact, the Big Bang is a model rather than a theory. As a model, it is self-consistent mathematically and when compared with observations, but it is not complete, as discussed above. This means that it contains free parameters (of which Ω and H_0 are the two most obvious examples). To put it another way, the mathematical equations of the Big Bang theory describe the evolution of the Universe, but in order to calculate a specific example we need to supply a set of initial conditions to act as a starting point. Since the mathematics on which it is based break down at the very beginning, we have no way of fixing the initial conditions theoretically. The Friedmann equation is well defined whatever the values of Ω and H_0, but our Universe happens to have been set up with one particular numerical combination of these quantities. All we can do, therefore, is to use observational data to make inferences about the cosmological parameters: these parameters cannot, at least with the knowledge presently available and within the framework of the standard Big Bang, be deduced by reason alone. On the other hand, there is the opportunity to use present-day cosmological observations to learn about the very early Universe, and perhaps even to penetrate the **quantum gravity** era.

Although the value of Ω is still not known to any great accuracy, progress has been made towards establishing its value, particularly with the discovery of overwhelming evidence that the Universe contains large quantities of **dark matter**. We still do not know precisely how much of it there is, or even what it is, but several independent arguments suggest that it is the dominant form of matter in the Universe.

Most cosmologists accept that the value of Ω cannot be smaller than 0.1. Even this minimum value requires the existence of dark matter in the form of exotic **elementary particles**, usually known as **weakly interacting massive particles**, or WIMPs. Many cosmologists favour a somewhat higher value of Ω, of around 0.3, which seems to be consistent with most of the observational evidence. Some claim that the evidence supports a value of the density close to the critical value, so that Ω can be very close to 1. The controversy over the value of Ω has arisen partly because it is difficult to assess the reliability and accuracy of the (sometimes conflicting) observational evidence. The most vocal arguments in favour of a high value for Ω (i.e. close to 1) are based on theoretical grounds, particularly in the resolution of the cosmological **flatness problem** by the idea of an **inflationary Universe**, which seems to require the density to be finely poised at the critical value. Ultimately, however, theoretical arguments will have to bow to empirical evidence, and, as technological developments continue, the question of the amount of dark matter may well be resolved in the early years of the new millennium.

FURTHER READING: Coles, P. and Ellis, G. F. R., *Is the Universe Open or Closed?* (Cambridge University Press, Cambridge, 1997).

DE SITTER, WILLEM (1872–1934) Dutch astronomer and mathematician. He was one of the first to derive solutions of the equations of **general relativity** that describe an expanding **cosmological model**. The solution he obtained is still used as the basis of the **inflationary Universe**: it corresponds to a universe dominated by vacuum energy or, equivalently, by a **cosmological constant**.

DE VAUCOULEURS, GÉRARD HENRI (1918–95) French-born astronomer, who worked in Australia and, from 1957, in the USA. He is known for his thorough survey of the brighter galaxies, from which he deduced the existence of a local supercluster of galaxies. This work provided a cornerstone for modern studies of the **large-scale structure** of the Universe. He also made significant contributions to establishing the **extragalactic distance scale**.

DICKE, ROBERT HENRY (1916–97) US physicist. Unaware of the earlier work by Ralph **Alpher**, Robert **Herman** and George **Gamow**, he predicted the existence of the **cosmic microwave background radiation** in 1964 on the basis of the **Big Bang theory** shortly, before its actual discovery by Arno **Penzias** and Robert **Wilson**. He also worked on a theory of gravity that explicitly takes account of **Mach's principle**, now known as the **Brans–Dicke theory**, and presented ideas that relate to the weak version of the **anthropic principle**.

DIGGES, THOMAS (*c.* 1546–95) English mathematician. In his 1576 exposition of the heliocentric model proposed by Copernicus, he depicted the sphere of fixed stars as infinite in extent – a significant innovation in post-medieval astronomy – and was one of the first to correctly formulate what is now called **Olbers' paradox**.

DILATON see **Brans–Dicke theory**.

DIMENSION see **Kaluza–Klein theory**, **string theory**.

DIRAC, PAUL ADRIEN MAURICE (1902–84) English physicist. He made many important contributions to particle physics and **quantum theory**, including the prediction of antimatter, and the equation now bearing his name which combines quantum mechanics with **special relativity**. He also developed some interesting ideas on cosmology (see **Dirac cosmology**).

DIRAC COSMOLOGY A novel and imaginative approach to the study of cosmology introduced in 1937 by Paul **Dirac**. It was built upon the so-called *large number hypothesis*, which relates physical constants measuring the strength of the **fundamental interactions** to macroscopic properties of the Universe at large such as its age and mean density. In particular, Dirac was struck by the properties of dimensionless numbers (quantities whose numerical value does not depend on any particular choice of units) and the relations between them.

For example, we can construct a dimensionless ratio measuring the relative strengths of the electrostatic force between a proton and neutron and the gravitational force between the same two particles. If the charge on the electron is e, the Newtonian gravitational constant is G and the masses of the proton and electron are m_p and m_e, then the required combination is $e^2/Gm_p m_e$, which turns out to be around 10^{40}. Strikingly, other dimensionless ratios which appear to be completely unrelated also turn out to be of the order of 10^{40}. For example, the ratio of the Compton wavelength (see **quantum physics**) to the Schwarzschild radius (see **black holes**) of a proton, also a dimensionless quantity, is given by hc/Gm_p^2, where h is the Planck constant (see **quantum physics**) and c is the velocity of light (see **special relativity**). This quantity also turns out to have around the same value. Another example: if we calculate the scale of the cosmological **horizon**, given roughly by c/H_0 where H_0 is the **Hubble constant**, and compare it with the classical electron radius, which is $e^2/m_e c^2$, the ratio between these two quantities again turns out to be similar to 10^{40}.

These coincidences deeply impressed Dirac, who decided that the fundamental constants from which they are constructed had to evolve with time. In particular, his theory has a value of the gravitational constant G that varies inversely with time in such a way that the dimensionless ratios mentioned above appear large simply because the Universe is so old. Unfortunately, the time variation in G that Dirac's theory predicted appears to make it impossible for stars to remain stable for billions of years, as they are required to in standard models of **stellar evolution**.

For this and other reasons Dirac's theory has never been widely accepted, but it was historically important because it introduced cosmologists to new ways of thinking about the relationship between cosmology and fundamental physics. The resolution of the large number coincidences which so puzzled Dirac is now thought to reside, at least partly, in an application of the weak version of the **anthropic principle**.

FURTHER READING: Dirac, P. A. M., 'The cosmological constants', *Nature*, 1937, **139**, 323; Narlikar, J. V., *Introduction to Cosmology*, 2nd edition (Cambridge University Press, Cambridge, 1993), Chapter 8.

DOPPLER EFFECT, DOPPLER SHIFT The Doppler effect, explained and formulated by Christian Johann Doppler (1803–53), was introduced to physics with a fanfare in the 1840s (literally, because several trumpeters on a steam train took part in its first experimental demonstration). Doppler used it to explain the properties of sound waves when there is relative motion between the source of the sound and the receiver. We are all familiar with the effect from everyday experience: an approaching police siren has a higher pitch than a receding one. The easiest way to understand the effect is to remember that the pitch of sound depends on the wavelength of the acoustic waves that carry the sound energy. High pitch means short wavelength. If a source is travelling close to the speed of sound, it tends to approach the waves it has emitted in front of it, thus reducing their apparent wavelength. Likewise, it tends to rush ahead of the waves it emits behind it, thus lowering their apparent pitch.

In astronomical contexts the Doppler effect applies to light waves, and it becomes appreciable if the velocity of a source is a significant fraction of the velocity of light. A moving source emitting **electromagnetic radiation** tends to produce light of shorter wavelength if it is approaching the observer, and light of a longer wavelength if it is receding. The light is thus shifted towards the blue and red parts of the spectrum, respectively. In other words, there is a blueshift (approaching source) or a **redshift** (receding source). If the source is emitting white light, however, we are not able to see any kind of shift. Suppose each line were redshifted by an amount $\Delta\lambda$ in wavelength. Then light which is emitted at a wavelength λ would be observed at wavelength $\lambda + \Delta\lambda$. But the same amount of light would still be observed at the original wavelength λ, because light originally emitted at wavelength $\lambda - \Delta\lambda$ would be shifted there. White light therefore still looks white, regardless of the *Doppler shift*. To see an effect, one has to look at **absorption lines** or **emission lines**, which occur at discrete wavelengths so that no such compensation can occur. A whole set of lines will be shifted one way or the other in the spectrum, but because the lines keep their relative spacing it is quite easy to identify how far they have shifted relative to a source at rest (e.g. in a laboratory).

The Doppler effect is useful for measuring properties of astronomical objects, such as **galaxies**. In a spiral galaxy, for example, **spectroscopy** can be used to map how different parts of the galaxy are moving, and hence to calculate the amount of **dark matter** from its gravitational effect on these motions. This task is simplified by the fact the rotation of such an object is relatively ordered and well behaved. In more chaotic systems, motions may be either towards or away from the observer at any place, so what happens is that lines are simply made thicker, a phenomenon known as *Doppler broadening*.

The most important application of this effect in cosmology is, however, as an indicator of the **expansion of the Universe** as described by **Hubble's law**. Hubble's original discovery was a measured correlation between the **redshift** of galaxies and their estimated distance from the Milky Way. At the time there was some confusion over whether this was a Doppler effect or not, particular as **general relativity** predicted another form of redshift produced solely by gravitational effects. To this day it is often stated (even in textbooks) that the Doppler effect is not responsible for the phenomenon of cosmological redshift which, it is claimed, is the result of space itself expanding rather than the galaxies moving through space. It is indeed often helpful to regard light as being stretched by the expansion of the Universe in this manner, but there is actually no difference between this and the Doppler effect. Space is expanding because galaxies are moving apart, and vice versa.

Doppler peaks see **Sakharov oscillations**.

E

EDDINGTON, SIR ARTHUR STANLEY (1882–1944) English astrophysicist. He pioneered the study of stellar structure and **stellar evolution**, and was responsible for bringing the work of Georges **Lemaître** to the attention of a wider audience. He also led the famous solar eclipse expedition in 1919 that measured the bending of light predicted by the theory of **general relativity**; this was the first recorded example of **gravitational lensing**.

EINSTEIN, ALBERT (1879–1955) German physicist, who took Swiss and later US citizenship. He made immense contributions to all branches of physics, formulating the theories of **special relativity** and **general relativity**, and making fundamental discoveries in **quantum theory**. The gravitational field equations of general relativity form the basis of most modern cosmological models, and provide the foundation of the **Big Bang theory**. Einstein was awarded the Nobel Prize for Physics in 1921, but it is curious that the citation did not mention his work on relativity: the award was in recognition of a paper he had published in 1905 on the photoelectric effect, in which he showed that light was quantised. He never really accepted the inherently indeterministic nature of quantum theory, leading to his famous remark that 'God does not play dice with the Universe'.

EINSTEIN–DE SITTER UNIVERSE see **cosmological models**.

EINSTEIN (FIELD) EQUATIONS The mathematical theory of **general relativity is constructed around the idea that **spacetime** must be treated as curved, rather than considered as being flat as in **special relativity**. This is because special relativity applies only locally in the general theory. The neighbourhood of a single point can be described by the flat spacetime of special relativity, but this does not work for an extended object because a different transformation is needed for each point in the object. For example, consider a set of particles arranged in a line. At a perpendicular distance d from some point on this line is a large mass M which exerts a gravitational field. Now consider the point P on the line nearest to the mass M. This point experiences a gravitational attraction directly towards M. To overcome this force we would need to accelerate in the same direction as the line from M to P. But another point Q farther along the line would also feel a force directed towards M. To overcome this force we would have to accelerate along the line from M to Q. The first transformation will not remove the gravitational force acting at Q as well as that acting at P because the direction needed is different. We therefore have to use a 'curved' transformation that is different for every point on the line.

A more precise version of this idea is to consider the properties of light rays. In special relativity these paths are described by straight lines with $ds = 0$ in the Minkowski **metric**:

$$ds^2 = c^2 dt^2 - (dx^2 + dy^2 + dz^2)$$

However, in an accelerated **frame of reference** light is not seen to travel in a straight line. Imagine, for example, a laser beam shining from one side of a lift to the other. In the absence of acceleration the beam shines in a straight line across the lift. But if the lift is accelerated upwards, then in the time it takes a photon to travel across the lift, the lift will have moved slightly upwards. The beam therefore strikes the opposite wall slightly below the point it would illuminate in the stationary lift. The light path is therefore deflected downwards in the lift's frame of reference. Because of the equivalence principle, the same thing happens if there is a gravitational field: light gets curved in a gravitational field, which is the principle behind **gravitational lensing**. Gravitational fields also cause time-dilation and **redshift** effects. Clocks run at different speeds in gravitational fields of different strengths; light travelling upwards against a gravitational field loses energy.

In general, ds in the above metric is invariant under changes of coordinate system, and for the path of a light ray it is always the case that $ds = 0$. If we accept that light rays in the general theory also have to correspond to intervals with $ds = 0$, then there is no choice but to use a different metric, one that describes a curved spacetime. Since gravitation exerts the same force per unit mass on all bodies, the essence of general relativity is to change gravitation from being a force to being a property of spacetime. In Einstein's theory, spacetime is described in mathematical terms as having a *Riemannian geometry* (formally, it is a four-dimensional Riemannian manifold). In general the metric of this geometry is a **tensor** denoted by g_{ij}, and the interval between two events is written as

$$ds^2 = g_{ij} dx^i dx^j$$

where repeated suffixes imply summation, and i and j both run from 0 to 3; $x^0 = ct$ is the time coordinate, and x^1, x^2, x^3 are the space coordinates. Particles acted on by no gravitational forces move along paths which are no longer straight because of the effects of curvature which are contained in g_{ij}. Free particles move on *geodesics* in the spacetime, but the metric g_{ij} is itself determined by the distribution of matter. The key factor in Einstein's equations is the relationship between the matter and the metric.

In Newtonian and special relativistic physics a key role is played by conservation laws of mass, energy and momentum. With the equivalence of mass and energy brought about by special relativity, these laws can be rewritten as

$$(\partial T_{ik}/\partial x^k) = 0$$

where T_{ik}, the *energy–momentum tensor*, describes the matter distribution; in cosmology this is usually assumed to be a perfect fluid with pressure p and density ρ. The form of T_{ik} is then

$$T_{ik} = (p + \rho c^2)u_i u_k - pg_{ik}$$

in which u_i is the fluid 4-velocity

$$u_i = g_{ik}u^k = g_{ik}\,\mathrm{d}x^k/\mathrm{d}s$$

where $x^k(s)$ is the *world line* of a particle in the fluid; that is, the trajectory the particle follows in spacetime. (This equation is a special case of the general rule for raising or lowering suffixes using the metric tensor.)

There is a problem with the conservation law as expressed above in that it is not a tensor equation, since the derivative of a tensor is not itself necessarily a tensor. This was unacceptable to Albert **Einstein** because he wanted to be sure that all the laws of physics had the same form regardless of the coordinates used to describe them. He was therefore forced to adopt a more complicated form of derivative called the *covariant derivative* (for details, see the FURTHER READING list on p. 168). The conservation law can therefore be written in fully covariant form as $T^k_{i\ ;k} = 0$. (A covariant derivative is usually written as a semicolon in the subscript; ordinary derivatives are usually written as a comma, so that the original equation would be written as $T^k_{i\ ,k} = 0$.)

Einstein wished to find a relation between the energy–momentum tensor for the matter and the metric tensor for the spacetime geometry. This was the really difficult part of the theory for Einstein to construct. In particular, he needed to find how to handle the curvature of spacetime embodied in the metric g_{ik} so that his theory could reproduce the well-understood and tested behaviour of Newton's laws of **gravity** when the gravitational fields were weak. This was the vital clue, because in order to reproduce Poisson's equation (see **gravity**) for gravity, the theory had to contain at most second-order derivatives of g_{ik}.

The properties of curved spaces were already quite well known to mathematicians when Einstein was working on this theory. In particular there was known to be a tensor called the *Riemann–Christoffel tensor*, constructed from complicated second-order derivatives of the metric and from quantities called the Christoffel symbols, which could be used to tell unambiguously whether a Riemannian manifold is flat or curved. This tensor, which is too complicated to write here, has four indices. However, it can be simplified by summing over two of the indices to construct the *Ricci tensor*, which has only two indices:

$$R_{ik} = R^l_{\ ilk}$$

From this we can form a scalar quantity describing the curvature, called the *Ricci scalar*:

$$R = g^{ik}R_{ik}$$

Einstein used these complicated mathematical forms to construct a new tensor, which is now known as the *Einstein tensor*:

$$G_{ik} = R_{ik} - \tfrac{1}{2}g_{ik}R$$

Einstein showed that the conservation law given above is equivalent to $G^k_{i\ ;k} = 0$. He therefore decided that his fundamental theory must make these two different tensors – one describing the matter distribution, the other describing the spacetime geometry – proportional to each other. The constant of proportionality can be fixed by requiring the resulting equations to match Poisson's equation, given above, in the limit of weak fields. What results are the famous Einstein field equations in the form:

$$G_{ik} = (8\pi G/c^4)T_{ik}$$

He subsequently proposed a modification of the original form of the Einstein tensor, so that it became

$$G'_{ik} = R_{ik} - \tfrac{1}{2}g_{ik}R - \Lambda g_{ik}$$

in which Λ is the so-called **cosmological constant**. He did this to ensure that static cosmological solutions could be obtained (see **cosmological models**).

The final form of the Einstein equations is deceptively simple because of the very compact tensor notation used. In the general case there are ten simultaneous nonlinear partial differential equations to solve, which is not an easy task compared with the single equation of Newtonian gravity. It is important to stress the essential nonlinearity of Einstein's theory, which is what sets it aside from the Newtonian formulation. Because of the equivalence between mass and energy embodied in special relativity through the relation $E = mc^2$, all forms of energy gravitate. The gravitational field produced by a body is itself a form of energy which also gravitates, so there is a great deal of complexity in any physical situation where these equations are applied. Note also that because of the form of the energy–momentum tensor, pressure itself also gravitates. Objects with zero rest mass (e.g. photons of **electromagnetic radiation**) can exert a gravitational force that corresponds to them having a total energy density which is nonzero, even though their mass density is zero. This has important consequences for the rate of expansion during the early stages of the **thermal history of the Universe**, at which time the pressure of **radiation** was what dominated the expansion of the Universe.

FURTHER READING: Berry, M. V., *Principles of Cosmology and Gravitation* (Adam Hilger, Bristol, 1989); Schutz, B. F., *A First Course in General Relativity* (Cambridge University Press, Cambridge, 1985); Rindler, W., *Essential Relativity: Special, General, and Cosmological*, revised 2nd edition (Springer-Verlag, New York, 1979); Misner, C. W., Thorne, K. S. and Wheeler, J. A., *Gravitation* (W. H. Freeman, San Francisco, 1972); Weinberg, S., *Gravitation and Cosmology: Principles and Applications of the General Theory of Relativity* (John Wiley, New York, 1972).

ELECTROMAGNETIC FORCE see **fundamental interactions**.

ELECTROMAGNETIC RADIATION Oscillating electric and magnetic fields propagating through space. Electromagnetic radiation is generally associated with the motion of charged particles. It can be either produced or absorbed, for example by electrons moving along a wire or by electrons changing orbits within an atom or molecule. The first theory of this form of radiation was encapsulated in *Maxwell's equations* of electromagnetism. James Clerk **Maxwell** showed that one solution of his equations corresponded to a wave travelling through space at the speed of light, c. It was subsequently realised that visible light is merely one form of electromagnetic radiation, corresponding to oscillations at a particular frequency.

Maxwell's theory of electromagnetism was a classical theory. With developments in **quantum physics**, it became clear that electromagnetic radiation was quantized: it was not emitted continuously, but in small packets now called *photons*. Each photon of a particular frequency ν carries a certain amount of energy E given by the formula $E = h\nu$, where h is the Planck constant. (The associated wavelength is simply c/ν, where c is the velocity of light.) Considered as **elementary particles**, photons are the bosons responsible for carrying the electromagnetic interaction between charged particles (see **fundamental interactions** for more details).

The spectrum of electromagnetic radiation spans many orders of magnitude in frequency, from very low frequency (and therefore low energy) radio waves to phenomenally energetic gamma-rays. Before the 20th century, astronomers were restricted to the use of optical wavelengths spanning the range from red (the lowest optical frequency) to blue colours (see **optical astronomy**). Nowadays a vast array of different devices, both on the ground and in space, can be used to obtain astronomical images and spectra in virtually all parts of the electromagnetic spectrum (see **gamma-ray astronomy, X-ray astronomy, ultraviolet astronomy, infrared astronomy, radio astronomy**; see also Essay 4).

The particular form of radiation emitted by an object depends on its chemical composition and the physical conditions within it. In general an object will emit different amounts of energy at different frequencies: the dependence of the energy radiated on the frequency of emission is usually called the *spectrum*. The simplest form of emission is the **black-body** radiation emitted by an object in a state of **thermal equilibrium**. In this case the spectrum has a characteristic curve predicted by quantum physics. Such radiation has a characteristic peak at a frequency that depends on the temperature of the source. At temperatures of several thousand degrees the peak lies in the optical part of the spectrum, as is the case for a star like the Sun. Hotter objects produce most of their radiation at higher frequencies, and cooler objects at lower frequencies. The best example of a black-body spectrum known to physics is the spectrum of the **cosmic microwave background radiation**, which has a temperature of 2.73 K (see Essay 5).

More complicated spectra are produced when non-thermal processes are involved. In particular, spectral lines can be produced at definite frequencies when electrons absorb or emit energy in order to move between one quantum state of an atom and another (see **absorption lines, emission lines**). The observational study of spectra and their interpretation in terms of the physics of the emitting or absorbing object is called **spectroscopy**.

ELECTROMAGNETISM see **fundamental interactions**.

ELECTRON see **elementary particles**.

Elementary particles The basic subatomic building blocks of all matter, which interact with each other by the various **fundamental interactions**. Strictly speaking, an elementary particle is one that cannot be broken down into any smaller components, but the term is used for some kinds of particle that do not meet this criterion. **Quantum physics** recognises two basic kinds of particle: *fermions* (which all have half-integer spin) and *bosons* (which have integer spin). Particles can also be classified in a different way, according to which of the fundamental interactions the particle experiences.

Particles that experience only the weak nuclear interaction are called *leptons*. The lepton family includes electrons and neutrinos and their respective antiparticles (see **antimatter**). These are, at least in current theories, genuinely fundamental particles which cannot be divided any further. All leptons are *fermions*. Particles that experience the strong nuclear interaction are called *hadrons*. This family is divided further into the *baryons* (which are fermions, and which include the proton and neutron) and the *mesons* (which are bosons). These particles, however, are not genuinely elementary because they are made up of smaller particles called *quarks*. Quarks come in six different varieties: up, down, strange, charm, top and bottom. These are usually abbreviated to the set of initials (u, d, s, c, t, b). The quarks have fractional electrical charge. Each hadron species is made up of a different combination of quarks: the baryons are quark triplets, and the mesons are doublets. In terrestrial laboratories, matter has to be accelerated to very high energies to smash hadrons into their constituent quarks, but suitable conditions would have existed in the inferno of the Big Bang.

Alongside these fundamental constituents are the particles responsible for mediating the fundamental interactions between them. These particles are all examples of gauge bosons (see **gauge theory**). The most familiar of these is the *photon*, the quantum of **electromagnetic radiation**, which carries the electromagnetic force between charged particles. The weak force is mediated by the *W* and *Z bosons* which, unlike the photon, are massive and therefore of short range. The strong force that binds quarks together is mediated by bosons called *gluons*.

The cosmological relevance of this plethora of possible particles is that, in the early stages of the **thermal history of the Universe**, the temperature was so high that there was enough energy in the **cosmic microwave background radiation** to create particle–antiparticle pairs out of the radiation. This led to cosmic epochs where lepton–antilepton pairs and hadron–antihadron pairs came to dominate the behaviour of the Universe. Many of the particles described above are highly unstable (unlike the proton and the electron) and do not survive for long, but they still had an impact on the evolution of the Universe. Outside the nucleus, the neutron (a baryon) is unstable, with a half-life of about 10 minutes, and its rate of decay helps to determine the amount of helium that was made by cosmological **nucleosynthesis**.

The families of particles discussed

above contain between them all the species whose existence is either known or can be confidently inferred from experiments carried out in particle accelerators. It is possible, however, that at energies beyond our current reach other particles can be created. In particular, theories of physics based on the idea of **supersymmetry** suggest that every bosonic particle has a fermionic partner. Thus, the photon has a supersymmetric partner called the photino, and so on. It is possible that at least one such particle might actually be stable, and not simply have had a fleeting existence in the early stages of the Big Bang. It has been speculated that the most stable supersymmetric fermion might be a good candidate for a **weakly interacting massive particle**, capable of forming the apparently ubiquitous **dark matter**.

SEE ALSO: **grand unified theory**, **theory of everything**.

FURTHER READING: Weinberg, S., *The First Three Minutes* (Fontana, London, 1983); Close, F., *The Cosmic Onion* (Heinemann, London, 1983).

EMISSION LINE A very sharp 'spike' in the spectrum of **electromagnetic radiation** emitted by an object, usually corresponding to a particular change in the energy state of an electron. For example, the so-called *recombination lines* arise when a free electron combines with an ion. As the electron falls through the different atomic energy levels, it emits photons at frequencies corresponding to the gaps between the levels. Since there are usually many energy states and hence many possible transitions, emission lines can be produced at a wide range of frequencies, from the radio band all the way up to X-rays (for certain elements). The pattern of lines produced is therefore a sensitive indicator of both the chemical makeup of the source and the physical conditions within it.

Broad emission lines are often very prominent in the spectra of **active galaxies**, indicating the presence of matter in a very excited state. Although such lines are naturally very narrow, they are broadened because the emitting material is moving very quickly either towards or away from the observer. This spreads out the line in frequency as a result of the **Doppler shift**. **Quasars** are often identified by their extremely strong Lyman-alpha emission, which is produced in the ultraviolet part of the electromagnetic spectrum. As a result of the extreme cosmological **redshifts** of these objects, this line is often shifted into the optical spectrum, where it is easily seen. Measurements of spectral emission lines in normal **galaxies** are the basis of the usual method for determining galaxy redshifts, and were what led to the discovery of **Hubble's law** for the **expansion of the Universe**.

FURTHER READING: Aller, L. H., *Atoms, Stars, and Nebulae* (Harvard University Press, Cambridge, MA, 1971).

EQUIVALENCE PRINCIPLE see **general relativity**.

ETHER see **aether**.

EUCLIDEAN SPACE see **curvature of space-time**, **flat universe**.

Evolution of galaxies Any systematic change in the properties of individual galaxies, or of populations of galaxies, with cosmic time. The properties of **galaxies** are expected to have been different in the past to what they are now. It may be, for example, that galaxies were systematically brighter in the past, since they may have contained more young, bright stars. This would be an example of *luminosity evolution*. It may also have been the case that galaxies were formerly more numerous, for example if the galaxies existing at present were formed by smaller objects merging together. This would be an example of *number evolution*.

Because light travels at a finite speed, any observation of a distant galaxy is actually an observation of how that galaxy looked in the past. In particular, the observed **redshift** z of a galaxy is a measures not only of its distance but also of the cosmic epoch at which it is seen. The most convenient way to describe this is in terms of the *lookback time* – the time taken for light from a distant object to reach the observer. Light from nearby galaxies typically takes several million years to reach us, but the lookback time to very distant galaxies may be very large indeed: galaxies with redshifts much greater than 1 may have emitted the light we now see more than 80% of the **age of the Universe** in the past. The fact that a telescope can be used as a time machine in this way allows astronomers to see how the stellar populations, gas content and physical size of galaxies have varied with time. With advances like the imaging of the Hubble Deep Field (see **Hubble Space Telescope**), history has become a central part of the astronomers' curriculum.

But galaxy evolution has not always been regarded with such interest and excitement. In the early days of **classical cosmology**, it was viewed as a major inconvenience. The aim of the classical cosmological tests was to measure directly the **curvature of spacetime** and the **deceleration parameter** of the Universe. But such tests require the use of standard rods and standard candles: objects whose physical size or luminosity do not change with time. It was soon realised that the effects of evolution on all known classes of object were so large that they completely swamped any differences in geometry or deceleration rate between the various **cosmological models**. The effect of evolution is particularly evident in the variation of **source counts** with magnitude. On the other hand, evolution in the properties of radio galaxies (see **radio astronomy**) did prove that the Universe at large was systematically changing with time, which ruled out the so-called perfect **cosmological principle** and, with it, the **steady state theory**.

There is a very neat and unambiguous way of testing for evolution which does not depend on the choice of cosmological model (evolution, geometry and deceleration cannot usually be separated out in the classical tests, but interact together in a complicated way). In this test, called the *luminosity–volume test*, we calculate the volume of the sphere centred on the observer which has a given source (i.e. a galaxy, or perhaps a **quasar**) at its edge. We then divide this number by the maximum possible such volume, which occurs when the source is so distant as to be just undetectable. The calculation is repeated for a large number of sources, and all the individual ratios are averaged. This average value should be 0.5 if there is no evolution in the properties of the sources. Surveys of quasars, for example, yield a value of about 0.7, which indicates that there are many more quasars at high redshift than we would expect without evolution.

According to the **Big Bang theory**, physical conditions would not always have allowed galaxies to exist. While the Universe was so hot that all matter was ionised, stars and galaxies could not have formed and survived. At some point, therefore, during the matter era of the **thermal history of the Universe**, galaxies must have formed by some process whose details are not yet fully understood (see **structure formation**). At sufficiently large redshifts, astronomers should be able to see *primordial galaxies*, objects which have only just formed and which may be related to the Lyman-alpha systems present in the **absorption line** spectra of quasars. Farther still we should see *protogalaxies* – objects representing the earliest stage of galaxy formation, when a massive gas cloud begins to condense out from the **intergalactic medium**. Such objects are yet to be unambiguously identified observationally, suggesting that galaxy formation must happen at redshifts of about 2 or more. With sufficient improvements in the sensitivity of detectors, we ought to be able to see so far out into space, and so far back in time, that no galaxies had yet formed.

FURTHER READING: Tayler, R. J., *Galaxies: Structure and Evolution* (Wiley-Praxis, Chichester, 1997).

EXPANSION OF THE UNIVERSE In the framework of the standard **Friedmann models** of the **Big Bang theory**, the increase with cosmic time of spatial separations between observers at all different spatial locations. The expansion of the Universe is described by **Hubble's law**, which relates the apparent velocity of recession of distant galaxies to their distance from the observer, and is sometimes known alternatively as the *Hubble expansion*. It is important, however, to stress that not everything takes part in this 'universal' expansion. Objects that are held together by forces other than gravity – **elementary particles**, atoms, molecules and crystals, for example – do not participate: they remain at a fixed physical size as the Universe swells around them. Likewise, objects in which the force of **gravity** is dominant also resist the expansion: planets, stars and galaxies are bound so strongly by gravitational forces that they are not expanding with the rest of the Universe. On scales even larger than galaxies, not all objects are moving away from one another either. The Local Group of galaxies is not expanding, for example: the Andromeda Galaxy is actually approaching the Milky Way because these two objects are held together by their mutual gravitational attraction. Some massive clusters of galaxies (see **large-scale structure**) are similarly held together against the cosmic flow. Objects larger than this may not necessarily be bound (as individual galaxies are), but their gravity may still be strong enough to cause a distortion of Hubble's law by generating **peculiar motions**. All these departures from the law discovered by Hubble are due to the fact that the Universe on these scales is not exactly homogeneous. But on larger scales still, no objects possess self-gravitational forces that are strong enough to counteract the overall tendency of the Universe to grow with time. In a broad sense, therefore, ignoring all these relatively local perturbations, all matter is rushing apart from all other matter at a speed described by Hubble's law. Observers moving with respect to one another in this way are sometimes called *fundamental observers*.

The expansion of the Universe is comfortably accommodated by the standard **cosmological models**, and it is normally viewed as a consequence of Albert **Einstein**'s theory of **general relativity**. It is true that, without the introduction of a **cosmological constant**, it is impossible to construct relativistic cosmological models that do not either expand or contract with time. Ironically, however, Einstein's original cosmological theory was explicitly constructed to be static, and it was not until the work by Georges **Lemaître** and Alexander **Friedmann** that expanding models were considered. Moreover, there is nothing about the expansion of the Universe that is inherently relativistic: all cosmological solutions of Newton's law of **gravity** must also expand or contract.

The fact that galaxies are moving away from the observer suggests that the observer must be at the centre of the expansion. But any other observer would also see everything moving away. In fact, every point in the Universe is equivalent in the expansion. Moreover, it can be shown quite easily that in a homogeneous and isotropic expanding universe – one in which the **cosmological principle** holds – Hubble's law must apply: it is the only mathematically possible description of such a universe.

It is traditional to visualise the expansion by an analogy in which the three dimensions of space are replaced by the two dimensions of the surface of a balloon (this would be a **closed universe**, but the geometry does not matter for this illustration). We paints dots on the surface of the balloon. If we imagine ourselves as an observer on one of the dots, and then blow the balloon up, we will see all the other dots moving away from our one as if we were the centre of expansion – no matter which dot we have chosen. The problem with this analogy is that it is difficult not to be aware that the two-dimensional surface of the balloon is embedded in the three dimensions of our ordinary space. We therefore see the centre of the space inside the balloon as the real centre of expansion. Instead, we must think of the balloon's surface as being the entire **Universe**. It is not embedded in another space, and there is nothing that corresponds to the centre inside it. Every point on the balloon is the centre. This difficulty is often also confused with the question of where the Big Bang actually happened: are we not moving away from the site of the original explosion? Where was this explosion situated? The answer to this is that the explosion happened everywhere and everything is moving away from it. But in the beginning, at the Big Bang **singularity**, everywhere and everything was in the same place.

The recessional velocities of galaxies are measured from their observed **redshift**. While the redshift is usually thought of as a **Doppler shift**, caused by the relative motion of source and observer, there is also another way of picturing the cosmological redshift which may be easier to understand. Light travels with a finite speed, c. Light arriving now from a distant source must therefore have set out at some time in the past, when the Universe was younger and consequently smaller. The size of the Universe is described mathematically by the cosmic scale factor $a(t)$ (see **Robertson–Walker metric, Friedmann models**). The scale factor can be thought of as a time-dependent magnification factor. In the expanding Universe, separations between points increase uniformly such that a regular grid at some particular time looks like a blown-up version of the same grid at an earlier time. Because the symmetry of the situation is preserved, we only need to know the factor by which the grid has been expanded in order to recover the past grid from the later one. Likewise, since a homogeneous and isotropic Universe remains so as it expands, we need only the scale factor in order to reconstruct a picture of the past physical conditions from present data.

If light from a distant source was emitted at time t_e and is observed now at time t_0 (see **age of the Universe**), then the Universe has expanded by a factor $a(t_0)/a(t_e)$ between then and now. Light emitted at some wavelength λ_e is therefore stretched by this factor as it travels through the Universe to be observed at wavelength λ_o. The **redshift** z of the source is then given by

$$1 + z = \lambda_o/\lambda_e = a(t_0)/a(t_e)$$

It is often stated in textbooks that this expansion effect is not really a Doppler shift, but there is only one difference between the cosmological redshift and the normal Doppler effect: the distances are so large that we are seeing the velocity the object had when the light was emitted (i.e. in the past).

Considering the expansion of the Universe in reverse leads to the conclusion that the density of matter must have been higher in the past than it is now. Also, since the wavelength of all light decreased (and thus its energy increased) in the past, the temperature of all **electromagnetic radiation** must have been higher. These two extrapolations are what led to the **Big Bang theory**, and detailed calculations allow the thermal history of the Universe to be followed as the Universe cooled and expanded after the initial singularity.

FURTHER READING: Hubble, E., *The Realm of the Nebulae* (Yale University Press, Newhaven, CT, 1936); Narlikar, J. V., *Introduction to Cosmology*, 2nd edition (Cambridge University Press, Cambridge, 1993).

EXTRAGALACTIC DISTANCE SCALE A complex array of observational techniques designed to determine the precise numerical value of the **Hubble constant**, H_0; sometimes called the *extragalactic distance ladder*. Before describing some of the rungs on the distance ladder, it is worth explaining what the essence of the problem is. In **Hubble's law**, the velocity of recession of a galaxy is proportional to its distance from the observer: $v = H_0 d$. Clearly, then, the task of determining H_0 boils down to measuring distances and velocities for an appropriate set of objects (i.e. distant **galaxies**). Measuring velocities is not a problem: using **spectroscopy**, we can measure the **Doppler shift** of **emission lines** in a galaxy's spectrum. Once the **redshift** is known, the velocity can be deduced straightforwardly. The real problem lies not with determining velocities, but with measuring distances.

Suppose you were in a large, dark room in which there is a lamp at an unknown distance from you. How could you determine its distance? One way would be to attempt to use a method based on triangulation. You could use a surveying device like a theodolite, moving around in the room, measuring angles to the lamp from different positions, and using trigonometry to work out the distance. An alternative approach would be to measure distances using the properties of the light emitted by the lamp. Suppose you knew that the lamp contained, say, a 100-watt bulb. Suppose also that you were equipped with a light meter. By measuring the amount of light received by the light meter, and remembering that the intensity of light falls off as the square of the distance, you could infer the distance to the lamp. If you did not know the power of the bulb, however, this method would not work. On the other hand, if there were two identical lamps in the room with bulbs of an unknown but identical wattage, then you could tell the relative distances between them quite easily. For example, if one lamp produced a reading on your light meter that was four times the reading produced by the other lamp, then the second lamp must be twice as far away as the first one. But still you would not know in absolute terms the distance of either lamp.

The problem of determining the distance scale of the Universe is very similar. Triangulation is difficult because it is not feasible to move very much relative to the distances concerned, except in special situations (see below). Measuring absolute distances using stars or other sources is also difficult unless we can find some way of determining their intrinsic luminosity. A feeble star nearby looks much the same as a very bright star far away, since stars, in general, cannot be resolved into disks even by the most powerful telescopes. If we know that two stars (or other sources) are identical, however, then measuring relative distances is not so difficult. It is the calibration of these relative distance measures that forms the central task of work on the extragalactic distance scale.

To put these difficulties into perspective, we should remember that it was not until the 1920s that there was even a rough understanding of the scale of the Universe. Before Edwin **Hubble**'s discovery that the spiral nebulae (as spiral galaxies were then called) are outside the Milky Way, the consensus was that the Universe was actually quite small. These nebulae were usually thought to represent the early stages of formation of structures like our Solar System. When Hubble announced the discovery of his eponymous law, the value of H_0 he obtained was about 500 kilometres per second per megaparsec, about ten times larger than current estimates. Hubble had made a mistake in identifying a kind of star to use as a distance indicator (see below) and, when his error was corrected in the 1950s by Walter **Baade**, the value dropped to about 250 km/s/Mpc. In 1958 Allan **Sandage** revised the value down still further, to between 50 and 100 km/s/Mpc, and present observational estimates still lie in this range. Modern measurements of H_0 use a battery of distance indicators, each taking one step upwards in scale, starting with local estimates of distances to stars within the Milky Way, and ending at the most distant galaxies and clusters of galaxies. The basic idea, however, is still the same as that pioneered by Hubble and Sandage.

First, we exploit local kinematic distance measurements to establish the scale of the Milky. Kinematic methods do not rely on knowledge of the absolute luminosity of a source, and they are analogous to the idea of triangulation mentioned above. To start with, distances to relatively nearby stars can be gauged using the *trigonometric parallax* of a star, i.e. the change in the star's position on the sky, as viewed against a backdrop of more distant stars, in the course of a year as a result of the Earth's motion in space. The usual astronomers' unit of distance – the parsec (pc) – stems from this method: a star 1 parsec away produces a parallax of 1 second of arc. (In other words, the parsec is the distance at which the radius of the Earth's orbit around the Sun subtends an angle of 1 second of arc.) Until recently, this method was limited to measuring distances out to about 30 pc, but the Hipparcos satellite has improved upon this dramatically, increasing the range by tenfold.

Other versions of this idea can also be used. The *secular parallax* of nearby stars is caused by the motion of the Sun with respect to them. For binary stars we can derive distances using the *dynamical parallax*, based on measurements of the angular size of the semi-major axis of the orbital ellipse, and other orbital elements of the binary system. Another related method is based on the properties of a moving open cluster of stars. Such a cluster is a group of stars which move through the Milky Way with the same speed and on parallel trajectories; a perspective effect makes these stars appear to converge to a point on the sky. The position of this point and the proper motion of the stars lead us to their distance. This method can be used on scales up to a few hundred parsecs; the Hyades cluster is a good example of a suitable cluster. With the method of *statistical parallax* we can derive distances of 500 pc or so; this technique is based on the statistical analysis of the proper motions and radial velocities of a group of stars. Taken together, all these kinematic methods allow us to establish distances up to a few hundred parsecs; however, this is much smaller than the size of our Galaxy.

Once we have determined the distances of nearby stars by kinematic methods, we can calculate their absolute luminosity from their apparent luminosity and their distances. In this way it was learned that most stars, the so-called main-sequence stars (see

stellar evolution), follow a strict relationship between spectral type (an indicator of surface temperature) and absolute luminosity: this is usually represented as the plot known as the Hertzsprung–Russell (HR) diagram. By using the properties of this diagram we can measure the distances of main-sequence stars of known apparent luminosity and spectral type by a process known as *spectroscopic parallax*. We can thus measure distances up to around 30 kpc (30,000 pc).

Another important class of distance indicator are variable stars of various kinds, including RR Lyrae and Cepheid variables. The variability of these objects gives clues about their intrinsic luminosity. The RR Lyrae stars all have a similar (mean) absolute luminosity; a simple measurement of the apparent luminosity suffices to provide a distance estimate for this type of star. These stars are typically rather bright, so this can extend the distance scale to around 300 kpc. The classical Cepheids, another group of bright variable stars, display a very tight relationship between their period of variation and their absolute luminosity (see the Figure). Measurement of the period of a distant Cepheid thus gives its luminosity, which allows us to estimate its distance. These stars are so bright that they can be seen in galaxies outside our own, and they extend the distance scale to around 4 Mpc (4,000,000 pc). Errors in the Cepheid distance scale, caused by interstellar absorption, galactic rotation and, above all, a confusion between Cepheids and another type of variable star, called W Virginis variables, were responsible for Edwin Hubble's large original value for H_0. Other distance indicators based on blue and red supergiants and also novae allow the scale to be extended slightly farther, to around 10 Mpc. Collectively, these methods are given the name *primary distance indicators*.

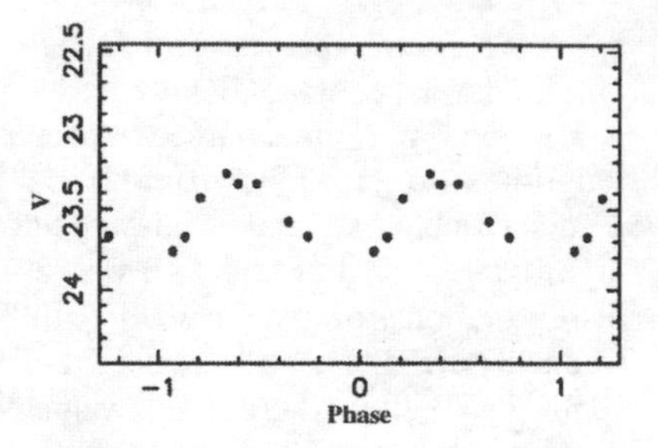

Extragalactic distance scale An example of the light curve of a Cepheid variable star, showing a regular pattern of variation that can be used to construct an indicator of the star's distance.

The *secondary distance indicators* include HII regions (large clouds of ionised hydrogen surrounding very hot stars) and **globular clusters** (clusters of around one hundred thousand to ten million stars). HII regions tend all to have similar diameters, and globular clusters similar luminosities With such relative indicators, calibrated using the primary methods, we can extend the distance scale out to about 100 Mpc.

The *tertiary distance indicators* include the brightest cluster galaxies and **supernovae**. Clusters of galaxies can contain up to about a thousand galaxies. The brightest elliptical galaxy in a rich cluster has a very standard total luminosity, probably because these objects are known to be formed in special way by *cannibalising* other galaxies. With the brightest galaxies we can reach distances of several hundred megaparsecs. Supernovae are stars that explode, producing a luminosity roughly equal to that of an entire galaxy. These stars are therefore easily seen in distant galaxies. These objects are particularly important for various cosmological reasons, so they are discussed in their own entry.

Much recent attention has been paid to the use of observed correlations between the intrinsic properties of galaxies themselves as distance indicators. In spiral galaxies, for example, one can use the empirical *Tully–Fisher relationship* between the absolute luminosity of the galaxy and the fourth power of its circular rotation speed: $L = kv^4$, where k is a constant. The measured correlation is tight enough for a measurement of v to allow L to be determined to an accuracy of about 40%. Since the apparent flux can be measured accurately, and this depends on the square of the distance to the galaxy, the resulting distance error is about 20%. This can be reduced further by applying the method to a number of spirals in the same cluster. A similar indicator can be constructed from properties of elliptical galaxies; this empirical correlation is called the *Faber–Jackson relationship*.

So there seems to be no shortage of techniques for measuring H_0. Why is it, then, that the value of H_0 is still known so poorly? One problem is that a small error in one rung of the distance ladder also affects higher levels of the ladder in a cumulative way. At each level there are also many corrections to be made: the effect of galactic rotation in the Milky Way; telescope aperture variations; the **K-correction**; absorption and obscuration in the Milky Way; the **Malmquist bias**; and the uncertainty introduced by the **evolution of galaxies**. Given this large number of uncertain corrections, it comes as no surprise that we are not yet in a position to determine H_0 with any great precision. Recently, however, methods (such as one based on the **Sunyaev–Zel'dovich effect**) have been proposed for determining the distance scale directly, without the need for this complicated ladder.

Controversy has surrounded the distance scale ever since Hubble's day. An end to this controversy seems to be in sight, however, because of the latest developments in technology. In particular, the **Hubble Space Telescope** (HST) is also be able to image stars directly in galaxies within the nearby Virgo Cluster of galaxies, an ability which bypasses the main sources of uncertainty in the calibration of traditional steps in the distance scale. The HST key programme on the distance scale is expected to fix the value of Hubble's constant to an accuracy of about 10%. This programme is not yet complete, but preliminary estimates suggest a value of H_0 in the range 60 to 70 km/s/Mpc.

FURTHER READING: Sandage, A. R., 'Distances to galaxies: The Hubble constant, the Friedmann time and the edge of the Universe', *Quarterly Journal of the Royal Astronomical Society*, 1972, **13**, 282; Rowan-Robinson, M. G., *The Cosmological Distance Ladder* (W. H. Freeman, New York, 1985); Freeman, W. *et al.*, 'Distance to the Virgo Cluster galaxy M100 from Hubble Space Telescope observations of Cepheids', *Nature*, 1994, **371**, 757.

F

Fermi, Enrico (1901–54) Italian physicist, who did most of his work in the USA. He specialised in nuclear physics, and was responsible for pioneering work on beta-decay that led ultimately to the theory of the weak nuclear interaction (see **fundamental interactions**). He was awarded the Nobel Prize for Physics in 1938. He put forward the so-called *Fermi paradox*, which purports to show that there are not many advanced civilisations in our Galaxy (see **life in the Universe**).

Fermion see **elementary particles**.

Feynman, Richard Phillips (1918–88) US theoretical physicist. He was responsible for the theory of quantum electrodynamics that unified electromagnetism (see **fundamental interactions**) with **quantum theory**. He also devised the 'sum-over-histories' approach to quantum mechanics. He was a brilliant lecturer and researcher, and his three volumes of *Feynman Lectures* are still used by students throughout the world. He was awarded the Nobel Prize for Physics in 1965.

Flatness problem The problem, left unresolved in the standard version of the **Big Bang theory**, stemming from the impossibility of predicting *a priori* the value of the **density parameter** Ω which determines whether the Universe will expand for ever or will ultimately recollapse. This shortcoming is ultimately a result of the breakdown of the **laws of physics** at the initial **singularity** in the Big Bang model.

To understand the nature of the mystery of cosmological flatness, imagine you are in the following situation. You are standing outside a sealed room. The contents are hidden from you, except for a small window covered by a small door. You are told that you can open the door at any time you wish, but only once, and only briefly. You are told that the room is bare, except for a tightrope suspended in the middle about two metres in the air, and a man who, at some indeterminate time in the past, began to walk the tightrope. You know also that if the man falls, he will stay on the floor until you open the door. If he does not fall, he will continue walking the tightrope until you look in.

What do you expect to see when you open the door? One thing is obvious: if the man falls, it will take him a very short time to fall from the rope to the floor. You would be very surprised, therefore, if your peep through the window happened to catch the man in transit from rope to floor. Whether you expect the man to be on the rope depends on information you do not have. If he is a circus artist, he might well be able to walk to and fro along the rope for hours on end without falling. If, on the other hand, he is (like most of us) not a specialist in this area, his time on the rope would be relatively brief. Either way, we would not expect to catch him in mid-air. It is reasonable, on the grounds of what we know about this situation, to expect the man to be either on the rope or on the floor when we look.

This may not seem to have much to do with Ω, but the analogy can be recognised when we realise that Ω does not have a constant value as time goes by in the Big Bang theory. In fact, in the standard **Friedmann models** Ω evolves in a very peculiar way. At times arbitrarily close to the Big Bang, these models are all described by a value of Ω arbitrarily close to 1. To put this another way, consider the Figure under **density parameter**. Regardless of the behaviour at later times, all three curves shown get closer and closer near the beginning, and in particular they approach the **flat universe** line. As time goes by, models with Ω just a little greater than 1 in the early stages develop larger and larger values of Ω, reaching values far greater than 1 when recollapse begins. Universes that start out with values of Ω just less than 1 eventually expand much faster than the flat model, and reach values of Ω very close to 0. In the latter case, which is probably more relevant given the contemporary estimates of $\Omega < 1$, the transition from Ω near 1 to a value near 0 is very rapid.

Now we can see the analogy. If Ω is, say, 0.3, then in the very early stages of cosmic history it was very close to 1, but less than this value by a tiny amount. In fact, it really is a tiny amount indeed! At the **Planck time**, for example, Ω has to differ from 1 only in the sixtieth decimal place. As time went by, Ω hovered close to the critical density value for most of the expansion, beginning to diverge rapidly only in the recent past. In the very near future it will be extremely close to 0. But now, it is as if we had caught the tightrope walker right in the middle of his fall. This seems very surprising, to put it mildly, and is the essence of the flatness problem.

The value of Ω determines the **curvature of spacetime**. It is helpful to think about the radius of spatial curvature – the characteristic scale over which the geometry appears to be non-Euclidean, like the radius of a balloon or of the Earth. The Earth looks flat if we make measurements on its surface over distances significantly less than its radius (about 6400 km). But on scales larger than this the effect of curvature appears. The curvature radius is inversely proportional to $1 - \Omega$ in such a way that the closer Ω is to unity, the larger is the radius. (A flat universe has a radius of infinite curvature.) If Ω is not too different from 1, the scale of curvature is similar to the scale of our cosmological **horizon**, something that again appears to be a coincidence.

There is another way of looking at this problem by focusing on the Planck time. At this epoch, where our knowledge of the relevant physical laws is scant, there seems to be only one natural timescale for evolution, and that is the Planck time itself. Likewise, there is only one relevant length scale: the **Planck length**. The characteristic scale of its spatial curvature would have been the Planck length. If spacetime was not flat, then it should either have recollapsed (if it were positively curved) or entered a phase of rapid undecelerated expansion (if it were negatively curved) on a timescale of order the Planck time. But the Universe has avoided going to either of these extremes for around 10^{60} Planck times.

These paradoxes are different ways of looking at what has become known as the cosmological flatness problem (or sometimes, because of the arguments that are set out in the preceding paragraph, the *age problem* or the *curvature problem*), and it arises from the incompleteness of the standard Big Bang theory. That it is such a big problem has convinced many scientists that it needs a big solution. The only thing that seemed likely to resolve the conundrum was that our Universe really is a professional circus artist, to stretch the above metaphor to breaking point. Obviously, Ω is not close to zero, as we have strong evidence of a lower limit to its value of around 0.1. This rules out the man-on-the-floor alternative. The argument then goes that Ω must be extremely close to 1, and that something must have happened in primordial times to single out this value very accurately.

The happening that did this is now believed to be cosmological inflation, a speculation by Alan **Guth** in 1981 about the very early stages of the Big Bang model. The **inflationary Universe** involves a curious change in the properties of matter at very high energies resulting from a **phase transition** involving a quantum phenomenon called a **scalar field**. Under certain conditions, the Universe begins to expand much more rapidly than it does in standard Friedmann models, which are based on properties of low-energy matter with which we are more familiar. This extravagant expansion – the inflation – actually reverses the kind of behaviour expected for Ω in the standard models. Ω is driven hard towards 1 when inflation starts, rather than drifting away from it as in the cases described above.

A clear way of thinking about this is to consider the connection between the value of Ω and the curvature of spacetime. If we take a highly curved balloon and blows it up to an enormous size, say the size of the Earth, then its surface will appear to be flat. In inflationary cosmology, the balloon starts off a tiny fraction of a centimetre across and ends up larger than the entire observable Universe. If the theory of inflation is correct, we should expect to be living in a Universe which is very flat indeed, with an enormous radius of curvature and in which Ω differs from 1 by no more than one part in a hundred thousand.

The reason why Ω cannot be assigned a value closer to 1 is that inflation generates a spectrum of **primordial density fluctuations** on all scales, from the microscopic to the scale of our observable Universe and beyond. The density fluctuations on the scale of our horizon correspond to an uncertainty in the mean density of matter, and hence to an uncertainty in the value of Ω.

One of the problems with inflation as a solution to the flatness problem is that, despite the evidence for the existence of **dark matter**, there is no really compelling evidence of enough such material to make the Universe closed. The question then is that if, as seems likely, Ω is significantly smaller than 1, do we have to abandon inflation? The answer is not necessarily, because some models of inflation have been constructed that can produce an **open universe**. We should also remember that inflation predicts a flat universe, and the flatness could be achieved with a low matter density if there were a **cosmological constant** or, in the language of particle physics, a nonzero vacuum energy density.

On the other hand, even if Ω were to turn out to be very close to 1, that would not necessarily prove that inflation happened either. Some other mechanism, perhaps associated with the epoch of **quantum gravity**, might have trained our Universe to walk the tightrope. It may be, for example, that for some reason quantum gravity favours a flat spatial geometry. Perhaps, then, we should not regard the flatness 'problem' as a problem: the real problem is that we do not have a theory of the very beginning in the Big Bang cosmology.

FURTHER READING: Coles, P. and Ellis, G. F. R., *Is the Universe Open or Closed?* (Cambridge University Press, Cambridge, 1997); Guth, A. H., 'Inflationary Universe: A possible solution to the horizon and flatness problems', *Physical Review* D, 1981, **23**, 347; Narlikar, J.V. and Padmanabhan, T., 'Inflation for astronomers', *Annual Reviews of Astronomy and Astrophysics*, 1991, **29**, 325.

FLAT UNIVERSE Any **cosmological model** in which the **curvature of spacetime** vanishes. In such a universe all the normal rules of Euclidean geometry hold. For example, the sum of the interior angles of a triangle is 180°, and parallel lines never meet. The particular **Friedmann model** that describes a flat universe is the one in which the **density parameter** Ω = 1 and the **deceleration parameter** q = 0.5. This model, the *Einstein–de Sitter universe*, is infinite in spatial extent. It also expands for ever, but only just: mathematically the deceleration generated by **gravity** does eventually cause the **expansion of the Universe** to cease, but only after an infinite time.

The Einstein–de Sitter model is, however, not the only **cosmological model** to be spatially flat. If we allow a **cosmological constant** term Λ into the equations of **general relativity**, then it is also possible for models with Ω different from 1 to be spatially flat. Such models can be either accelerating or decelerating in the present, so q can be either positive or negative.

Most theories of the **inflationary Universe** predict that the Universe should be flat, but this can be achieved either by having a critical density of matter or by postulating a nonzero vacuum energy density left over from the inflationary phase. To all intents and purposes, a vacuum energy plays exactly the same role as a cosmological constant.

FURTHER READING: Berry, M. V., *Principles of Cosmology and Gravitation* (Adam Hilger, Bristol, 1989); Narlikar, J. V., *Introduction to Cosmology*, 2nd edition (Cambridge University Press, Cambridge, 1993); Islam, J. N., *An Introduction to Mathematical Cosmology* (Cambridge University Press, Cambridge, 1992).

Fowler, William Alfred (1911–95) US physicist and astrophysicist, who spent most of his working life at Caltech. In the 1950s he collaborated with Fred **Hoyle** and Margaret and Geoffrey **Burbidge** on the 'B^2FH theory' of **nucleosynthesis** in stars, which led to the realisation that the **light element abundances** have a cosmological explanation.

***Fractal** A *fractal set* is a mathematical object that is often associated with the theory of chaotic dynamics. Some aspects of the **large-scale structure** of the distribution of galaxies in space can also be described in terms of fractals; indeed, some of the earliest models of this structure associated with the idea of a **hierarchical cosmology** made use of this concept.

A *fractal object* is one that possesses a fractional dimension. Geometrical objects with which we are familiar tend to have dimensions described by whole numbers (integers). A line has one dimension, and a plane has two, whereas space itself is three-dimensional. But it is quite possible to define objects mathematically for which the appropriate dimension is not an integer. Consider, for example, a straight line of unit length. This has dimension $d = 1$. Now remove the central one-third of this line. You now have two pieces of line, each one-third the length of the original, with a gap in between. Now remove the central one-third of each of the two remaining pieces. There are now four pieces. Now imagine carrying on this process of dividing each straight line into thirds and discarding the central part. The mathematical limit of this eternal editing process is a set which is, in some senses, a set of points with zero dimension but in other senses it retains some of the characteristics of a line. It is actually a fractal called the *Cantor ternary set*, which has a dimension (formally the *Hausdorff dimension*) given by $d = \ln 2/\ln 3 = 0.6309\ldots$. The dimension lies between that of a set of points and that of the original line.

What does this idea have to do with galaxy clustering? Imagine a distribution of galaxies in space such that all the galaxies are distributed uniformly throughout three-dimensional space. If you could sit on one galaxy and draw a sphere of radius R around it, then the amount of mass inside the sphere simply scales with its radius R to the third power: $M(R) \propto R^3$. Now imagine that, instead of filling space, galaxies are restricted to lines on two-dimensional sheets. In this case if we draw a sphere around a given galaxy, the mass contained scales as the area of the sheet contained within the sphere: $M(R) \propto R^2$. If, instead, galaxies were distributed along filaments like pieces of string, the behaviour of $M(R)$ would be proportional simply to R: it just scales with the length of the string that lies inside the sphere.

But measurements of the **correlation function** of galaxies can be used to determine how $M(R)$ behaves for the real Universe. It can be shown mathematically that if the correlation function is a power law (i.e. $\xi(R) \propto R^{-\gamma}$, as seems to be the case, then it follows that $M(R) \propto R^{3-\gamma}$. Since it appears that $\gamma \approx 1.8$ (see **correlation function**), the appropriate fractional dimension for galaxy clustering is around 1.2. This in turn indicates that the dimensionality of the distribution lies somewhere between that of one-dimensional filaments and two-dimensional sheets, in accordance with the qualitative picture of **large-scale structure**.

Some have argued that this fractal behaviour continues on larger and larger scales, so that we can never see a scale where the structure becomes homogeneous, as is required by the **cosmological principle**. If this were so, we would have to abandon the standard **cosmological models** and look for alternative theories, perhaps based on the old idea of a **hierarchical cosmology**. Recent galaxy **redshift surveys** do, however, support the view that there is a scale, larger than the characteristic sizes of voids, filaments and walls, above which large-scale homogeneity and isotropy is reached (see **large-scale structure**).

FURTHER READING: Mandelbrot, B. B., *The Fractal Geometry of Nature* (W. H. Freeman, San Francisco, 1982); Heck, A. and Perdang, J. M. (editors), *Applying Fractals in Astronomy* (Springer-Verlag, Berlin, 1991).

FRAME OF REFERENCE In mathematical terms, a system of coordinates used to chart the positions of objects and/or events. We can think of lines of latitude and longitude on the Earth's surface as defining a frame of reference, likewise with the coordinates of right ascension and declination used to indicate the positions of astronomical objects on the celestial sphere. In physics the idea of a frame of reference extends beyond mere cartography (description of positions in space) to incorporate also the location of events in **time**. In particular, the mathematical coordinates used in **special relativity** and **general relativity** describe the properties of **spacetime** as perceived by an observer in some particular state of motion.

We can think of a frame of reference in this context as consisting of spatial coordinates (say the usual Cartesian coordinates x, y and z, or the polar coordinates r, θ and ϕ) as well as a time coordinate, t. These coordinates are used in mathematical formulations of the **laws of physics** to describe the motions and interactions undergone by particles and fields. The spatial coordinates can be visualised in terms of a regular grid defining the positions of objects (like a three-dimensional sheet of graph paper). Suppose an object is at rest in the coordinate system of an observer A. Now, a different observer, B, moving with respect to A, has her own set of coordinates which moves with her: her sheet of graph paper moves with respect to that of A. The object which is not moving through A's frame will be moving through B's. The position of a particle at any time will also not be the same in terms of A's coordinates as it is for B's, unless the two sheets of graph paper just happen to lie exactly on top of each other at that time. Clearly, then, the spatial position of an object depends on the state of motion of the coordinate system used by the observer measuring it. In special relativity, all frames of reference in constant relative motion are equivalent. One of the consequences of this is that not only the spatial coordinates but also the time t must depend on the state of motion of the observer. This is why time as well as space must be included in the frame of reference.

General relativity extends these concepts to frames of reference undergoing accelerated motion. This has even deeper and more counter-intuitive consequences than those appearing in special relativity, such as the **curvature of spacetime**.

Curiously, although relativity theory forbids the existence of preferred frames of reference (because the laws of physics must be the same for all observers), **cosmological models** in which the **cosmological principle** applies do possess a preferred frame. According to this principle, for example, the **cosmic microwave background radiation** should have the same temperature in all directions on the sky. But this cannot be the case for observers regardless of their state of motion because of the **Doppler effect**: a moving observer should see a higher temperature in the direction of motion than in the opposite direction. This preferred frame does not violate the principle of relativity, however, because it is a consequence of the special property of overall homogeneity implied by the cosmological principle that allows observers to choose a preferred time coordinate: cosmological **proper time**.

SEE ALSO: **Robertson–Walker metric**.

* **FREE STREAMING** According to the standard theory of the gravitational **Jeans instability**, any structures that are so large that their self-gravity exceeds the restoring force provided by their internal pressure will collapse. Any that are not so large will oscillate like sound waves. The dividing line between the two is called the *Jeans length*, and it defines a characteristic size for structures that can form by condensing out of a cloud of gas or dust by gravitational processes.

This idea forms the basis of most theories of **structure formation** in cosmology, but there are several important additional physical processes which make it necessary to modify the standard theory. *Free streaming* is one such process, and it is important in theories which deal with some form of material other than a standard gas. In particular, cosmological structure formation is generally thought to involve the gravitational instability of some form of non-baryonic **dark matter** in the form of **weakly interacting massive particles** (WIMPs) (see also **elementary particles**). WIMPs do not constitute a gas in the normal sense of the word because they are collisionless particles (by virtue of their weakly interacting nature). This complicates the standard theory of the Jeans instability because a fluid of collisionless particles does not really exert a pressure. These particles merely stream around, seemingly oblivious to one another's presence except for any gravitational interactions that might occur – hence the term 'free streaming'.

However, there is a characteristic scale for WIMPs analogous to the Jeans length, defined by the distance over which the WIMPs are able to stream in the time it takes for structure to form. This depends on their velocity: fast-moving WIMPs – the so-called hot dark matter (HDM) – have a very large streaming length, while slow-moving ones (cold dark matter, CDM) do not stream very much at all. The free-streaming length, roughly speaking, plays the same role as the Jeans length, but the situation for WIMPs differs from the standard Jeans instability in that fluctuations on scales smaller than the free-streaming length cannot oscillate like acoustic waves. Because there is no restoring force to create oscillations, the WIMPs simply move out of dense regions into empty ones and smooth out the structure.

To help to understand this effect, we might ask why there are no very small sandhills in a desert, only large rolling dunes. The answer is that the high winds in desert climates cause sand particles to free-stream quite a large distance. If you build a sandcastle in the Sahara desert, by the next day the particles from which it was made will be smoothed out. All that remains are structures larger than the free-streaming length for grains of sand.

The effect of free-streaming is not particularly important for structure formation within a CDM model because the lengths in question are so small. But for the HDM model, which was a favourite of many scientists in the early 1980s, the effect is drastic: the only structures that can survive the effects of neutrino free-streaming are on the scale of giant superclusters, hundreds of millions of light years across (see **large-scale structure**). Individual galaxies must form in this model after much larger structures have formed; it is thus a 'top-down' model. Difficulties in understanding how galaxies might have formed by fragmentation led to the abandonment of the HDM theory by most cosmologists in favour of the CDM alternative.

FURTHER READING: Coles, P. and Lucchin, F., *Cosmology: The Origin and Evolution of Cosmic Structure* (John Wiley, Chichester, 1995), Chapter 10.

FRIEDMANN, ALEXANDER ALEXANDROVICH (1888–1925) Russian cosmologist. He developed the mathematical **Friedmann models** of an expanding Universe, which form the basis of the modern **Big Bang theory**. He performed his calculations in the midst of the siege of Petrograd in 1917.

* **FRIEDMANN MODELS** A special family of solutions to the **Einstein equations** of general relativity, obtained by assuming that the **cosmological principle** holds – that the Universe is the same in every place and looks the same in every direction. This considerably simplifies the otherwise complicated task of trying to solve the Einstein equations. Generally speaking, these equations mix together time and space in a complicated way through the **metric**. In the absence of any special symmetry there is no unambiguous way of defining **time** and space separately, and we have to deal with a truly four-dimensional **spacetime**.

In the Friedmann models this problem is avoided because there is a unique way of separating space and time. Since the Universe is supposed to be the same in every place, it follows that the density of matter must be the same everywhere. Hypothetical observers can then set their clocks according to the local density of matter, and a perfect synchronisation is thus achieved. The time coordinate that results is usually called the **proper time** (t). Application of the cosmological principle also reduces the amount of complexity in the spatial parts of the Einstein equations. In mathematical terms, the assumption of the **Robertson–Walker metric** fixes all the terms in the Einstein equations that deal with the spatial geometry, which is then completely described by a single number k, which distinguishes between an **open universe** ($k = -1$), a **closed universe** ($k = +1$) and a **flat universe** ($k = 0$).

All that is left is the overall **expansion of the Universe**. Because, according to the assumed property of isotropy, this must proceed at the same rate in all directions, we can assume that all **proper distances** between observers simply expand by a constant factor as time goes on. The expansion can therefore be described by a function $a(t)$, usually called the *cosmic scale factor*. If any two observers are expanding with the Universe in such a way that at times t_1 and t_2 they are separated by distances d_1 and d_2, then

$$d_1/a(t_1) = d_2/a(t_2)$$

All that is then needed to solve the Einstein equations is a description of the bulk properties of the matter on cosmological scales. In general this requires us to assume a so-called *equation of state* that relates the pressure p of the material to its density ρ. At least for the later stages of the **thermal history of the Universe**, this is also very simple and takes the form of pressureless ($p = 0$) matter (i.e. dust) which is simply described by its density ρ. Since the Universe is expanding, the density ρ falls off as the volume, i.e. as a^3. The density therefore decreases with increasing time in the expanding Universe.

It is then straightforward to derive the Friedmann equation:

$$(\mathrm{d}a/\mathrm{d}t)^2 - 8\pi G\rho a^2/3 = -kc^2$$

On the left-hand side, the first term represents the square of the rate of expansion, and is something like the kinetic energy per unit mass of the expanding Universe; the second term is analogous to the gravitational potential energy per unit mass. The term on the right-hand side is constant (c is the speed of light). The Friedmann equation is therefore nothing more than the law of conservation of energy. If the total energy is positive, then the Universe has $k = -1$ and will expand for ever; if it is negative, then $k = +1$ and the Universe is 'bound', so it will recollapse. We can obtain a very similar equation using only Newton's theory of **gravity**, but we cannot then identify the constant term with the **curvature of spacetime**, a meaningless concept in Newtonian physics. If the **cosmological constant** Λ is taken into account, the above equation is modified by the addition of an extra term to the left-hand side which describes an overall repulsion; this is rather like a modification of Newton's law of gravitation.

We can simplify matters further by defining two important parameters: the Hubble parameter (see **Hubble constant**), which is simply related to the expansion rate via $H = (\mathrm{d}a/\mathrm{d}t)/a$; and the **density parameter** $\Omega = 8\pi G\rho/3H^2$. The Friedmann equation then reduces to the constraint that $H^2(1 - \Omega)$ is a constant. Note, however, that H and Ω both vary with time. To specify their values at the present time, denoted by $t = t_0$, a zero subscript is added: the present value of the density parameter is written as Ω_0 and the present value of the Hubble parameter (or Hubble constant) as H_0. Since these two parameters are at least in principle measurable, it makes sense to write the equation in terms of them. On the other hand, they are not predicted by the Big Bang model, so they must be determined empirically from observations.

The Friedmann equation is the fundamental equation that governs the time-evolution of the Friedmann models, and as such it is the fundamental equation governing the evolution of the **Universe** in the standard **Big Bang theory**. In particular, it shows that the cosmological density becomes infinite when $t = 0$, signalling the presence of a **singularity**. Although they form the standard framework for modern cosmology, it is important to stress that the Friedmann models are not the only cosmological solutions that can be obtained from the theory of **general relativity** (see **cosmological models** for more details).

FURTHER READING: Berry, M. V., *Principles of Cosmology and Gravitation* (Adam Hilger, Bristol, 1989).

FUNDAMENTAL INTERACTIONS The four ways in which the various **elementary particles** interact with one another: electromagnetism, the weak nuclear force, the strong nuclear force and **gravity**. They vary in strength (gravity is the weakest, and the strong nuclear force is the strongest) and also in the kinds of elementary particle that take part.

The electromagnetic interaction is what causes particles of opposite charge to attract each other, and particles of the same charge to repel each other, according to the Coulomb law of electrostatics. Moving charges also generate **magnetic fields** which, in the early history of physics, were thought to be a different kind of phenomenon altogether, but which are now realised to be merely a different aspect of the electromagnetic force. James Clerk **Maxwell** was the first to elucidate the character of the electromagnetic interactions. In this sense, Maxwell's equations were the first unified physical theory, and the search for **laws of physics** that unify the other interactions is still continuing (see **grand unified theories**, **theory of everything**).

The theory of electromagnetism was also an important step in another direction. Maxwell's equations show that light can be regarded as a kind of **electromagnetic radiation**, and demonstrate that light should travel at a finite speed. The electromagnetic theory greatly impressed Albert **Einstein**, and his theory of **special relativity** was constructed specifically from the requirement that Maxwell's theory should hold for observers regardless of their velocity. In particular, the speed of light had to be identical for all observers, whatever the relative motion between emitter and receiver.

The *electromagnetic force* holds electrons in orbit around atomic nuclei, and is thus responsible for holding together all the material with which we are familiar. However, Maxwell's theory is a classical theory, and it was realised early in the 20th century that, in order to apply it in detail to atoms, ideas from **quantum physics** would have to be incorporated. It was not until the work of Richard **Feynman** that a full quantum theory of the electromagnetic force, called *quantum electrodynamics*, was developed. In this theory, which is usually abbreviated to QED, electromagnetic radiation in the form of photons is responsible for carrying the electromagnetic interaction between particles.

The next force to come under the spotlight was the *weak nuclear force*, which is responsible for the so-called beta decay of certain radioactive isotopes. It involves elementary particles belonging to the lepton family (which includes electrons). As with electromagnetism, weak forces between particles are mediated by other particles – not photons, in this case, but massive particles called the *W* and *Z bosons*. The fact that these particles have mass (unlike the photon) is the reason why the weak nuclear force has such a short range. The W and Z particles otherwise play the same role in this context as the photon does in QED: they are all examples of *gauge bosons* (see **gauge theory**). In this context, the particles that interact are always fermions, while the particles that carry the interaction are always bosons.

A theory that unifies the electromagnetic force with the weak nuclear force was developed in the 1960s by Sheldon Glashow, Abdus Salam and Steven Weinberg. Called the *electroweak theory*, it represents these two distinct forces as being the low-energy manifestations of a single force. At high enough energies, all the gauge bosons involved change character and become massless entities called *intermediate vector bosons*. That electromagnetism and the weak force appear so different at low energies is a consequence of **spontaneous symmetry-breaking**.

The *strong nuclear interaction* (or strong force) involves the hadron family of elementary particles, which includes the baryons (protons and neutrons). The theory of these interactions is called *quantum chromodynamics* (QCD) and it is built upon similar lines to the electroweak theory. In QCD there is another set of gauge bosons to mediate the force: these are called *gluons*. There are eight of them, and they are even more massive than the W and Z particles. The strong force is thus of even shorter range than the weak force. Playing the role of electric charge in QED is a property called 'colour'. The hadrons are represented as collections of particles called *quarks*, which have a fractional electrical charge and come in six different 'flavours': up, down, strange, charmed, top and bottom. Each distinct hadron species is a different combination of the quark flavours.

The electroweak and strong interactions coexist in a combined theory of the fundamental interactions called the *standard model*. This model is, however, not really a unified theory of all three interactions, and it leaves many questions unanswered. Physicists hope eventually to unify all three of the forces discussed so far in a single **grand unified theory**. There are many contenders for such a theory, but it is not known which (if any) is correct.

The fourth fundamental interaction is gravity, and the best theory of it is **general relativity**. This force has proved extremely resistant to efforts to make it fit into a unified scheme of things. The first step in doing so would involve incorporating quantum physics into the theory of gravity in order to produce a theory of **quantum gravity**. Despite strenuous efforts, this has not yet been achieved. If this is ever done, the next task will be to unify quantum gravity with the grand unified theory. The result of this endeavour would be a **theory of everything**. The difficulty of putting the theory of interactions between elementary particles (grand unified theories) together with the theory of space and time (general relativity) is the fundamental barrier to understanding the nature of the initial stages of the **Big Bang theory**.

FURTHER READING: Roos, M., *Introduction to Cosmology*, 2nd edition (John Wiley, Chichester, 1997), Chapter 6; Davies, P. C. W., *The Forces of Nature* (Cambridge University Press, Cambridge, 1979); Pagels, H. R., *Perfect Symmetry* (Penguin, London, 1992).

G

Galaxy A gravitationally bound agglomeration of stars, gas and dust. Observational **cosmology** is concerned with the distribution of matter on scales much larger than that of individual stars, or even individual galaxies. For many purposes, therefore, we can take the basic building block of cosmology to be the galaxy. Galaxies come in a rich variety of shapes and sizes, and it would take a whole book on its own to describe their properties in detail. Moreover, the way in which galaxies formed in the **expanding Universe** is still not completely understood, and neither is their detailed composition. All the evidence suggests that the visible parts of galaxies constitute a tiny fraction of their mass and that they are embedded in enormous *haloes* of **dark matter**, probably in the form of **weakly interacting massive particles** (WIMPs). Nevertheless there are some properties of galaxies that are worthy of mention as they must be accounted for by any theory of cosmological **structure formation**.

Galaxies come in three basic types: *spiral* (or *disk*), *elliptical* and *irregular*. As late as the early years of the 20th century controversy raged about what the spiral nebulae that had been observed by William **Herschel** in the 18th century actually were, and how far away they were (see Harlow **Shapley**). This was finally resolved in the 1920s by Edwin **Hubble**, who showed that they were much more distant than the stars in our own Galaxy, the Milky Way, now itself known to be a spiral galaxy. Before this, most astronomers believed that the spiral nebulae like the famous one in Andromeda (shown in Figure 1) might be relatively nearby objects, like our own Solar System seen in the process of formation. Hubble proposed a morphological classification, or taxonomy, for galaxies in which he envisaged the three basic types (spiral, elliptical and irregular) as forming a kind of evolutionary sequence. Although it is no longer thought that this evolutionary sequence is correct, Hubble's nomenclature, in which ellipticals are *early types* and spirals and irregulars are *late types*, is still sometimes used.

The elliptical galaxies (type E), which account for only around 10% of

Galaxy (1) The Andromeda Nebula (M31), the nearest large galaxy to our own, and a fine example of a spiral galaxy.

Galaxy (2) An example of a giant elliptical galaxy (NGC4881) in the centre of a cluster of galaxies known as the Coma cluster.

observed bright galaxies, are (not surprisingly) elliptical in shape and have no discernible spiral structure. They are usually reddish in colour, have very little dust and show no sign of active star formation. The average luminosity profile of an elliptical galaxy is of the form

$$I(r) = I_0(1 + r/R)^{-2}$$

where I_0 and R are constants, and r is the distance from the centre of the galaxy. The scale length R is typically around 1 kiloparsec (i.e. about 3000 light years). Elliptical galaxies are subclassified by adding to the 'E' a number n which depends on the ratio of the minor axis b to the major axis a of the ellipse: $n \approx 10(1 - b/a)$. Ellipticals show no significant rotational motions, and their shape is thought to be sustained by the anisotropic thermal motions of the stars within them (see **virial theorem**). Elliptical galaxies – *giant elliptical galaxies* in particular – occur preferentially in dense regions, inside clusters of galaxies (see Figure 2). This has prompted the idea that they might originally have been spiral galaxies which have lost their spiral structure through mergers or interactions with other galaxies.

Spiral galaxies account for more than half the galaxies observed in our neighbourhood. Hubble distinguished between *normal* (S) and *barred* (SB) spiral galaxies, according to whether the prominent spiral arms emerge directly from the central *nucleus*, or originate at the ends of a luminous bar projecting symmetrically on either side of the nucleus. Spirals often contain copious amounts of dust, and the spiral arms in particular show evidence of ongoing star formation (i.e. lots of young supergiant stars), generally giving the arms a noticeably blue colour. The nucleus of a spiral galaxy resembles an elliptical galaxy in morphology, luminosity profile and colour. Many spirals also demonstrate some kind of 'activity' (non-thermal emission processes) (see **active galaxies**). The intensity profile of spiral galaxies (outside the nucleus) does not follow the same law as an elliptical galaxy, but can instead be fitted by an exponential form:

$$I(r) = I_0 \exp(-r/R)$$

The normal and barred spirals S and SB are further subdivided into types a, b and c, depending on how tightly the spiral arms are wound (a being tightly wound, and c loosely wound). Spirals show ordered rotational motion which can be used to estimate their masses. This is the strongest evidence in favour of large amounts of **dark matter**. The above light profile falls away so rapidly that the total amount of light is much less than would be produced if all the mass responsible for generating the rotation were in the form of stars. Moreover, the galaxy continues to rotate even at such large *r* that the intensity profile is negligibly small. This is usually reconciled by appealing to a large extended halo of dark matter extending to ten or more times the scale of the light profile (*R*).

Lenticular, or S0 galaxies, were added later by Hubble to bridge the gap between normal spirals and ellipticals. Around 20% of galaxies we see have this morphology. They are more elongated than elliptical galaxies, but have neither bars nor a spiral structure.

Irregular galaxies have no apparent structure and no rotational symmetry. Bright irregular galaxies are relatively rare; most are faint and small, and consequently very hard to see. Their irregularity may stem from the fact that they are have such small masses that the material within them is relatively loosely bound, and may have been disturbed by the environment in which they sit.

The masses of elliptical galaxies vary widely, from 10^5 to 10^{12} solar masses, which range includes the mass scale of globular star clusters. Small elliptical galaxies appear to be very common: for example, 7 out of the 17 fairly bright galaxies in the Local Group (which also includes the Andromeda Galaxy) are of this type. Spiral galaxies seem to have a smaller spread in mass, with a typical value of around 10^{11} solar masses.

The problem of galaxy formation is one of the outstanding challenges facing modern cosmology. The main problem is that most models are based on the assumption that there is a dominant component of non-baryonic, weakly interacting dark matter in the Universe. The distribution of this material can be predicted using the theory of the gravitational **Jeans instability** because its evolution depends only on **gravity**: it does not interact in any other way. The baryonic matter that forms all the stars and gas in a galaxy is, as far as gravity is concerned, merely a small contaminating influence on this dark matter. But whether a concentration of WIMPs becomes a galaxy or not is very difficult to decide. The physics involved in the heating and cooling of gas, fragmentation processes leading to star formation, and the feedback of energy from **supernova** explosions into the **intergalactic medium** is very complicated. These processes are very difficult to simulate, even with the most powerful computers. We are therefore left with the problem that the distribution of visible galaxies is related in a very uncertain way to the distribution of the underlying dark matter, and theories are consequently difficult to test directly against observations of galaxies alone.

SEE ALSO: **evolution of galaxies, large-scale structure, luminosity function**.

FURTHER READING: Hubble, E., *The Realm of the Nebulae* (Yale University Press, Newhaven, CT, 1936); Tayler, R. J., *Galaxies: Structure and Evolution* (Wiley-Praxis, Chichester, 1997).

GALILEO GALILEI (1564–1642) Italian scientist. He was one of the first physicists in the modern sense of the word and also, because of his pioneering use of the telescope for observations, one of the first modern astronomers. He strongly advocated the Copernican heliocentric view of the Solar System, and was consequently condemned to house arrest by the Inquisition. He also advocated an early form of the principle of relativity, which was later generalised by Albert **Einstein**.

GAMMA-RAY ASTRONOMY The study of gamma-rays – photons of **electromagnetic radiation** with the very highest energies and the shortest wavelengths of all – from astrophysical sources. Because of their extremely high energies, gamma-rays are invariably produced in regions of extremely high temperature, which are associated with some of the most violent phenomena known.

All photons with wavelengths less than about 0.01 nm are usually called gamma-rays; luckily for terrestrial life, no significant flux of such extremely energetic radiation reaches the surface of the Earth. Early experiments in gamma-ray astronomy used balloons to carry instruments sufficiently high to escape the worst effects of atmospheric absorption. Some early observations were also made from spacecraft, during for example the Apollo lunar missions of the late 1960s and early 1970s. The first satellites to undertake systematic surveys of the gamma-ray sky were the second Small Astronomy Satellite (SAS-2) and COS-B, launched in 1972 and 1974 respectively. Later on, the High Energy Astrophysical Observatory HEAO-1 and HEAO-3 missions also carried gamma-ray telescopes. The 1990s saw the launch of the satellite Granat and the Compton Gamma Ray Observatory. In the future, various other missions are planned, including the European Space Agency's cornerstone mission, Integral.

The extremely high energies and short wavelengths of gamma-ray photons necessitate special kinds of detectors and observational methods, featuring special collimators or *coded masks* for relatively low energies, and *spark chambers* at higher energies. For the most energetic radiation, above 100 GeV, no suitable spaceborne detector has yet been devised, but it is possible to observe the passage of such photons through the Earth's atmosphere using the properties of *Cherenkov radiation* (a burst of optical light), which is emitted by electrons created as the gamma-rays ionise matter.

Among the most interesting sources of this radiation are the *gamma-ray bursters*, which produce intense pulses of gamma-rays that can last anything from a few milliseconds to several minutes. These were first detected accidentally during the course of a programme to monitor terrestrial nuclear explosions. There has been a long controversy about the nature of these sources, particularly focusing on the issue of whether they are local or cosmological. Their isotropic distribution on the sky suggests an extragalactic origin, but this idea requires the energy produced to be phenomenally large, leading some scientists to prefer models in which the sources are distributed in our Galaxy. This controversy has recently been resolved by observations that have identified the source of a gamma-ray burst with a galaxy at high **redshift**. The sources are therefore cosmological, and theorists must now account for the enormous energies released. Several models have been proposed, including the merger of two neutron stars, but none has yet emerged as the likely favourite.

As well as gamma-rays produced by individual objects, there also exists a cosmic background of gamma-rays with energies up to several hundred MeV. The origin of this background, which is distributed in an approximately isotropic fashion on the sky, is not yet known, but it could be produced by processes involving the annihilation of **antimatter** or by the evaporation of primordial **black holes** via **Hawking radiation**.
SEE ALSO: Essay 4.

GAMOW, GEORGE (1904–68) Russian physicist, who moved to the USA in 1934. He pioneered the investigation of **nucleosynthesis** in the **expanding Universe**, and thereby helped to establish the **Big Bang theory** in its modern form. With Ralph **Alpher** and Robert **Herman**, he predicted that the Big Bang would leave a residual radiation with a temperature of about 5 K.

** **GAUGE THEORY** The essential role of symmetry in the **laws of physics** describing the **fundamental interactions** between **elementary particles** is long-established. We know, for example, that electromagnetism is symmetrical with respect to changes in the sign of electrical charge: it is merely a convention that electrons are taken to have negative change, and protons positive charge. The mathematical laws describing electromagnetism would look the same if we changed the signs of all charges. Charge is an example of a *discrete symmetry*: there are only two possible states (+ and −) in this theory. Unified theories of the fundamental interactions, however, involve a more subtle kind of symmetry called *gauge symmetry*. These symmetries relate to quantities in the theory that do not change when the coordinate system is changed.

The simplest kind of gauge symmetry appears in the **quantum field theory** version of Maxwell's equations of electromagnetism, usually known as *quantum electrodynamics* (QED). In the simplest case of a single electron moving in an electromagnetic field, we needs two mathematical functions to describe the system. The first is the wavefunction for the electron, ψ, and the second is a vector field A representing the electromagnetic interaction (the quantum states of this field are simply the photons responsible for mediating the electromagnetic force). If we write the equations of QED in terms of these two functions, we find that they are unchanged if we add a term to the vector potential A that describes the gradient of a scalar potential and changes the phase of the wavefunction ψ. (All the forces emerge from taking the curl of A, and since the gradient of a scalar potential has zero curl, the physics is not changed.) In the language of group theory, this symmetry is called a *unitary symmetry of one dimension*, and the appropriate gauge group is given the symbol U(1); it basically corresponds to a symmetry under transformations of phase.

Theories that combine the fields in electrodynamics with those in the other fundamental interactions involve more complicate symmetries, because there are more field equations than in the simple case of QED. The symmetry groups therefore have higher dimensionality and correspondingly more complicated structures. The gauge group for the theory of the strong interactions, which is known as *quantum chromodynamics* (QCD), is denoted by SU(3); the three-dimensional character of this group arises from the fact that there are three distinct quark fields. The order of the symmetry group determines the number of gauge bosons responsible for mediating the interactions between elementary particles. The U(1) symmetry has only one (the photon); SU(2) has three (the W^+, W^- and Z bosons); the SU(3) group has eight (corresponding to eight different gluons).

The symmetry corresponding to the weak interaction is likewise denoted by SU(2), while the electroweak interaction which unifies the weak interaction with the U(1) symmetry of electromagnetism is denoted by SU(2)⊗U(1). The interesting thing about the weak interactions, however, is that they have an in-built chirality (or handedness): all neutrinos are left-handed. The symmetry group for the weak interactions is therefore sometimes denoted by $SU_L(2)$ to denote only the left-handed part.

We can trivially write a combination of strong, weak and electromagnetic interactions in terms of a gauge theory with a symmetry group SU(3)⊗SU(2)⊗U(1). This is essentially what is done in the standard model of particle physics, which seems to fit the results of experiments fairly well. This is not, however, a truly unified theory because there is no overriding reason for the different parts of the theory to behave in the way they do. A **grand unified theory** would combine all three of these forces in single gauge group, such as SU(5); the particular form of the lower-energy interactions would then hopefully emerge as a result of some kind of **spontaneous symmetry-breaking** process.

FURTHER READING: Davies, P. C. W., *The Forces of Nature* (Cambridge University Press, Cambridge, 1979); Narlikar, J. V., *Introduction to Cosmology*, 2nd edition (Cambridge University Press, Cambridge, 1993), Chapter 6; Roos, M., *Introduction to Cosmology*, 2nd edition (John Wiley, Chichester, 1997), Chapter 6; Collins, P. D. B., Martin, A. D. and Squires, E. J., *Particle Physics and Cosmology* (John Wiley, New York, 1989); Barrow, J. D. and Silk, J., *The Left Hand of Creation: The Origin and Evolution of the Expanding Universe* (Basic Books, New York, 1983).

GELLER, MARGARET JOAN (1947–) US astronomer. In 1985, at the Harvard-Smithsonian Center for Astrophysics, she and John **Huchra** began a survey of galaxies out to around 200 Mpc which revealed new features in the **large-scale structure** of the Universe, including voids, sheets and the Great Wall.

GENERAL RELATIVITY The strongest of the **fundamental interactions** on large scales is **gravity**, so the most important part of a physical description of the **Universe** as a whole is a consistent theory of gravitation. The best candidate we have for this is Albert **Einstein**'s theory of *general relativity*. This theory is mathematically extremely challenging (see **Einstein equations**), but it is founded on fairly straightforward physical ideas, which are described here.

Einstein's theory of **special relativity**, upon which the general theory is partly based, introduced the important idea that **time** and space are not absolutes, but depend to some extent on the state of motion of the observer. However, special relativity is restricted to so-called *inertial* motions – the motions of particles which are not acted on by any external forces. This means that special relativity cannot describe accelerated motion of any kind; in particular, it cannot describe motion under the influence of gravity.

Einstein had a number of deep insights into how to incorporate gravitation into relativity theory. For a start, consider Isaac Newton's theory of gravity (which is not a relativistic theory). In this theory the force exerted on a particle of mass m by another particle of mass M is given by the famous *inverse-square law*: $F = GMm/r^2$, where G is the Newtonian constant of universal gravitation. According to Newton's laws of motion, this force induces an acceleration a in the first particle, the magnitude of which is given by $F = ma$. The m in this second equation is called the *inertial mass* of the particle, and it determines the particle's resistance to being accelerated. In the first equation, however, the mass m measures the reaction to the gravitational field produced by the other particle; it is therefore called the *passive gravitational mass*. But Newton's laws of motion also state that if a body A exerts a force on a body B, then the body B exerts a force on the body A which is equal and opposite. This means that m must also be the *active gravitational mass* (if you like, the 'gravitational charge') produced by the particle. In Newton's theory all three of these masses – the inertial mass, and the active and passive gravitational masses – are equivalent. But there seems to be no reason in Newton's theory why this should be the case.

Einstein decided that this equivalence must be the consequence of a deeper principle called *equivalence principle*. In his own words, this means that 'all local, freely falling laboratories are equivalent for the performance of all physical experiments'. What this means is essentially that we can do away with gravity altogether and regard it instead as a consequence of moving between accelerated **frames of reference**. To see how this is possible, imagine a lift equipped with a physics laboratory. If the lift is at rest on the ground floor, experiments will reveal the presence of gravity to the occupants. For example, if we attach a weight on a spring fixed to the ceiling of the lift, the weight will extend the spring downwards. Now let us take the lift to the top of the building and let it fall freely. Inside the freely falling lift there is no perceptible gravity: the spring does not extend, as the weight is falling at the same rate as the rest of the lift. This is what would happen if we took the lift into space far away from the gravitational field of the Earth. The absence of gravity therefore looks very much like the state of free fall in a gravitational field. Now imagine that our lift is actually in space (and out of gravity's reach), but that it is mounted on a rocket. Firing the rocket would make the lift accelerate. There is no up or down in free space, but let us assume that the rocket is below the lift so that the lift would accelerate in the direction of its ceiling. What happens to the spring? The answer is that the acceleration makes the weight move in the reverse direction relative to the lift, thus extending the spring towards the floor. (This is similar to what happens when a car suddenly accelerates: the passenger's head is flung backwards.) But this is just like what happened when there was a gravitational field pulling the spring down. If the lift carried on accelerating, the spring would remain extended, just as if it were not accelerating but placed in a gravitational field.

Einstein's insight was that these situations do not merely appear to be similar: *they are completely indistinguishable*. Any experiment performed in an accelerated lift would give us exactly the same results as one performed in a lift upon which gravity is acting. This set Einstein the task of describing gravitational fields in terms of transformations between accelerated frames of reference (or coordinate systems). This is a difficult mathematical challenge, and it took him ten years from the publication of the theory of special relativity to arrive at a consistent formulation, now known as the Einstein field equations.

General relativity is the best theory of **gravity** that we have. Among its notable successes are the prediction of **gravitational lensing** and of the precession of the perihelion of Mercury, both of which have been tested and found to match the observations. On the other hand, all the tests of this theory that have so far been carried out concern relatively weak gravitational fields. It is possible that tests in strong fields might reveal departures from the theory. We should therefore bear in mind that, impressive though its successes undoubtedly have been, it may not be a complete theory of gravitation that works well under all circumstances.

FURTHER READING: Berry, M. V., *Principles of Cosmology and Gravitation* (Adam Hilger, Bristol, 1989); Schutz, B.F., *A First Course in General Relativity* (Cambridge University Press, Cambridge, 1985); Rindler, W., *Essential Relativity: Special, General, and Cosmological*, revised 2nd edition (Springer-Verlag, New York, 1979); Misner, C. W., Thorne, K. S. and Wheeler, J. A., *Gravitation* (W. H. Freeman, San Francisco, 1972); Weinberg, S., *Gravitation and Cosmology: Principles and Applications of the General Theory of Relativity* (John Wiley, New York, 1972).

GEODESIC see **space-time**.

GLOBULAR CLUSTERS Aggregations of stars found inside the halo of the Milky Way, and also in the haloes of other **galaxies**. They are small, about 10 parsecs across, and consist of typically a hundred thousand stars, usually of quite low mass and quite old. They are generally assumed to have formed during the very early stages of the formation of the galaxies themselves, and then to have evolved without much evidence of interaction with the rest of their host galaxy.

Globular clusters, as well as being interesting in their own right, provide an important check on **cosmological models**. Because they are thought to have formed early and in a relatively short period, it is reasonable to infer that their stars are all more or less of the same age. If we plot a colour–magnitude diagram for stars in such objects, the result looks very different from the Hertzsprung–Russell diagram obtained for stars in general (see the Figure). The strange appearance of the globular cluster HR diagram is, however, quite reasonable given our current knowledge of the origin and age of these objects.

According to the theory of **stellar evolution**, the lifetime of a star on the main sequence of the HR diagram is, roughly speaking, determined by its mass. More massive stars burn more quickly and leave the main sequence earlier than those with lower mass. If we could populate the main sequence with objects of different masses then, as time went on, we would find that objects of higher mass would move away from the main sequence before those of lower mass. At any particular time, therefore, there will be stars of some particular mass that are just leaving the main sequence and heading upwards to the right on the HR diagram. This is shown in the Figure by the series of *isochrones* (lines of constant age). Fitting isochrones to the observed HR diagram therefore provides a fairly direct measurement of the ages of these systems: the results are quite controversial, with ages of between 12 and 20 billion years having been claimed. These estimates can be compared with the time elapsed since the initial Big Bang in particular **Friedmann models** to provide a relatively simple test of theory against observation.

SEE ALSO: **age of the Universe**.

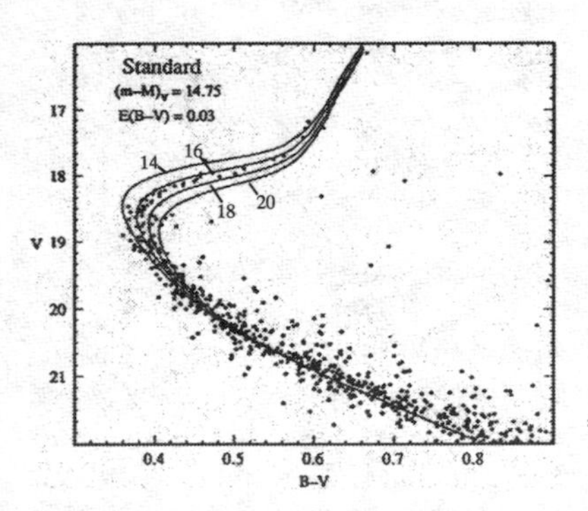

Globular clusters The observed colour–magnitude diagram (Hertzsprung–Russell diagram) for an example of a globular cluster, showing the characteristic main-sequence turnoff. The solid lines are computations for stars of the constant age (isochrones), with the chosen age ranging from 14 to 20 billion years.

GOLD, THOMAS (1920–) Austrian-born astronomer and physicist, who moved to Britain in the late 1930s, and to the USA in the mid-1950s. In 1948, with Hermann **Bondi** and Fred **Hoyle**, he developed the **steady-state theory**; though now discarded, it stimulated much important research.

***GRAND UNIFIED THEORY (GUT)** Any theory that attempts to describe three of the four **fundamental interactions** (electromagnetism, and the strong and weak nuclear forces) within a single mathematical formulation. Unification of the weak nuclear force with electromagnetism has already been satisfactorily achieved, by

Sheldon Glashow, Abdus Salam and Steven Weinberg in the 1960s in a theory called the *electroweak theory*. Attempts to incorporate the electroweak force with the strong nuclear force in a fully defined grand unified theory have been only partially successful, though experiments with particle accelerators do suggest that such a unification should be consistent with the data.

At present the best theory of particle physics available is called the *standard model*: it is a relatively *ad hoc* merging of three **gauge theories**: the U(1) symmetry of *quantum electrodynamics* (QED), the SU(2) symmetry of the weak interactions and the SU(3) symmetry of the strong interaction between quarks. The standard model is written as SU(3)⊗SU(2)⊗U(1). This, however, does not constitute a grand unified theory because the couplings between different parts of the theory are not explained.

A typical GUT theory has a much bigger gauge group, such as that corresponding to the SU(5) symmetry. Taking this as an example, we can make some comments about how a grand unified theory might behave. (In fact this simple GUT theory is excluded by experimental considerations, but it serves as a useful illustration nevertheless.) The number of distinct particles in a gauge theory is one less than the order of the group. The SU(5) theory is built from 5 × 5 matrices, and so has 24 arbitrary constants, corresponding to 24 different *gauge bosons*. In some sense four of these must correspond to the electroweak bosons (which are called the *intermediate vector bosons*), which become the photon and the W and Z bosons when the symmetry is broken at low energies. There will also be eight gauge bosons corresponding to the gluons. There are therefore 12 unidentified particles in the theory. For want of any better name, they are usually called the *X-bosons*. In a GUT theory, one of the six quark species can change into one of the six lepton species by exchanging an X-boson. This is one way in which **baryogenesis** might occur, because these processes do not necessarily conserve baryon number. It is likewise possible in such a theory for protons, for instance, to decay into leptons. The best estimates for the proton half-life are extremely large: about 10^{32} years. The masses of the X-bosons are also expected to be extremely large: about 10^{15} GeV.

In the **Big Bang theory**, energies appropriate to the electroweak unification are reached when the Universe has cooled to a temperature of about 10^{15} K (about 10^{-12} seconds after the Big Bang). Before then, the electromagnetic and weak interactions would have acted like a single physical force, whereas at lower energies they are distinct. The appropriate temperature for GUT unification is of the order of 10^{27} K, which occurs only 10^{-35} seconds after the Big Bang. Particles surviving to the present epoch as relics from this phase are possible candidates for non-baryonic **dark matter** in the form of **weakly interacting massive particles** (WIMPs). Phase transitions under which some of the fields present in a grand unified theory undergo **spontaneous symmetry-breaking** may also lead to a release of vacuum energy and a drastic acceleration in the cosmic expansion (see **inflationary Universe**).

Unification of the GUT interaction with gravity may take place at higher energies still, but there is no satisfactory theory that unifies all four physical forces in this way. Such a theory would be called a **theory of everything** (TOE).

FURTHER READING: Davies, P. C. W., The *Forces of Nature* (Cambridge University Press, Cambridge, 1979); Narlikar, J. V., *Introduction to Cosmology*, 2nd edition (Cambridge University Press, Cambridge, 1993), Chapter 6; Roos, M., *Introduction to Cosmology*, 2nd edition (John Wiley, Chichester, 1997), Chapter 6; Collins, P. D. B., Martin, A. D. and Squires, E. J., *Particle Physics and Cosmology* (John Wiley, New York, 1989).

Gravitational lensing According to **general relativity**, a deflection of the path followed by rays of light travelling near any relatively massive and compact object, such as a star or **galaxy**, in much the same way that an optical lens deflects light. This happens because the gravitating object induces a local distortion of the surrounding **spacetime**. Light paths in this distorted spacetime are not in general straight lines.

The possible bending of light rays by the gravitational fields around massive bodies was studied by Albert **Einstein** himself, and it furnished the first successful prediction made by the theory of general relativity. The first verification of the effect was made during Arthur **Eddington**'s famous solar eclipse expedition in 1919. The eclipse of the Sun made it possible to measure the positions of stars whose line of sight passed close to the solar surface. The expedition found the deflection of light to be consistent with the size predicted by Einstein's theory.

Since those early days, gravitational lensing phenomena have been observed in many different branches of astronomy and cosmology. The most common observational manifestation of this effect is the formation of double or multiple images by a galaxy (or other massive body) lying in the line of sight from the observer to a distant galaxy or **quasar**. A famous example of this kind of system is called the Einstein Cross or, sometimes in the USA, the Huchra Cross. Careful observations of the quasar G2237+0305 revealed four distinct images arranged symmetrically around the image of a nearby galaxy which is acting as the lens. In this case the lensed quasar is about 8 billion light years away, while the galaxy much closer, at a mere 400 million light years.

More complicated lensing effects can also occur. For example, an Einstein ring can be produced when a point-like mass lies exactly on the line of sight to a distant galaxy or quasar. In practice, we are unlikely to observe an idealised situation, but near-circular *arcs* have been detected in images of rich clusters of galaxies (see **large-scale structure**). The detection of such arcs makes it possible to estimate the mass of the intervening cluster and thus probe the gravitational effect of any **dark matter**. Weaker lensing phenomena include the formation of arcs and *arclets*, and the slight distortions of images of background galaxies in the presence of a foreground mass.

A different kind of phenomenon called *microlensing* takes place when the foreground object has insufficient mass to significantly bend the light rays around it and form multiple or distorted images. Instead, the lens manifests itself by causing a brightening of the distant source (in fact, multiple images *are* formed, but they are separated by too small an angle for them to be resolved separately; more light reaches the observer, however, than if the lens were not there). It is believed that this effect has been detected in the form of a brightening and fading of star images caused by the passage of low-mass stars or planets across the line of sight from us to the star. This kind of observation has provided the first clear evidence for the existence of **massive compact halo objects** (MACHOs) in our Galaxy. The application of gravitational lensing techniques to cosmology is now a boom industry.

SEE ALSO: Essay 6.

FURTHER READING: Kneib, J.-P. and Ellis, R. S., 'Einstein applied', *Astronomy Now*, May 1996, p. 435; Fort, B. and Mellier, Y., 'Arc(let)s in clusters of galaxies', *Astronomy and Astrophysics Review*, 1994, **5**, 239; Blandford, R. and Narayan, R., 'Cosmological applications of gravitational lensing', *Annual Reviews of Astronomy and Astrophysics*, 1992, **30**, 311; Walsh, D., Carswell, R. F. and Weymann, R. J., '0957+561A,B: Twin quasistellar objects or gravitational lens?', *Nature*, 1979, **279**, 381.

GRAVITATIONAL WAVES One of the important results to emerge from Maxwell's theory of electromagnetism was that it was possible to obtain solutions to Maxwell's equations that describe the propagation of an electromagnetic wave through a vacuum. Similar solutions can be obtained in Einstein's theory of **general relativity**, and these represent what are known as *gravitational waves* or, sometimes, gravitational radiation.

Gravitational waves represent distortions in the **metric** of **spacetime** in much the same way that fluctuations in the density of matter induce distortions of the metric in **perturbation theory**. The metric fluctuations that are induced by density fluctuations are usually called *scalar perturbations*, whereas those corresponding to gravitational waves are generally described as *tensor perturbations*. The reason behind this difference in nomenclature is that gravitational waves do not result in a local expansion or contraction of spacetime. Scalar perturbations can do this because they are *longitudinal waves*: the compression and rarefaction in different parts of the wave correspond to slight changes in the metric such that some bits of spacetime become bigger and some become smaller. Gravitational waves instead represent a distortion of the geometry that does not change the volume. Formally, they are transverse–traceless density fluctuations. (In fact, there is another possible kind of metric fluctuation, called a *vector perturbation*, which corresponds to vortical motions which are transverse but not traceless.) Gravitational waves are similar to the shear waves that can occur in elastic media: they involve a twisting distortion of spacetime rather than the compressive distortion associated with longitudinal scalar waves.

Gravitational waves are produced by accelerating masses and in rapidly changing tidal fields. The more violent the accelerations, the higher the amplitude of the resulting gravitational waves. Because general relativity is nonlinear, however, the waves become very complicated when the amplitude gets large: the wave begins to experience the gravitational effect produced by its own energy. These waves travel at the speed of light, just as **electromagnetic radiation** does. The problem with detecting gravitational waves, however, is that gravity is very weak. Even extremely violent events like a **supernova** explosion produce only a very slight signal.

Gravitational wave detectors have been built that attempt to look, for example, for changes in the length of large metal blocks when a wave passes through. Because the signal expected is much smaller than any thermal fluctuations or background noise, such experiments are extremely difficult to carry out. In fact, the typical fractional change in length associated with gravitational waves is less than 10^{-21}. Despite claims by Joseph Weber in the 1960s that he had detected signals that could be identified with gravitational radiation, no such waves have yet been unambiguously observed. The next generation of gravitational wave detectors such as GEO (a UK/German collaboration), Virgo (France/Italy) and LIGO (USA) should reach the desired sensitivity by using interferometry rather than solid metal bars. The LIGO experiment, for example, is built around an interferometer with arms 4 km long. Moreover, plans exist to launch satellites into space that should increase the baseline to millions of kilometres, thus increasing the sensitivity to a given fractional change in length. One such proposal, called LISA, is pencilled in for launch by the European Space Agency sometime before 2020.

Although these experiments have not yet detected gravitational radiation, there is very strong indirect evidence for its existence. The period of the binary pulsar 1913+16 is gradually decreasing at a rate which matches to a great precision relativistic calculations of the expected motion of a pair of neutron stars. In these calculations the dominant form of energy loss from the system is via gravitational radiation, so the observation of the spin-up in this system is tantamount to an observation of the gravitational waves themselves. Russell Hulse and Joseph Taylor were awarded the 1993 Nobel Prize for Physics for studies of this system. It is also possible that gravitational waves have already been seen directly. The temperature fluctuations in the **cosmic microwave background radiation** (described in Essay 5) are usually attributed to the **Sachs–Wolfe effect** produced by scalar density perturbations (see **primordial density fluctuations**). But if these fluctuations were generated in the **inflationary Universe** phase by quantum fluctuations in a **scalar field**, they are expected to have been accompanied by gravitational waves which in some cases could contribute an observable Sachs–Wolfe effect of their own. It could well be that the famous **ripples** detected by the **Cosmic Background Explorer** (COBE) satellite are at least partly caused by gravitational waves with wavelengths of the same order as the cosmological **horizon**.

It can be speculated that, in a theory

of **quantum gravity**, the quantum states of the gravitational field would be identified with gravitational waves in much the same way that the quantum states of the electromagnetic field are identified with photons. The hypothetical quanta of gravitation are thus called *gravitons*.

FURTHER READING: Schutz, B. F., *A First Course in General Relativity* (Cambridge University Press, Cambridge, 1985).

* **GRAVITY** The weakest of the **fundamental interactions**, representing the universal tendency of all matter to attract all other matter. Unlike electromagnetism, which has two possible charges so that both attraction (between unlike charges) and repulsion (between like charges) can occur, gravity is always attractive. The force of gravity is, however, extremely weak. For example, the electrostatic force between a proton and an electron is about 10^{40} times stronger than the gravitational force between them. Despite its weakness, though, gravity is more important than electromagnetism in astronomical situations because there is no large-scale separation of electrostatic charge, which there would have to be to produce electrostatic forces on large scales.

One of the first great achievements of theoretical physics was Isaac **Newton**'s theory of universal gravitation. This law unified what, at the time, seemed to be disparate physical phenomena. Newton's theory of mechanics is embodied in three simple laws:

> Every body continues in a state of rest or uniform motion in a straight line unless it is compelled to change that state by forces acting on it.
>
> The rate of change of momentum is proportional to the acting force, and is in the direction in which this force acts.
>
> To every action, there is always opposed an equal reaction.

These three laws of motion are general, and Newton set about using them to explain the motions of the heavenly bodies. He realised that a body orbiting in a circle, like the Moon going around the Earth, is experiencing a force acting in the direction of the centre of motion (just as a weight tied to the end of a piece of string does when you whirl it around your head). Gravity could cause this motion in the same way as it could cause an object (like an apple) to fall to the Earth when dropped. In both cases the force has to be directed towards the centre of the Earth. These two situations can be described in terms of a simple mathematical law, stating that the size of the attractive force F between any two bodies depends on the product of the masses M and m of the bodies and on the square of the distance r between them:

$$F = GMm/r^2$$

The constant G is called the Newtonian constant of gravitation. It was a triumph of Newton's theory that it could explain the laws of planetary motion obtained over half a century before by Johannes **Kepler**. In particular, the elliptical orbits that were introduced by Kepler to match Tycho Brahe's observations come naturally out of this theory. So spectacular was this success that the idea of a Universe guided by Newton's laws of motion was to dominate scientific thinking for more than two centuries.

Albert **Einstein**'s early work on **special relativity**, however, convinced him that there were serious problems with Newton's laws. For example, the gravitational force described by the inverse-square law has to involve instantaneous action at a distance, which is at odds with the idea that no signal can travel faster than light. Einstein therefore abandoned Newtonian mechanics, and in particular Newtonian gravity, in favour of an entirely different formulation – the theory of **general relativity** – upon which modern **cosmological models** are based and which forms the basis of the standard **Big Bang theory**.

While it is undoubtedly true that general relativity is a more complete theory that Newtonian gravity, there are nevertheless many situations, even in cosmology, where Newtonian calculations are perfectly adequate. While Newton's theory cannot predict the bending of light by a gravitational field (**gravitational lensing**) or the existence of **black holes**, it is possible, for example, to obtain the equations governing the evolution of the **Friedmann models** by using entirely Newtonian arguments. Newton's theory is what is obtained by taking the limit of weak gravitational fields in the Einstein theory. The enormous computations involved in simulating the evolution of the Universe in models **of structure formation** based on the idea of gravitational **Jeans instability** are similarly entirely Newtonian. In fact, the latter calculations simply involve solving *Poisson's equation*:

$$\nabla^2\phi(\boldsymbol{x}) = 4\pi G\rho(\boldsymbol{x})$$

This is a slightly more sophisticated formulation of Newton's theory which allows us to deal with simpler quantities, like the gravitational potential ϕ, rather than forces. The force (per unit mass) on an object at position $\boldsymbol{x}$ is simply the gradient of ϕ at that point, and ρ is the density of matter (see ***N*-body simulations**).

FURTHER READING: Symon, K. R., *Mechanics*, 3rd edition (Addison-Wesley, Reading MA, 1980); Hawking, S. W. and Israel, W. (editors), *300 Years of Gravitation* (Cambridge University Press, Cambridge, 1987).

GREAT ATTRACTOR see **peculiar motions**.

GUTH, ALAN HARVEY (1947–) US astrophysicist and cosmologist, who works at the Massachusetts Institute of Technology. He developed much of the theory of cosmological **phase transitions**, and was largely responsible for the ideas behind the **inflationary Universe**.

H

Hadron see **elementary particles**.

Hadron era see **thermal history of the Universe**.

Hawking, Stephen William (1942–) English theoretical physicist. He has made significant contributions to the theory of **black holes,** the nature of the cosmological **singularity** and ideas associated with **quantum cosmology**. He established that black holes, far from being black, actually radiate with a temperature inversely proportional to their mass, a phenomenon known as **Hawking radiation**. He has been confined to a wheelchair for many years, and is unable to speak without a voice synthesiser, because of the debilitating effects of motor neurone disease from which he has suffered since his student days.

Hawking radiation It seems that **black holes** are inevitably surrounded by an event **horizon** from which no **electromagnetic radiation** can escape. Their name therefore seems entirely appropriate. But this picture of black holes was shattered in the 1970s by calculations performed by the (then) young British physicist Stephen **Hawking** which showed that, under certain circumstances, not only could black holes emit radiation, but they could emit so much radiation that they might evaporate entirely. The radiation emitted by black holes is now called *Hawking radiation*.

How can radiation be emitted by a black hole when the hole is surrounded by a horizon? The reason is that Hawking radiation is essentially a quantum process. Nothing described by classical physics can leave the horizon of a black hole, but this prohibition does not arise with **quantum physics**. The violation of classical restrictions occurs in many other instances when quantum phenomena lead to *tunnelling*, such as when **elementary particles** are able to escape from situations where they appear to be trapped by electromagnetic forces (see **fundamental interactions**).

The conditions around the horizon of a black hole are discussed briefly in Essay 3. Basically, the **spacetime** around a black hole may be represented as a vacuum, but this vacuum is not entirely empty. Tiny quantum fluctuations are continually forming and decaying according to *Heisenberg's uncertainty principle* from quantum physics. This has the effect of continually filling the vacuum with *virtual particles* which form and decay on a very short timescale. Such processes usually create particle–antiparticle pairs out of nothing, but the two members of each pair never separate very far, and the two particles annihilate each other (see **antimatter**). On the edge of a horizon, however, even a small separation can be crucial. If one particle of the pair happens to move inside the horizon it is lost for ever, while the other particle escapes. To all intents and purposes this looks as if the black hole is radiating from its event horizon. Taking account of the energy states of the ingoing and outgoing particles leads inevitably to the conclusion that the mass of the black hole decreases. The process is more efficient for smaller black holes, and it only really has consequences which are observable for black holes which are very small indeed.

A black hole emits a spectrum of **electromagnetic radiation** like that from a **black body**. The typical energy of photons emitted approximates to kT, where k is the Boltzmann constant and T is the characteristic temperature of the Planck spectrum (see **black body**). The temperature turns out to be inversely proportional to the mass M of the hole. The time needed for such a black hole to completely evaporate (i.e. to lose all its rest-mass energy Mc^2) turns out to depend on the cube of the mass. Black holes of smaller mass therefore evaporate more quickly than large ones, and produce a much higher temperature. But evaporation is the fate of all black holes: they glow dimly at first, but as they fritter away their mass they glow more brightly. The less massive they get the hotter they get, and the more quickly they lose mass. Eventually, when are very small indeed, they explode in a shower of high-energy particles.

This effect is particularly important for very small black holes which, in some theories, form in the very early Universe: the so-called *primordial black holes*. Any such objects with a mass less than about 10^{15} grams would have evaporated by now, and the radiation they produced may well be detectable by techniques developed in **gamma-ray astronomy**. The fact that this radiation is not observed places strong constraints on theories that involve these primordial objects.

There is another interesting slant on the phenomenon of Hawking radiation which is also directly relevant to cosmology. In the **inflationary Universe** scenario, the Universe undergoes a period of rapid accelerated expansion described by the de Sitter solution (see **cosmological models**). This solution also has an event horizon. There is an important difference, however, because the observer is usually outside the black hole horizon, whereas in inflation the horizon forms around the observer. Putting this difference to one side and thinking of the Universe as an inside-out black hole, Hawking's calculations show that the horizon should also have a finite temperature. The temperature of this radiation is given by the expansion rate, as determined by the value of the **Hubble parameter** H, and this turns out to depend on the potential energy of the vacuum associated with the **scalar field** that drives the inflation. This temperature manifests itself by the creation of quantum fluctuations in the form of both **primordial density fluctuations** and **gravitational waves**.

FURTHER READING: Hawking, S. W., 'Black hole explosions?', *Nature*, 1974, **248**, 30; Thorne, K. S., *Black Holes and Time Warps* (Norton, New York, 1994).

Heat death of the Universe The great advances made in science during the Renaissance and after were founded on Isaac **Newton**'s laws of motion. These **laws of physics** were assumed to describe all the workings of Nature in such a way that if you knew the positions and velocities of all existing particles at the present time, their positions and velocities at some later time could be calculated exactly. In the mechanistic view of the Universe engendered by these laws, the Universe was seen as a giant clockwork device, elaborate yet built upon relatively simple principles.

Newton's laws of motion do not prescribe a preferred direction of **time**: any solution of the equations of motion is still a solution if we reverse the time dependence. There was therefore a temptation to view the Universe as essentially timeless. As there was no reason to single out a 'forward' or 'backward' sense of time, it seemed a logical conclusion that the cosmic clockwork machine had ticked for ever in the past and would continue to tick for ever in the future. The planets had always travelled in their orbits around the Sun, and would continue to do so *ad infinitum*. This view of creation as an eternal machine was brought into question by developments in the theory of **thermodynamics** that were stimulated by the Industrial Revolution. These showed that things cannot run just as easily backwards in time as they can forwards, and that nothing goes on for ever. Steam engines can never be perfect, perpetual motion (which had been thought possible) is impossible and, so it seemed, the Universe cannot carry on the same for ever.

The reason lies in the second law of thermodynamics, which stipulates that the entropy of any closed system must increase with time. *Entropy* is a measure of the degree of disorder of a system. Whatever happens to the system, the degree of disorder in it increases with time. A ball bouncing on a table-top gradually comes to rest when the energy stored in its bounces has been transformed into disordered motions of the molecules of the table. The table heats up slightly, so that the total energy is conserved, but this heat cannot be extracted to start the ball bouncing again. This may at first seem to be at odds with the fact that ordered systems (such as those associated with **life in the Universe**) can arise naturally, but this is because these systems are supplied by an external source of energy. Life on Earth, for example, is powered by energy from the Sun: if this source were to be extinguished, the inevitable result would be the gradual cooling of our planet and the extinction of all life.

The consequences of the second law of thermodynamics for cosmology became known as *the heat death of the Universe*, a phrase probably first used in print by the German physicist Hermann von Helmholtz in 1854. It seemed inevitable that Newton's stately machine would gradually wind down as all the energy of all the ordered motions in all the cosmos gradually became disordered and incoherent. These ideas were at large before the discoveries of **quantum physics** and **general relativity** in the early years of the 20th century, but they still dominated popular discussions of cosmology in the 1920s. A famous example is the influential book *The Nature of the Physical World* by Arthur **Eddington**, first published in 1927. The idea of a Universe that gradually winds down and fizzles out provided a powerful corrective to pseudo-Darwinian notions of continuous evolution towards higher and better life forms.

The idea of a universal heat death is still valid in modern **cosmological models**, especially those built around an **open universe**. The gradual increase in entropy also prevents a **closed universe** from undergoing infinite identical oscillations. But there is one major issue that remains unresolved: it is still not known exactly how to define the entropy associated with the gravitational field in general relativity. It would appear that the final state of the evolution of a self-gravitating system might well be a **black hole**. (Newton himself speculated that since gravity exerts a universal attraction, all matter in the Universe could end up in isolated lumps with vast areas of empty space between them.) But black holes themselves do not last for ever because they lose energy by **Hawking radiation**. If it is true that all matter in the Universe eventually ends up in black holes, then it must ultimately all be recycled as radiation. The late stages of evolution of an open universe would therefore contain no matter, only a gradually cooling sea of low-energy photons

Those who seek *teleological* explanations for the laws of physics (i.e. those who try to explain the existence of life in the Universe by means of arguments from design – see **anthropic principle**) reject this somewhat pessimistic conclusion. There seems little point in creating a Universe specifically for life, only to wipe it out later on. Some physicists, including Frank Tipler, have argued that intelligent beings could prevent the heat death and thus ensure eternal life. But most scientists hold a more pragmatic view, similar to the thoughts of the British philosopher Bertrand Russell:

> Therefore, although it is of course a gloomy view to suppose that life will die out . . . it is not such as to render life miserable. It merely makes you turn your attention to other things.
>
> *Why I Am Not a Christian*
> (Allen & Unwin, New York, 1957)

SEE ALSO: Essay 1.

FURTHER READING: Barrow, J. D. and Tipler, F. J., *The Anthropic Cosmological Principle* (Oxford University Press, Oxford, 1986); Dyson, F. J., 'Time without end: Physics and biology in an open universe', *Reviews of Modern Physics*, 1979, **51**, 447; Eddington, A. S., *The Nature of the Physical World* (Cambridge University Press, Cambridge, 1928); Tipler, F. J., *The Physics of Immortality* (Macmillan, London, 1995).

HEISENBERG, WERNER KARL (1901–76) German physicist. He worked on the theory of **quantum mechanics**, and formulated the *uncertainty principle* with which his name is now associated. The **primordial density fluctuations** that led to the process of cosmological **structure formation** are thought to have been generated according to this principle.

HERMAN, ROBERT (1914–) US physicist. In the late 1940s, with George **Gamow** and Ralph **Alpher**, he predicted that the event now called the Big Bang would leave a residual radiation with a temperature of about 5 K.

Herschel, (Frederick) William (1738–1822) German-born English astronomer. Among his numerous contributions to astronomy, he pioneered the use of statistics to probe the structure of the known Universe (which in his day was delimited by the Milky Way). Like others before him, including Immanuel **Kant**, he realised that the Milky Way is the plane of a disk-shaped stellar system, as viewed from our vantage point inside it. Herschel estimated its extent in different directions by counting the numbers of stars visible through his telescope in different directions. He also observed and catalogued nebulae, many of which were later found to other galaxies.

Hierarchical cosmology Before the discovery of the **expansion of the Universe** described by **Hubble's law** and its subsequent incorporation into the **Friedmann models** founded on Albert **Einstein**'s theory of **general relativity**, most astronomers imagined the **Universe** to be infinite, eternal and static. They also would have thought, along with Newton, that **time** was absolute and that space was necessarily Euclidean. The distribution of matter within the Universe was likewise assumed to be more or less homogeneous and static. (The discovery that **galaxies** were actually external to the Milky Way and comparable to it in size was made only a few years before Hubble's discovery of the **expansion of the Universe**.)

Nevertheless, from the beginning of the 19th century there were a number of prominent supporters of the *hierarchical cosmology*, according to which the material contents of the Universe are distributed in a manner reminiscent of the modern concept of a **fractal**. In such a scenario, galaxies occur in clusters which, in turn, occur in superclusters, and so on without end. Each level of the hierarchy is supposed to look like a photographic blow-up of the lower levels. A hierarchical cosmology does not possess the property of homogeneity on large scales and is therefore not consistent with the **cosmological principle**. Indeed, in such a model the mean density of matter ρ on a scale r varies as $\rho \propto r^{-\gamma}$, where γ is some constant. In this way the mean density of the Universe tends to zero on larger and larger scales.

The idea of a perfectly fractal Universe still has its adherents today, despite the good evidence we have from the extreme isotropy of the **cosmic microwave background radiation** that the Universe is close to being homogeneous and isotropic on scales greater than about a million light years. Analysis of the pattern of galaxy clustering seen in large-scale **redshift surveys** also suggests that the Universe is approximately hierarchical on scales up to perhaps a hundred million light years, but thereafter becomes homogeneous.

FURTHER READING: Mandelbrot, B. B., *The Fractal Geometry of Nature* (W. H. Freeman, San Francisco, 1982).

Homogeneity see **cosmological principle**.

* **Horizon** In **general relativity**, the boundary between regions of **spacetime** across which no signals can cross. An example is the horizon that forms around a black hole: light that has entered the region surrounded by this horizon cannot escape from it at any time in the future, at least not without the aid of the quantum processes of **Hawking radiation**.

There are different kinds of horizon relevant to **cosmological models**. We can understand the existence of some sort of horizon by looking at the form of **Hubble's law**, $v = H_0 d$. In the standard **Friedmann models** this relationship is exact between the observed recessional velocity v of an object moving with the **expansion of the Universe** and the **proper distance** d of that object from the observer. The constant of proportionality is the **Hubble constant**. This is an exact expression for all values of d. As d increases, the recessional velocity increases without bound. If $d = c/H_0$ the object is receding at the speed of light. Objects at distances farther than this value of d are said to be receding at *superluminal velocities*. This does not violate any physical laws because no light signals can be exchanged between such objects and the observer. In other words, some kind of horizon must form. To look at it another way, the redshift associated with an object moving away at the velocity of light is infinite, and the object would therefore be infinitely faint.

The distance $d = c/H_0$ is usually called the *Hubble distance* or *Hubble radius*, or sometimes the *cosmological horizon*; it roughly defines the scale of the observable Universe. It is not, however, a horizon in the exact sense as defined above. The Hubble distance defines those points that are moving away now at the velocity of light. But light signals from them could not be observed now even if they were not receding from us, because light travels at the finite speed c. What really defines a horizon is not the instantaneous properties of objects at a single **time**, but the entire history of behaviour in **spacetime**.

Perhaps a better way of thinking about this problem is to start from the realisation that, if **the age of the Universe** t_0 is finite, there might exist objects so far away now that light from them would have taken more than time t_0 to reach us. Such objects could obviously not be observed. This suggests that the size of the horizon should be ct_0. Again, this is roughly the right order of magnitude, but is not really accurate because the expansion of the Universe has not always been proceeding at the same rate at all times in the past.

Now consider the question of how to find the set of points in **spacetime** that are capable of sending light signals which can be received by an observer at any time in the observer's past. For simplicity, we can assume that the observer is at the origin O of a system of spatial coordinates. If the observer is asking this question at time t, then any light signal received at O must have set out from its source at some time between $t = 0$ (the Big Bang **singularity**) and t. Light signals travel from source to observer through spacetime, which is described by the **Robertson–Walker metric**. Let us assume for further simplicity that we are dealing with a

flat universe, so that the required metric is of the form

$$ds^2 = c^2\,dt^2 - a(t)^2[dr^2 + r^2(d\theta^2 + \sin^2\theta\,d\phi^2)]$$

(for which $k = 0$). Since it does not matter what the direction of the source is, we can ignore the dependence on the angular coordinates θ and ϕ: we simply point the angular coordinates directly at the object.

Light travels along so-called *null geodesics* in spacetime: these trajectories have $ds = 0$ in the **metric**. It is then clear that the set of points that can have communicated with O at any point must be contained within a sphere whose *proper size* at time t is given by

$$R(t) = a(t)\int dr = a(t)\int c\,dt'/a(t')$$

where the second integration is from 0 to t. The size of R depends on the expansion rate as a function of time, $a(t)$. It is possible that $a(t)$ tends to zero for small t sufficiently quickly that the integral diverges. This shows that the observer can have communicated with the entire Universe up to the time t. If, on the other hand, the integral converges, then $R(t)$ represents the size of the region around O at time t that contains all the points that can have sent signals to O. If it is finite, then R is called the radius of the *particle horizon*. It divides all the points in spacetime (up to and including t) into two classes: those inside the horizon that can have communicated with O, and those outside that cannot have.

The integral that determines R can be evaluated for particular cosmological models. The result, for example, for the special case of a flat, matter-dominated Friedmann model is that $R = 3ct$. The size of the particle horizon at the present epoch is then $3ct_0$, somewhat larger than the ct_0 calculated above. Since the Hubble constant in this model is related to the age of the Universe by $H_0=2/3t_0$, then the horizon is $2c/H_0$ and not the simple c/H_0 inferred from Hubble's law. Note, in particular, that objects can be outside our Hubble radius now, but still inside our particle horizon. The behaviour of the particle horizon in standard Friedmann models gives rise to the perplexing issue of the cosmological **horizon problem**.

The particle horizon refers to the past, but the event horizon mentioned at the beginning in the context of **black holes** refers to the future ability of light to escape. There is an analogous event horizon in cosmology which can be obtained simply by changing the limits of integration in the calculation that led to the particle horizon. Instead of asking where the points are that can send signals to O between the beginning and some arbitrary time t, we ask where the points are that can communicate with O between t and the end of the universe. In a **closed universe** the end of the Universe means the *Big Crunch* wherein the second singularity forms; in an **open universe** we have to take the upper limit to infinity. In the latter case, the event horizon may or may not exist, depending on the behaviour of $a(t)$. For example, in the de Sitter solution describing the exponential expansion that occurs in the **steady state theory** with constant $H(t) = H_0$, the event horizon lies at a distance c/H_0.

FURTHER READING: Rindler, W., *Essential Relativity: Special, General, and Cosmological*, revised 2nd edition (Springer-Verlag, New York, 1979).

* **Horizon problem** The standard **Big Bang theory** is based on the assumption that the Universe is, on sufficiently large scales, homogeneous and isotropic. This assumption goes under the grand-sounding name of the **cosmological principle**. The name, however, belies the rather pragmatic motivations that led the early relativistic cosmologists to introduce it. Having virtually no data to go on, Albert **Einstein**, Alexander **Friedmann**, Georges **Lemaître** and the rest simply chose to explore the simplest cosmological models they could find. Somewhat fortuitously, it seems that the Universe is reasonably compatible with these simple models.

More recently, cosmologists started to ask whether homogeneity could be *explained* within the Big Bang theory rather than simply being assumed at the start. The prospects appear fairly promising: there are many physical processes that we can imagine having smoothed out any fluctuations in the early Universe, in much the same way that inhomogeneous media reach a homogeneous state, for example by diffusion. But there is a fundamental problem that arises when we appeal to these processes in cosmology: diffusion or other physical homogenisation mechanisms take time. And in the early stages of the rapid **expansion of the Universe** there does not seem to have been enough time for these mechanisms to come into play.

This shortage of time is indicated by the presence of cosmological (particle) **horizons**. Even the most rapid process for smoothing out fluctuations cannot occur more quickly over a scale L than the time it takes light to traverse that scale. Therefore, assuming that the initial state of the Universe was not homogeneous, we should expect it to remain inhomogeneous on a scale L unless the horizon is large enough to encompass L. Roughly speaking, this means that $L > ct$ for homogenisation to occur at some time t. But the cosmological particle horizon grows in proportion to time t in the standard Friedmann models, while the **proper distance** between two points moving with the expansion scales with t more slowly than this. (For example, in the Friedmann model describing a **flat universe** – the Einstein–de Sitter solution – the proper distance between points scales as $t^{2/3}$.)

The existence of a cosmological horizon makes it difficult to accept that the cosmological principle results from a physical process. This principle requires that there should be a very strong correlation between the physical conditions in regions which are outside each other's particle horizons and which, therefore, have never been able to communicate by causal processes. For example, the observed isotropy of the **cosmic microwave background radiation** implies that this radiation was homogeneous and isotropic in regions on the **last scattering surface** (i.e. the spherical surface centred upon us, here on Earth, which is at a distance corresponding to the lookback time to the era at which this radiation was last scattered by matter). The last scattering probably took place at a cosmic epoch characterised by some time t_{ls} corresponding to a redshift of $z_{ls} \approx 1000$. The distance of the last scattering surface is now roughly ct_0, since the time of last scattering was very soon after the Big Bang singularity. Picture a sphere delimited by this surface. The size of the sphere at the epoch when the last scattering occurred was actually smaller than its present size because it has been participating since then in the expansion of the Universe. At the epoch of last scattering the sphere had a radius given roughly by $ct_0/(1 + z_{ls})$. This is about one-tenth the size of the particle horizon at the same epoch. But our last scattering sphere seems smooth and uniform. How did this happen, when different parts of it have never been able to exchange signals with each other in order to cause homogenisation?

Various avenues have been explored in attempts to find a resolution of this problem. Some homogeneous but anisotropic **cosmological models** do not have a particle horizon at all. One famous example is the *mixmaster universe* model proposed by Charles Misner. Other possibilities are to invoke some kind of modification of Einstein's equations to remove the horizon, or some process connected with the creation of particles at the Planck epoch of **quantum gravity** that might lead to a suppression of fluctuations. Indeed, we might wonder whether it makes sense to talk about a horizon at all during the era governed by **quantum cosmology**. It is generally accepted that the distinct *causal* structure of spacetime that is responsible for the behaviour of light signals (described by the signature of the **metric**) might break down entirely, so the idea of a horizon becomes entirely meaningless (see e.g. **imaginary time**).

The most favoured way of ironing out any fluctuations in the early Universe, however, is generally accepted to be the **inflationary Universe** scenario. The horizon problem in the standard models stems from the fact that the expansion is invariably decelerating in the usual Friedmann models. This means that when we look at the early Universe the horizon is always smaller, compared with the distance between two points moving with the expansion, than it is now. Points simply do not get closer together quickly enough, as we turn the clock back, to be forced into a situation where they can communicate. Inflation causes the expansion of the Universe to accelerate. Regions of a given size now come from much smaller initial regions in these models than they do in the standard, decelerating models. This difference is illustrated in the Figure by the convex curves showing expansion in the

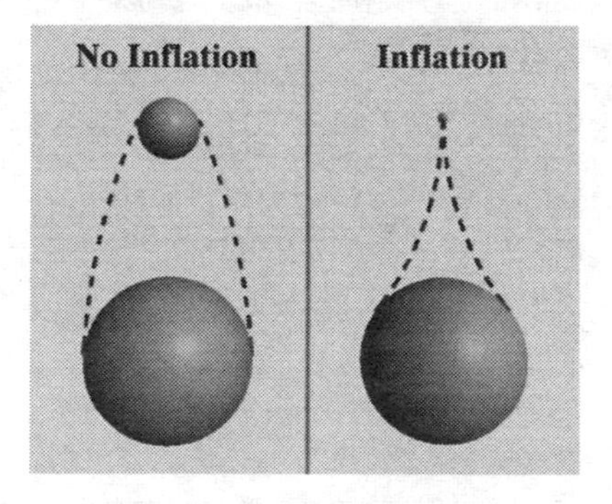

Horizon problem Our observable patch of the Universe grows from a much smaller initial patch in the inflationary Universe (right) than it does in the standard Friedmann models (left).

inflationary model, and the concave curves with no inflation.

With the aid of inflation, we can make models in which the present-day Universe comes from a patch of the initial Universe that is sufficiently small to have been smoothed out by physics rather than by cosmological decree. Interestingly, though, having smoothed away any fluctuations in this way, inflation puts some other fluctuations in their place. These are the so-called **primordial density fluctuations** which might be responsible for cosmological **structure formation**. The difference with these fluctuations, however, is that they are small – only one part in a hundred thousand or so – whereas we might have expected the initial pre-inflation state of the Universe to be arbitrarily large.
SEE ALSO: Essays 3 and 5.

FURTHER READING: Guth, A. H., 'Inflationary universe: A possible solution to the horizon and flatness problems', *Physical Review* D, 1981, **23**, 347; Narlikar, J. V. and Padmanabhan, T., 'Inflation for astronomers', *Annual Reviews of Astronomy and Astrophysics*, 1991, **29**, 325.

HOYLE, SIR FRED (1915–) English astronomer and physicist. He is best known for his advocacy of the **steady state theory** of cosmology, but he also made outstanding contributions to the theories of **stellar evolution** and **nucleosynthesis**.

HUBBLE, EDWIN POWELL (1889–1953) US astronomer, the founding-father of observational **cosmology**. He was the first to prove unambiguously that the 'spiral nebulae' were extragalactic, constructed the first systematic classification of the **galaxies**, made the first attempts to calibrate the **extragalactic distance scale** and established the **expansion of the Universe**. NASA's Hubble Space Telescope, launched in 1990, was named in his honour.

HUBBLE CONSTANT, HUBBLE PARAMETER In the standard **Friedmann models** upon which the **Big Bang theory** is based, the **expansion of the Universe** is described mathematically by a global scaling of distances with (proper) **time**: all distances between objects moving with the expanding Universe are simply proportional to a function of time called the *scale factor* and usually given the symbol $a(t)$. The rate of expansion is then simply the derivative of this function with respect to the time coordinate t: $\mathrm{d}a/\mathrm{d}t$. It is convenient to discuss the rate of expansion by defining the *Hubble parameter* $H(t)$ to be equal not to $\mathrm{d}a/\mathrm{d}t$, but to the logarithmic derivative of a, written as $(1/a)(\mathrm{d}a/\mathrm{d}t)$. Since the expansion of the Universe is not uniform in time, the Hubble parameter is itself a function of time (unless the Universe is completely empty). The particular value that the Hubble parameter takes *now* (at $t = t_0$) is called the *Hubble constant*. As with other, present-day values of quantities such as the density parameter and the deceleration parameter, the present-day value of the Hubble parameter is usually distinguished by adding a subscript '0': $H(t_0) = H_0$. This much sought-after number indicates the present rate of the expansion of the Universe.

The Hubble constant H_0 is one of the most important numbers in cosmology because it defines the size of the observable **horizon** and **the age of the Universe**. If its value were known exactly, the Hubble constant could be used to determine a number of interesting things, such as the intrinsic brightness and masses of stars in nearby **galaxies**. Those same properties could then be examined in more distant galaxies and galaxy clusters. The amount of **dark matter** present in the Universe could then be deduced, and we could also obtain the characteristic size of **large-scale structure** in the Universe, to serve as a test for theoretical **cosmological models**.

In 1929, Edwin **Hubble** announced his discovery that galaxies in all directions appear to be moving away from us according to what is now called **Hubble's law**. This phenomenon was observed as a displacement of known spectral lines towards the red end of a galaxy's spectrum (when compared with the same spectral lines from a source here on Earth). This **redshift** appeared to have a larger displacement for faint, presumably more distant galaxies. Hence, the more distant a galaxy, the faster it is receding from the Earth. The Hubble constant can therefore be defined by a simple mathematical expression: $H_0 = v/d$, where v is the galaxy's velocity of recession (in other words, motion along our line of sight) and d is the galaxy's distance.

However, obtaining a true value for H_0 is a very complicated business. Astronomers need two measurements. First, spectroscopic observations will reveal a galaxy's redshift, indicating its recessional velocity. This part is relatively straightforward. The second measurement, and the one most difficult to carry out, is of the galaxy's precise distance from the Earth. Reliable 'distance indicators', such as variable stars and **supernovae**, must be found in galaxies in order to calibrate the **extragalactic distance scale**. The value of H_0 itself must also be derived from a sample of galaxies that are sufficiently far away for their **peculiar motions** due to local gravitational influences to be negligibly small.

The Hubble constant is normally measured in units of kilometres per second per megaparsec (km/s/Mpc). In other words, for each megaparsec of distance, the velocity of a distant object appears to increase by some value in kilometres per second. For example, if the Hubble constant were determined to be 50 km/s/Mpc, a galaxy at 10 Mpc would have a redshift corresponding to a radial velocity of 500 kilometres per second.

The value of the Hubble constant initially obtained by Hubble was around 500 km/s/Mpc, and has since been radically revised downwards because the assumptions originally made about stars yielded underestimated distances. Since the 1960s there have been two major lines of investigation into the Hubble constant. One team, associated with Allan **Sandage** of the Carnegie Institutions, has derived a value for H_0 of around 50 km/s/Mpc. The other team, associated with Gérard De Vaucouleurs of the University of Texas, has obtained values that indicate H_0 to be around 100 km/s/Mpc. One of the long-term key programmes of the **Hubble Space Telescope** has been to improve upon these widely discrepant estimates; preliminary results indicate a value of

60–70 km/s/Mpc, between the Sandage and De Vaucouleurs estimates (see **extragalactic distance scale**).
SEE ALSO: Essay 4.

FURTHER READING: Sandage, A. R., 'Distances to galaxies: The Hubble constant, the Friedmann time and the edge of the Universe', *Quarterly Journal of the Royal Astronomical Society*, 1972, **13**, 282; Rowan-Robinson, M. G., *The Cosmological Distance Ladder* (W. H. Freeman, New York, 1985); Freeman, W. *et al.*, 'Distance to the Virgo Cluster galaxy M100 from Hubble Space Telescope observations of Cepheids', *Nature*, 1994, **371**, 757.

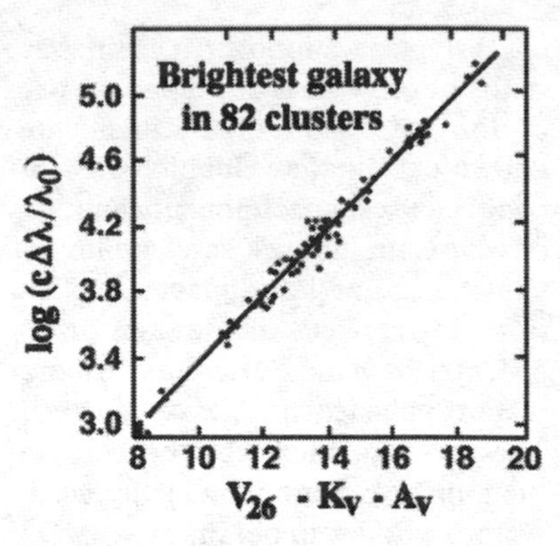

Hubble's law The linearity of Hubble's law has been demonstrated observationally using a variety of distance indicators, including the brightest cluster galaxies shown in this plot. An estimate of the distance is plotted along the *x*-axis with the redshift shown vertically.

* **HUBBLE'S LAW** The statement that the apparent recession velocity of a galaxy v is proportional to its distance d from the observer: $v = H_0 d$, where the constant of proportionality H_0 is known as the **Hubble constant**.

The law was first published in 1929 by Edwin **Hubble**, who had noticed a linear relationship between the **redshift** of **emission lines** of a sample of **galaxies** and their estimated distance from the Milky Way. Interpreting the redshift as a **Doppler shift**, he was able to relate the redshift to a velocity and hence derive the law by plotting a diagram called the *Hubble diagram*, like one shown in the Figure. However, the American astronomer Vesto **Slipher** should be given a large part of the credit for the discovery of Hubble's law. By 1914 Slipher had obtained the spectra of a group of nebulae that also displayed this relationship, and presented his results at the 17th Meeting of the American Astronomical Association; they were published the following year. In giving most of the credit to Hubble, history has tended to overlook Slipher's immense contribution to the development of cosmology.

Hubble's law is now assumed to represent the **expansion of the Universe**. Hubble's original paper, however, does not claim this origin for the empirical correlations he measured. Georges **Lemaître** was probably the first theorist to present a theoretical cosmological model in which Hubble's law is explained in this way, by objects moving with a global expansion of **spacetime**. Lemaître's paper, published in 1927 and so prefiguring Hubble's classic paper of 1929, made little impression as it was written in French and published in an obscure Belgian journal. It was not until 1931 that Arthur **Eddington** had Lemaître's paper published (in English) in the more influential *Monthly Notices of the Royal Astronomical Society*. The identification of Hubble's law with the expansion of the Universe is one of the main supporting pillars of the **Big Bang theory**, so Lemaître too should be given due credit for making this important connection.

There are some ambiguities in the phrasing of Hubble's law. It is sometimes stated to be a linear relationship between redshift z and distance d, rather than a relationship between recessional velocity v and d. If the velocities concerned are much smaller than the speed of light c, then $z \approx v/c$. If the relationship between v and d is linear, then so is the relationship between z and d, but for large redshifts this relationship breaks down.

There is also a potential problem with what exactly is meant by d. Astronomers cannot measure directly the present **proper distance** of an object (i.e. the distance the object has at the present cosmological **proper time**) because they have to make measurements using light emitted by the object. Since light travels at a finite speed and, as we know thanks to Hubble, the Universe is expanding, objects are not at the same position now as they were when their light set out. Proper distances are therefore not amenable to direct measurement, and astronomers have to use indirect distance measures like the **luminosity distance** or **angular-diameter distance**. These alternative distance measures are close to the proper distance if the object is not so distant that light has taken a time comparable to the Hubble time ($1/H_0$) to reach the telescope. Hubble's law is therefore approximately true for any of these distance measurements as long as the object is not too distant. The exact form of the law, however, holds only for proper distance. We should expect deviations from linearity if, for example, we plot redshift z against luminosity distance for objects at high z because of the **curvature of spacetime** and the effect of deceleration; this is one of the tests in **classical cosmology**.

Hubble's law, in the precise form of a linear relationship between recessional velocity and proper distance, is an exact property of all **cosmological models** in which the **cosmological principle** holds. This can be shown quite easily to hold as long as v is much greater than c. (It is true for larger velocities, but the proof is more complicated because the effects of special relativity need to be taken into account.)

The cosmological principle requires that space be homogeneous (i.e. that all points should be equivalent). Consider a triangle formed by three points O, O′ and P. Let the (vector) distance from O to O′ be $\boldsymbol{d}$, and the corresponding vector from O to P be $\boldsymbol{r}$. The distance between O′ and P is then $\boldsymbol{s} = \boldsymbol{r} - \boldsymbol{d}$. Suppose the law relating velocity $\boldsymbol{v}$ to distance $\boldsymbol{x}$ is of some mathematical form $\boldsymbol{v}(\boldsymbol{x})$. If the velocity vectors of P and O′ (as measured from O) are then written as $\boldsymbol{v}(\boldsymbol{r})$ and $\boldsymbol{v}(\boldsymbol{d})$ respectively, then by simple vector addition the velocity of P with respect to O′ is found to be

$$\boldsymbol{u}(\boldsymbol{s}) = \boldsymbol{v}(\boldsymbol{r}) - \boldsymbol{v}(\boldsymbol{d})$$

But the function $\boldsymbol{u}$ must be the same as the function $\boldsymbol{v}$ if all points are equivalent. This means that

$$\boldsymbol{v}(\boldsymbol{r} - \boldsymbol{d}) = \boldsymbol{u}(\boldsymbol{r} - \boldsymbol{d}) = \boldsymbol{v}(\boldsymbol{r}) - \boldsymbol{v}(\boldsymbol{d})$$

The cosmological principle also requires the velocity field to be isotropic. From this requirement we can deduce straightforwardly that the function $v(x)$ must be linear in x. Writing the constant as H, we derive the Hubble law as $v = Hx$. Note that any point on the triangle OO′P can be regarded as the centre for this analysis: all observers will see objects receding from them.

The linearity of Hubble's law is well established out to quite large distances. The *Hubble diagram* shown in the Figure on p. 218 is based on data from a paper by Allan Sandage. The distances used are estimated distances of brightest cluster galaxies (see **extragalactic distance scale**). The black data point at the bottom left-hand corner of this plot indicates where Hubble's (1929) data set resided. The considerable scatter about the mean relationship is partly caused by statistical errors in the measurement of the relevant distances, but this is not the whole story. Hubble's law is true only for objects moving in an idealised homogenous and isotropic universe. Although our Universe may be roughly like this on large enough scales, it is not exactly homogeneous: there is a wealth of **large-scale structure** in the distribution of galaxies. This structure induces fluctuations in the gravitational field of the Universe, which in turn generate **peculiar motions**. While these motions show themselves as scatter in the Hubble diagram, they can also be used (at least in principle) to figure out how much **dark matter** is responsible for inducing them.

SEE ALSO: **supernova**.

FURTHER READING: Slipher, V. M., 'Spectrographic observations of nebulae', *Popular Astronomy*, 1915, **23**, 21; Hubble, E., 'A relation between distance and radial velocity among extragalactic nebulae', *Proceedings of the National Academy of Sciences*, 1929, **15**, 168; Sandage, A. R., 1972 'Distances to galaxies: The Hubble constant, the Friedmann time and the edge of the Universe', *Quarterly Journal of the Royal Astronomical Society*, **13**, 282; Lemaître, G., 'A homogeneous universe of constant mass and increasing radius accounting for the radial velocity of the extragalactic nebulae', *Monthly Notices of the Royal Astronomical Society*, 1931, **91**, 483.

HUBBLE SPACE TELESCOPE (HST) A space telescope, jointly built by NASA and the European Space Agency, launched in 1990 and orbiting at an altitude of about 600 km. It has revolutionised many branches of astronomy, and its impact on cosmology has been immense. Its technical specifications are relatively modest: it is a 2.4-metre reflecting telescope, which makes it quite small by terrestrial standards. Aside from the telescope itself, with its famously misconfigured mirror, the HST is equipped with an impressive set of instruments. The original instrumentation was:

Wide Field and Planetary Camera (WFPC);

Faint Object Camera (FOC);

Faint Object Spectrograph (FOS);

Goddard High Resolution Spectrograph (GHRS);

High Speed Photometer (HSP).

During the first servicing mission in December 1993, astronauts replaced the original WFPC1 with an improved instrument, imaginatively called WFPC2. They also installed a

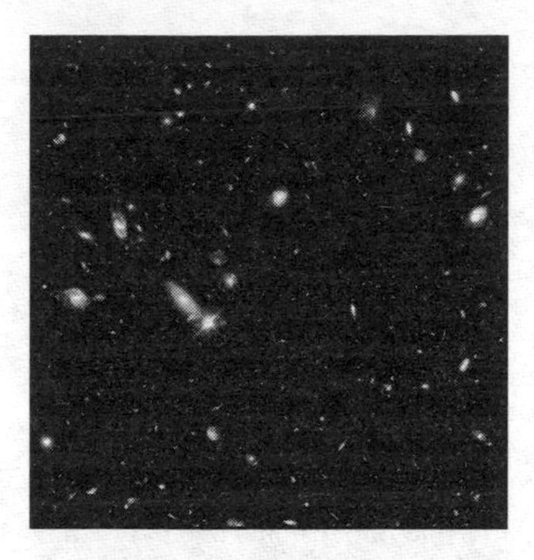

Hubble Space Telescope The Hubble Deep Field, the result of a prolonged pointing of the telescope at an unremarkable piece of the sky, shows hundreds of galaxies, many of them extremely distant.

device called the Corrective Optics Space Telescope Axial Replacement (COSTAR) to correct the faulty optics; this was at the expense of the HSP. In the second servicing mission in February 1997, the FOS and GHRS were removed to make way for new instruments called the Space Telescope Imaging Spectrograph (STIS) and the Near-Infrared Camera and Multi-Object Spectrometer (NICMOS). A third servicing mission is planned for 1999 which will see the installation of the Hubble Advanced Camera for Exploration (HACE).

Although the actual telescope is relatively small, being in space gives it the chance to avoid atmospheric absorption or seeing effects, and also to keep pointing at a given object for longer than is possible on Earth. Among the cosmological applications, HST has been used to image distant **quasars** in order to look for their putative host galaxies. Perhaps the HST's most impressive achievement so far has been to photograph what is called the Hubble Deep Field (HDF), which is the longest, deepest exposure ever taken in the optical part of the electromagnetic spectrum (see the Figure). Some of the images in this exposure are of sources that are so distant that it has taken light more than 90% of the **age of the Universe** to reach us. The HDF allows cosmologists to gauge the extent of the **evolution of galaxies** over this period.

Perhaps the most important role of HST is now to calibrate the **extragalactic distance scale** using Cepheid variable stars that are too distant to be resolved from the ground. The HST key programme of measuring distances to galaxies in the Virgo Cluster of galaxies should bring the uncertainty in the value of the **Hubble constant** down to around 10%.

SEE ALSO: Essay 4.

FURTHER READING: Petersen, C. C. and Brandt, J. C., *Hubble Vision: Astronomy with the Hubble Space Telescope* (Cambridge University Press, 1995); Barbree, J. and Caidin, M., *A Journey Through Time: Exploring the Universe with the Hubble Space Telescope* (Penguin, London, 1995); Fischer, D. and Duerbeck, H., *Hubble: A New Window to the Universe* (Springer-Verlag, New York, 1996); Gribbin, J. and Goodwin, S., *Origins: Our Place in Hubble's Universe* (Constable, London, 1997); Freeman, W. *et al.*, 'Distance to the Virgo Cluster galaxy M100 from Hubble Space Telescope observations of Cepheids', *Nature*, 1994, **371**, 757.

HUCHRA, JOHN PETER (1948–) US astronomer. In 1985, at the Harvard-Smithsonian Center for Astrophysics, he and Margaret **Geller** began a survey of galaxies out to around 200 Mpc which revealed new features in the **large-scale structure** of the Universe, including voids, sheets and the Great Wall. Huchra has investigated **gravitational lensing**, and discovered the multiple image of a distant quasar known as the Einstein Cross or Huchra Lens.

HUMASON, MILTON LASSELL (1891–1972) US astronomer. In the 1920s he became the foremost photographer of the spectra of faint galaxies, first with the 100-inch telescope at Mount Wilson and later with the 200-inch at Mount Palomar. His hundreds of measurements of the radial velocities of galaxies provided Edwin **Hubble** a flow of data to support his law relating redshift to distance.

HYPERBOLIC SPACE see **curvature of spacetime**.

** **HYPERSURFACE** Einstein's theory of **general relativity** relates the geometry of four-dimensional **spacetime**, as described by the **metric** g_{ij}, to the properties of matter. The metric contains all the information required to specify the geometrical properties of **spacetime**. In general, any lower-dimensional part of this four-dimensional space is called a *hypersurface*.

General relativity is explicitly four-dimensional, so in general there is no unambiguous way of defining time and space separately. But in those **cosmological models** in which the **cosmological principle** of overall homogeneity and isotropy applies, the metric has a simple form called the **Robertson–Walker metric** where there is a preferred time coordinate t (**proper time**) and in which the **tensor** form of the metric appears diagonal. This illustrates the different possible kinds of hypersurface in a simple way. At a fixed spatial location, only time changes: we have in effect selected a region of spacetime that corresponds to a one-dimensional line in the time direction. This is called a *time-like hypersurface*. On the other hand, if we fix the time coordinate then what is left is a three-dimensional hypersurface corresponding to a snapshot of space taken at the particular time chosen. This is called a *space-like hypersurface*. A region corresponding to the path taken by a photon of **electromagnetic radiation** (called a *geodesic*) is a one-dimensional line called a *null hypersurface*.

In more complicated metrics this classification of slices of spacetime into time-like, space-like and null requires more mathematical subtlety, but it can be achieved. This is because, though time and space are to some extent unified in general relativity, spacetime always has a causal structure defined by the ability of photons to carry information along null geodesics.

FURTHER READING: Schutz, B. F., *A First Course in General Relativity* (Cambridge University Press, Cambridge, 1985).

I

* **Imaginary time** In **general relativity** there is an intimate connection between the effect of **gravity** and the **curvature of spacetime**. Einstein's theory, however, does not include any effects of **quantum physics**, which are thought to be important in the earliest stages of the Big Bang. These effects need to be included in any theory of **quantum cosmology**.

One aspect of general relativity that might change when quantum effects become important is the structure of **spacetime** itself. In classical relativity theory, spacetime is a four-dimensional construction in which the three dimensions of space and one dimension of **time** are welded together. But space and time are not equivalent. The easiest way of seeing the differences between space and time is to look at the simple **metric** of **special relativity** which is called the *Minkowski metric*. This describes a flat space in which gravity and acceleration are absent; the metric in general relativity may be more complicated than this, but the crucial points of the argument remain the same in this simplified case.

Spacetime intervals (s) in the Minkowski space are represented in terms of the three Cartesian space coordinates (x, y, z) and the time coordinate t by an expression of the form

$$s^2 = (x^2 + y^2 + z^2) - c^2t^2$$

The terms in brackets, according to Pythagoras's theorem, simply add up to the square of the spatial distance l, through $l^2 = x^2 + y^2 + z^2$. The term in t also has the dimensions of distance squared, but it has a different sign. This is a consequence of the *signature* of the metric, which singles out the special nature of time, guaranteeing that faster-than-light signals cannot be exchanged between observers. The simple metric shown above is a particular example of the general set of *Lorentz metrics*, all of which have the same signature but whose spatial components might be more complicated than the simple Euclidean behaviour of Minkowski space.

One idea associated with quantum cosmology is that this signature may change when the gravitational field is very strong. The idea is based on the properties of imaginary numbers. Imaginary numbers are all multiples of the number i, which is defined to be the square root of minus one: $i^2 = \sqrt{-1}$. If we were to replace t by it in the equation above, it would simply become

$$s^2 = (x^2 + y^2 + z^2) + c^2t^2$$

Note that there is now no difference at all between space and time in this theory: the signature of the new metric is the same as that of a four-dimensional Euclidean space rather than the Minkowski signature we started with. This replacement is sometimes called the *Euclideanisation of space*, and it is part of the *no-boundary* hypothesis in quantum cosmology proposed by James Hartle and Stephen **Hawking**. Since, in this theory, time loses the characteristics that separate it from the spatial parts of the metric, the concept of a beginning in time becomes meaningless. Spacetimes with this signature therefore have no boundary.

FURTHER READING: Hartle, J. B. and Hawking, S. W., 'The wave function of the Universe', *Physical Review* D, 1983, **28**, 2960; Hawking, S.W., *A Brief History of Time* (Bantam, New York, 1988).

* **INFLATIONARY UNIVERSE** A modern variation of the standard **Big Bang theory** that includes a finite period of accelerated expansion (*inflation*) in the early stages of its evolution. Inflation is the mechanism by which various outstanding problems of the standard **cosmological models** might be addressed, providing a possible resolution of the **horizon problem** and the **flatness problem**, as well as generating the **primordial density fluctuations** that are required for **structure formation** to occur and which appear to have produced the famous **ripples** in the **cosmic microwave background radiation** (see also Essay 5).

Assuming that we accept that an epoch of inflation is desirable for these reasons, how can we achieve the accelerated expansion physically, when the standard **Friedmann models** are always decelerating? The idea that lies at the foundation of most models of inflation is that there was an epoch in the early stages of the evolution of the Universe in which the energy density of the vacuum state of a **scalar field** Φ, perhaps associated with one of the **fundamental interactions**, provided the dominant contribution to the energy density. In the ensuing phase the scale factor $a(t)$ describing the **expansion of the Universe** grows in an accelerated fashion, and is in fact very nearly an exponential function of time if the energy density of the scalar field is somehow held constant. In the inflationary epoch that follows, a microscopically small region, perhaps no larger than the **Planck length**, can grow – or *inflate* – to such a size that it exceeds the size of our present observable **Universe** (see **horizon**).

There exist many different versions of the inflationary universe. The first was formulated by Alan **Guth** in 1981, although many of his ideas had been presented by Alexei Starobinsky in 1979. In Guth's model inflation is assumed to occur while the Universe undergoes a first-order **phase transition**, which is predicted to occur in some **grand unified theories**. The next generation of inflationary models shared the characteristics of a model called the *new inflationary universe*, which was suggested in 1982 independently by Andrei Linde, and by Andreas Albrecht and Paul Steinhardt. In models of this type, inflation occurs during a phase in which the region that will grow to include our observable patch evolves more gradually from a 'false vacuum' to a 'true vacuum'. It was later realised that this kind of inflation could also be achieved in many different contexts, not necessarily requiring the existence of a phase transition or a **spontaneous symmetry-breaking**. This model is based on a certain choice of parameters of a particular grand unified theory which, in the absence of any other experimental evidence, appears a little arbitrary. This problem also arises in other inflationary models based on theories like **supersymmetry** or **string theories**, which are yet to receive any experimental confirmation or, indeed, are likely to in the foreseeable future. It is fair to say that the inflationary model has become a sort of paradigm for resolving some of the difficulties with the standard model, but no particular version of it has so far received any strong physical support from particle physics theories.

Let us concentrate for a while on the physics of generic inflationary models involving symmetry-breaking during a phase transition. In general, **gauge theories** of elementary particle interactions involve an *order parameter* which we can identify with the scalar field Φ determining the breaking of the symmetry. The behaviour of the scalar field is controlled by a quantity called its *Lagrangian action*, which has two components: one (denoted by U) concerns time-derivatives of Φ and is therefore called the kinetic term, and the other (V) describes the interactions of the scalar field and is called the potential term. A scalar field behaves roughly like a strange form of matter with a total energy density given by $U + V$ and a pressure given by $U - V$. Note that if V is much larger than U, then the density and pressure are equal but of opposite sign. This is what is needed to generate inflation, but the way in which the potential comes to dominate is quite complicated.

The potential function V changes with the temperature of the Universe, and it is this that induces the phase transition, as it becomes energetically favourable for the state of the field to change when the Universe cools sufficiently. In the language of thermodynamics, the potential $V(\Phi)$ plays the role of the free energy of the system. A graph of $V(\Phi)$ will typically have a minimum somewhere, and that minimum value determines the value of Φ which is stable at a given temperature. Imagine an inverted parabola with its minimum value at $\Phi = 0$; the configuration of the field can be represented as the position of a ball rolling on this curve. In the stable configuration it nestles in the bottom of the potential well at $\Phi = 0$. This might represent the potential $V(\Phi)$ at very high temperatures, way above the phase transition. The vacuum is then in its most symmetrical state. What happens as the phase transition proceeds is that the shape of $V(\Phi)$ changes so that it develops additional minima. Initially these 'false' minima may be at higher values of $V(\Phi)$ than the original, but as the temperature continues to fall and the shape of the curve changes further, the new minima can be at lower values of V than the original one. This happens at a critical temperature T_c at which the vacuum state of the Universe begins to prefer one of the alternative minima to the original one.

The transition does not occur instantaneously. How it proceeds depends on the shape of the potential, and this in turn determines whether the transition is first or second order. If the phase transition is second order it moves rather smoothly, and fairly large 'domains' of the new phase are generated (much like the Weiss domains in a ferromagnet). One such region (bubble or domain) eventually ends up including our local patch of the Universe. If the potential is such that the transition is first order, the new phase appears as bubbles nucleating within the false vacuum background; these then grow and coalesce so as to fill space with the new phase when the transition is complete.

Inflation arises when the potential term V greatly exceeds the kinetic term U in the action of the scalar field. In a phase transition this usually means that the vacuum must move relatively slowly from its original state into the final state. In fact, the equations governing the evolution of Φ are mathematically identical to those describing a ball moving under the action of the force $-\mathrm{d}V/\mathrm{d}\Phi$, just as in standard Newtonian dynamics. But there is also a frictional force, caused by the expansion of the Universe, that tends to slow down the rolling of Φ from one state into another. This provides a natural self-regulation of the speed of the transition. As long as the Universe is expanding at the start, the kinetic term U is quickly reduced by the action of this friction (or viscosity); the motion of the field then resembles the behaviour of particles during sedimentation.

In order to have inflation we must assume that, at some time, the Universe contains some expanding regions in **thermal equilibrium** at a temperature $T > T_c$ which can eventually cool below T_c before they recollapse. Let us assume that such a region, initially trapped in the false vacuum phase, is sufficiently homogeneous and isotropic to be described by a **Robertson–Walker metric**. In this case the evolution of the patch is described by a Friedmann model, except that the density of the Universe is not the density of matter, but the effective density of the scalar field, i.e. the sum $U + V$ mentioned above. If the field Φ is evolving slowly then the U component is negligibly small. The Friedmann equation then looks exactly like the equation describing a Friedmann model incorporating a **cosmological constant** term but containing no matter. The cosmological model that results is the well-known de Sitter solution in which the scale factor $a(t) \propto \exp(Ht)$, with H (the Hubble parameter) roughly constant at a value given by $H^2 = 8\pi GV/3$. Since Φ does not change very much as the transition proceeds, V is roughly constant.

The de Sitter solution is the same as that used in the **steady state theory**, except that the scalar field in that theory is the so-called C-field responsible for the continuous creation of matter. In the inflationary Universe, however, the expansion timescale is much more rapid than in the steady state. The inverse of the Hubble expansion parameter, $1/H$, is about 10^{-34} seconds in inflation. This quantity has to be fixed at the inverse of the present-day value of the Hubble parameter (i.e. at the reciprocal of the **Hubble constant**, $1/H_0$) in the steady state theory, which is around 10^{17} seconds.

This rapid expansion is a way of solving some riddles which are not explained in the standard Big Bang theory. For example, a region which is the same order of size as the horizon before inflation, and which might therefore be reasonably assumed to be smooth, would then become enormously large, encompassing the entire observable Universe today. Any inhomogeneity and anisotropy present at the initial time will be smoothed out so that the region loses all memory of its initial structure. Inflation therefore provides a mechanism for avoiding the horizon problem. This effect is, in fact, a general property of inflationary universes and it is described by the so-called *cosmic no-hair theorem* (see also **black hole**).

Another interesting outcome of the inflationary Universe is that the characteristic scale of the **curvature of spacetime**, which is intimately related to the value of the **density parameter** Ω, is expected to become enormously large. A balloon is perceptibly curved because its radius of curvature is only a few centimetres, but it would appear very flat if it were blown up to the radius of the Earth. The same happens in inflation: the curvature scale may be very small indeed initially, but it ends up greater than the size of our observable Universe. The important consequence of this is that the **density parameter** Ω should be very close to 1. More precisely, the total energy density of the Universe (including the matter and any vacuum energy associated with a cosmological constant) should be very close to the critical density required to make a **flat universe**.

Because of the large expansion, a small initial patch also becomes practically devoid of matter in the form of **elementary particles**. This also solves problems associated with the formation of monopoles and other **topological defects** in the early Universe, because any defects formed during the transition will be drastically diluted as the Universe expands, so that their present density will be negligible.

After the slow rolling phase is complete, the field Φ falls rapidly into its chosen minimum and then undergoes oscillations about the minimum value. While this happens there is a rapid liberation of energy which was trapped in the potential term V while the transition was in progress. This energy is basically the *latent heat* of the transition. The oscillations are damped by the creation of elementary particles coupled to the scalar field, and the liberation of the latent heat raises the temperature again – a phenomenon called *reheating*. The small patch of the Universe we have been talking about thus acquires virtually all the energy and entropy that originally resided in the quantum vacuum by this process of particle creation. Once reheating has occurred, the evolution of the patch again takes on the character of the usual Friedmann models, but this time it has the normal form of matter. If $V(\Phi_0) = 0$, then the vacuum energy remaining after inflation is zero and there will be no remaining cosmological constant Λ.

One of the problems left unsolved by inflation is that there is no real reason to suppose that the minimum of V is exactly at zero, so we would expect a nonzero Λ-term to appear at the end. Attempts to calculate the size of the cosmological constant Λ

induced by phase transitions in this way produce enormously large values. It is important that the inflationary model should predict a reheating temperature sufficiently high that processes which violate the conservation of baryon number can take place so as to allow the creation of an asymmetry between matter and **antimatter** (see **baryogenesis**).

As far as its overall properties are concerned, our Universe was reborn into a new life after reheating. Even if before it was highly lumpy and curved, it was now highly homogeneous, and had negligible curvature. This latter prediction may be a problem because, as we have seen, there is little strong evidence that the density parameter is indeed close to 1, as required.

Another general property of inflationary models, which we shall not go into here, is that fluctuations in the quantum scalar field driving inflation can, in principle, generate a spectrum of primordial density fluctuations capable of initiating the process of cosmological structure formation. They may also produce an observable spectrum of primordial gravitational waves.

There are many versions of the basic inflationary model which are based on slightly different assumptions about the nature of the scalar field and the form of the phase transition. Some of the most important are described below.

Old inflation The name now usually given to the first inflationary model, suggested by Guth in 1981. This model is based on a scalar field theory which undergoes a first-order phase transition. The problem with this is that, being a first-order transition, it occurs by a process of bubble nucleation. It turns out that these bubbles would be too small to be identified with our observable Universe, and they would be carried apart by the expanding phase too quickly for them to coalesce and produce a large bubble which we could identify in this way. The end state of this model would therefore be a highly chaotic universe, quite the opposite of what is intended. This model was therefore abandoned soon after it was suggested.

New inflation The successor to old inflation: again, a theory based on a scalar field, but this time the potential has no potential barrier, so the phase transition is second order. The process which accompanies a second-order phase transition, known as *spinodal decomposition*, usually leaves larger coherent domains, providing a natural way out of the problem of old inflation. The problem with new inflation is that it suffers from severe fine-tuning difficulties. One is that the potential must be very flat near the origin to produce enough inflation and to avoid excessive fluctuations due to the quantum field. Another is that the field Φ is assumed to be in thermal equilibrium with the other matter fields before the onset of inflation; this requires Φ to be coupled fairly strongly to the other fields that might exist at this time. But this coupling would induce corrections to the potential which would violate the previous constraint. It seems unlikely, therefore, that thermal equilibrium can be attained in a self-consistent way before inflation starts and under the conditions necessary for inflation to happen.

Chaotic inflation One of the most popular inflationary models, devised by Linde in 1983; again, it is based on a scalar field but it does not require any phase transitions at all. The idea behind this model is that, whatever the detailed shape of the effective potential V, a patch of the Universe in which Φ is large, uniform and relatively static will automatically lead to inflation. In chaotic inflation we simply assume that at some initial time, perhaps as early as the Planck time, the field varied from place to place in an arbitrary chaotic manner. If any region is uniform and static, it will inflate and eventually encompass our observable Universe. While the end result of chaotic inflation is locally flat and homogeneous in our observable patch, on scales larger than the horizon the Universe is highly curved and inhomogeneous. Chaotic inflation is therefore very different from both the old and new inflationary models. This difference is reinforced by the fact that no mention of supersymmetry or grand unified theories appears in the description. The field Φ that describes chaotic inflation at the Planck time is completely decoupled from all other physics.

Stochastic inflation A natural extension of Linde's chaotic inflation, sometimes called *eternal inflation*; as with chaotic inflation, the Universe is extremely inhomogeneous overall, but quantum fluctuations during the evolution of Φ are taken into account. In stochastic inflation the Universe will at any time contain regions which are just entering an inflationary phase. We can picture the Universe as a continuous 'branching' process in which new 'mini-universes' expand to produce locally smooth patches within a highly chaotic background Universe. This model is like a Big Bang on the scale of each mini-universe, but overall it is reminiscent of the steady state model. The continual birth and rebirth of these mini-universes is often called, rather poetically, the *phoenix universe*. This model has the interesting feature that the laws of physics may be different in different mini-universe, which brings the **anthropic principle** very much into play.

Modified gravity There are versions of the inflationary Universe model that do not require a scalar field associated with the fundamental interaction to drive the expansion. We can, for example, obtain inflation by modifying the laws of gravity. Usually this is done by adding extra terms in the curvature of spacetime to the equations of general relativity. For certain kinds of modification, the resulting equations are mathematically equivalent to ordinary general relativity in the presence of a scalar field with some particular action. This effective scalar field can drive inflation in the same way as a genuine physical field can. An alternative way to modify gravity might be to adopt the **Brans–Dicke theory** of gravity. The crucial point with this kind of model is that the scalar field does not generate an exponential expansion, but one in which the expansion is some power of time: $a(t) \propto t^{\alpha}$. This modification even allows old inflation to succeed: the bubbles that nucleate the new phase can be made to merge and fill space if inflation proceeds as a power law in time rather than an exponential. Theories based on Brans–Dicke modified gravity are usually called *extended inflation*.

There are many other possibilities: models with more than one scalar field, models with modified gravity and a scalar field, models based on more complicated potentials, models based on supersymmetry, on grand unified theories, and so on. Indeed, inflation has led to an almost exponential increase in the number of cosmological models!

It should now be clear that the inflationary Universe model provides a conceptual explanation of the horizon problem and the flatness problem. It also may rescue grand unified theories which predict a large present-day abundance of monopoles or other topological defects. Inflationary models have gradually evolved to avoid problems with earlier versions. Some models are intrinsically flawed (e.g. old inflation) but can be salvaged in some modified form (e.g. extended inflation). The magnitude of the primordial density fluctuations and gravitational waves they produce may also be too high for some particular models. There are, however, much more serious problems associated with these scenarios. Perhaps the most important is intimately connected with one of the successes. Most inflationary models predict that the observable Universe at the present epoch should be almost flat. In the absence of a cosmological constant this means that $\Omega \approx 1$. While this possibility is not excluded by observations, it cannot be said that there is compelling evidence for it and, if anything, observations of dark matter in galaxies and clusters of galaxies favour an open universe with a lower density than this. It is possible to produce a low-density universe after inflation, but it requires very particular models. To engineer an inflationary model that produces $\Omega \approx 0.2$ at the present epoch requires a considerable degree of unpleasant fine-tuning of the conditions before inflation. On the other hand, we could reconcile a low-density universe with apparently more natural inflationary models by appealing to a relic cosmological constant: the requirement that space should be (almost) flat simply translates into $\Omega_0 + (\Lambda c^2/3H_0^2) \approx 1$. It has been suggested that this is a potentially successful model of structure formation, but recent developments in **classical cosmology** put pressure on this alternative (see also Essay 6 for constraints on the cosmological constant from observations of **gravitational lensing**).

The status of inflation as a physical theory is also of some concern. To what extent is inflation predictive? Is it testable? We could argue that inflation does predict that we live in a flat universe. This may be true, but a flat universe might emerge naturally at the very beginning if some process connected with quantum gravity can arrange it. Likewise, an open universe appears to be possible either with or without inflation. Inflationary models also produce primordial density fluctuations and gravitational waves. Observations showing that these phenomena had the correct properties may eventually constitute a test of inflation, but this is not possible at the present. All we can say is that the observed properties of fluctuations in the cosmic microwave background radiation described in Essay 5 do indeed seem to be consistent with the usual inflationary models. At the moment, therefore, inflation has a status somewhere between a theory and a paradigm, but we are still far from sure that inflation ever took place.

SEE ALSO: Essays 1 and 3.

FURTHER READING: Guth, A. H., 'Inflationary Universe: A possible solution to the horizon and flatness problems', *Physical Review* D, 1981, **23**, 347; Albrecht, A. and Steinhardt, P. J., 'Cosmology for grand unified theories with radiatively induced symmetry breaking', *Physical Review Letters*, 1982, **48**, 1220; Linde, A. D., 'Scalar field fluctuations in the expanding Universe and the new inflationary Universe scenario', *Physics Letters* B, 1982, **116**, 335; Narlikar, J. V. and Padmanabhan, T., 'Inflation for astronomers', *Annual Reviews of Astronomy and Astrophysics*, 1991, **29**, 325.

INFRARED ASTRONOMY The branch of astronomy that concerns itself with observations made in the region of the spectrum of **electromagnetic radiation** with wavelengths between about 1 and 300 μm (1 μm is 1 millionth of a metre). Because the energies E of infrared photons are related to a temperature T via $E = kT$, with T not far from room temperature, the temperature of infrared telescopes poses problems for observers. The telescopes are usually cooled considerably using liquid nitrogen, or even by liquid helium in some cases.

Terrestrial infrared observations are also hampered by the intervention of the Earth's atmosphere, which both absorbs and emits radiation in various parts of the relevant wavelength region. Ground-based observations are therefore restricted to a relatively small number of atmospheric windows where these effects are small, particularly in the near-infrared region with wavelengths of a few micrometres. Even in these windows the effects of atmospheric water have to be avoided as much as possible, since water vapour absorbs in the infrared and produces spurious lines in infrared spectra. The solution is to site infrared telescopes at high altitudes and in very dry locations. The United Kingdom Infrared Telescope (UKIRT) and the Infra Red Telescope Facility (IRTF) run by NASA are both situated at Mauna Kea in Hawaii, which is by far the world's best site for such observations.

Attempts to perform observations at wavelengths outside the atmospheric infrared windows began with experiments flown on balloons and aircraft, but since the 1980s the emphasis has shifted to satellite missions, notably the Infrared Astronomical Satellite (IRAS) and the Infrared Space Observatory (ISO). The infrared region encroaches significantly into the region of the spectrum where the **cosmic microwave background radiation** is observed. The Cosmic Microwave Background Explorer (COBE), for example, carried a number of infrared experiments as well as the experiment that detected the famous ripples (see **black body** and Essay 6). Many future infrared missions are also scheduled, including the Far

Infrared and Submillimetre Space Telescope (FIRST), a cornerstone of the European Space Agency's science programme.

Sources of infrared radiation often contain significant quantities of dust. Many **active galaxies** emit the bulk of their radiation in the infrared region; this is probably radiation that has been reprocessed by dust in the region around the active nucleus. Ordinary spiral galaxies are also quite luminous in the infrared region, and it is in this part of the spectrum that the Tully–Fisher measurements used in constructing the **extragalactic distance scale** are usually made. One of the major successes of IRAS was that, contrary to expectations, it detected a large number of normal **galaxies** as well as **active galaxies**; a follow-up **redshift survey** of these galaxies has helped to refine theoretical models of **structure formation**.

SEE ALSO: Essay 4.

INFRARED BACKGROUND The existence of the **cosmic microwave background radiation** is an important piece of evidence in favour of the standard **Big Bang theory**. But various other possible sources of background radiation exist that might produce a kind of foreground to the microwave background. In particular, any sources which emit in the infrared or submillimetre regions of the spectrum of **electromagnetic radiation** might swamp the 2.73 K **black-body** spectrum of the microwave background radiation. A cosmological infrared background is very difficult to detect, even in principle, because of the many local sources of radiation at these frequencies. Nevertheless, the current upper limits on the flux in various wavelength regions can narrow down considerably the numbers of 'pregalactic' objects that we might expect to have emitted light which is now observable in this region.

For simplicity, we can characterise these sources by the wavelength at which their emission is strongest; the wavelength at which the peak of the black-body spectrum of the cosmic microwave background radiation occurs is around 1400 μm (1 μm is 1 millionth of a metre). For example, the lack of any distortions of the observed spectrum reported by the FIRAS experiment on board the **Cosmic Background Explorer** satellite (COBE) suggests that an excess background at wavelengths between 500 and 5000 μm have contributed no more than 0.03% of the total energy in the microwave background. Other rocket and ground-based experiments yield similar limits for other parts of the infrared spectrum.

One obvious potential source of infrared background radiation is galaxies. To estimate their contribution is rather difficult and requires complicated modelling because the observed near-infrared background would be generated by redshifted optical emission from normal galaxies. We therefore need to start with the spectrum of emission as a function of time for a single galaxy, which requires knowledge of the initial mass function of stars, the star formation rate and the laws of **stellar evolution**. To get the total background we need to examine different types of galaxy, taking into account redshift and the effect of the **density parameter** on the rate of expansion of the Universe. If galaxies are extremely dusty, then radiation from them will appear in the far-infrared region. Such radiation can emanate from dusty disks, star-forming clouds (perhaps associated with the starburst phenomenon), **active galaxies** and **quasars**. The evolution of these phenomena is very complex and poorly understood at present.

More interesting are the possible *pregalactic* sources of infrared radiation. Most of these sources produce an approximately black-body spectrum. For example, if gas clouds collapse to produce a system with typical random motions of about 300 km/s (see **virial theorem**), at a redshift of $z = 4$ they produce infrared radiation as an approximate black body but with a wavelength of about 0.1 μm. This kind of calculation could, in principle, be of some use in theories of **structure formation**, but the number of objects forming as a function of redshift is difficult to estimate in all but the simplest clustering models. On the other hand, pregalactic *explosions*, sometimes suggested as an alternative to the standard theories of galaxy formation, would produce a much larger background, but limits on the spectral distortions obtained by COBE appear to rule this model out. Constraints can also be placed on the number of black holes in the galactic halo, on the numbers of halo brown dwarfs and on the possibility of a decaying particle ionising the **intergalactic medium**.

The interpretation of results is complicated if the radiation from the source does not propagate freely to the observer without absorption or scattering. Many sources of radiation observed at the present epoch in the infrared and submillimetre regions were initially produced in the optical or ultraviolet, and have been redshifted by the **expansion of the Universe**. The radiation may therefore have been reprocessed if there was any dust in the vicinity of the source. Dust grains are associated with star formation, and may consequently be confined to galaxies; however, if there was a cosmological population of pregalactic stars, dust could be smoothly distributed throughout space. Interestingly, the peak wavelength that tends to be produced by such reprocessing is around 700 μm, not far into the infrared side of the peak in the black-body spectrum of the cosmic microwave background (CMB) radiation. In 1987, before the launch of COBE, a rocket experiment by the Nagoya–Berkeley collaboration claimed to have detected an excess in the microwave background spectrum in this wavelength region. Unfortunately, it was later shown that the experiment had detected hot exhaust fumes from the parent rocket!

A dust background would also be expected to be anisotropic on the sky if it were produced by galaxies or a clumpy distribution of pregalactic dust. We can study the predicted anisotropy by allowing the dust to cluster like galaxies, for example, and computing the resulting statistical fluctuations. Various experiments have been devised, along the lines of the CMB anisotropy experiments, to detect such fluctuations. At present, however, no such fluctuations have yet been observed.

FURTHER READING: Bond, J. R., Carr, B. J. and Hogan, C. J., 'The spectrum and anisotropy of the cosmic infrared background', *Astrophysical Journal*, 1986, **306**, 428.

INHOMOGENEITY According to the **cosmological principle**, the Universe is roughly homogeneous (it has the same properties in every place) and isotropic (it looks the same in every direction). These mathematical features are built into the standard **cosmological models** used to describe the bulk properties of the cosmos in the standard **Big Bang theory**.

But the Universe is not exactly homogeneous. There are considerable fluctuations in density from place to place on small scales. For example, the density of water is about 10^{29} times greater than the density of the **intergalactic medium**. However, if we average the density of material over sufficiently large volumes, the fluctuations become smaller and smaller as the volumes get larger. Imagine having a vast cubical box of a particular volume, placing it in a large number of different positions and weighing the amount of matter contained in the box. The average mass of the box will be some quantity M (which depends on the volume), but sometimes the box will contain more mass than M, and sometimes less. Suppose, for example, that the side of the box is about 100 kiloparsecs – about the size of a galaxy. If the box is placed exactly around a galaxy, the mass it contains is very large. But most of space is not occupied by galaxies, so the average amount of mass calculated when we have placed the box at random in different places is much smaller than that contained in a galaxy. In fact, fluctuations of the amount of matter in regions of this size are typically of the order of 100,000 times the average for this volume. Volumes about ten megaparsecs across have typical box-to-box mass fluctuations about the same as the mean value.

Large-scale structure corresponds to much smaller fluctuations, in terms of deviations from the mean, but within much larger volumes. For example, volumes about a hundred megaparsecs across vary by no more than a few per cent of the mean from box to box. On very large scales, the fluctuation must be negligibly small for the cosmological principle to hold. If, on the other hand, the Universe has a **fractal** structure, as has been claimed by some researchers, then no mean density can be defined and no scale of homogeneity is ever reached. Such a Universe could not be described by the standard **Friedmann models**.

SEE ALSO: **anisotropy**, **structure formation**.

INTERGALACTIC MEDIUM (IGM) Much of cosmology is concerned with studying the properties of particular objects (such as **galaxies** and **quasars**) that can be observed directly by the **electromagnetic radiation** they emit. But with the increasing realisation that the Universe might contain copious quantities of **dark matter**, the properties of whatever medium lies between these objects are being studied as well. This medium, whatever it consists of, is usually called the *intergalactic medium*. To be precise, this term applies only to cosmic material that is in the form of baryons (see **elementary particles**), so it specifically excludes dark matter in the form of **weakly interacting massive particles** (WIMPs), whose presence can be inferred only from dynamical arguments. Although the IGM has not condensed into definite structures, it is possible to perform observations that can test its properties and indicate its possible forms.

Prominent among such methods are those based on **spectroscopy**. Although radiation may not be emitted directly by the IGM, light from distant objects has to pass through it and may consequently be absorbed. For example, observations of quasar spectra allow us to probe a line of sight from our Galaxy to the quasar. Absorption or scattering of light during its journey to us can, in principle, be detected by its effect upon this spectrum. This in turn can be used to investigate the number and properties of whatever absorbers or scatterers are associated with the baryonic content of the IGM.

One of the most important techniques is called the *Gunn–Peterson test*. Neutral hydrogen has a peculiar physical property, known as *resonant scattering*, which is associated with the Lyman-alpha atomic transition (corresponding, according to the rules of **quantum mechanics**, to the transition between the two lowest energy levels of the hydrogen atom). This resonance is so strong that it is possible for clouds of relatively low density to produce a significant **absorption line** at the corresponding wavelength (which lies in the far ultraviolet – see **ultraviolet astronomy**). But since quasars are at quite high **redshifts**, and the medium between us and the quasar is spread out over a wide range of redshifts lower than that of the quasar, there are not absorption lines as such but an *absorption trough*: a huge number of closely spaced lines all merging together. The Gunn–Peterson test relies on the fact that in observed quasar spectra there is no apparent drop between the long-wavelength side of the Lyman-alpha **emission line** produced by the quasar and the short-wavelength side, where the effects of this resonant scattering might be expected.

The lack of any observed difference translates into a very tight limit on the amount of neutral hydrogen gas in the spaces between quasars. The conclusion is that either there is very little baryonic material in these gaps, or it is so hot that it is not neutral but ionised. If it were too hot, however, this gas would produce some of the cosmic **X-ray background** and also generate distortions of the **black-body** spectrum of the **cosmic microwave background radiation** via the **Sunyaev–Zel'dovich effect**. We are therefore left only with the first option, so the density of neutral hydrogen in the IGM must be very low indeed – much less than the total amount of hydrogen predicted by calculations of primordial **nucleosynthesis**. This means that some source of radiation must have almost completely ionised the IGM around the redshifts of $z = 4$ or so sampled by quasars.

Although quasar spectra do not exhibit any general absorption consistent with a smoothly distributed hydrogen component, there are many absorption lines in such spectra which are interpreted as originating in clouds of gas between the quasar and the observer absorbing at the

Lyman-alpha resonant frequency. These clouds are divided into three categories, depending on the strength of the absorption line they produce. The strongest absorbers contain about as much gas as there is in a present-day spiral galaxy. This is enough to produce a very wide absorption trough at the Lyman-alpha frequency, and these systems are usually called *damped Lyman-alpha systems*. They are relatively rare, and are usually interpreted as being protogalaxies of some kind. They occur at redshifts up to around $z = 3$, and their presence indicates that the process of structure formation was already well under way at these redshifts. More abundant are the *Lyman limit systems*. These are dense enough to block radiation at wavelengths near the photoionisation edge of the Lyman series of lines (see **ultraviolet astronomy**). The importance of the Lyman limit is that material at the centre of the cloud is shielded from ionising radiation by the material at its edge; at low densities this cannot happen. Smaller clouds appear as sharp absorption lines at the Lyman-alpha wavelength. These are very common, and reveal themselves as a mass of lines in the spectra of quasars; they are usually called the *Lyman-alpha forest*.

The Lyman-alpha forest clouds have a number of interesting properties. For a start, they provide evidence that quasars are capable of ionising the IGM. The numbers of such systems along lines of sight to different quasars are similar, which strengthens the impression that they are intervening objects not connected with the quasar. At redshifts near that of the quasar their numbers decrease markedly, an effect known as the *proximity effect*. The idea here is that radiation from the quasar substantially reduces the neutral hydrogen fraction in the clouds by ionisation, thus inhibiting absorption at the Lyman-alpha resonance. Secondly, the total mass in the clouds appears to be close to that in the damped systems or that seen in present-day galaxies. Thirdly, the number of such systems changes strongly with redshift, indicating, perhaps, that the clouds are undergoing dissipation. Finally, and most interestingly from the point of view of **structure formation**, the absorption systems seem to be unclustered, in contrast to the distribution of galaxies (see **large-scale structure**). How these smaller Lyman-alpha systems fit into the picture of galaxy formation is presently unclear.

FURTHER READING: Peebles, P. J. E., *Principles of Physical Cosmology* (Princeton University Press, Princeton, 1993); Coles, P. and Lucchin, F., *Cosmology: The Origin and Evolution of Cosmic Structure* (John Wiley, Chichester, 1995).

ISOTROPY see **cosmological principle**.

J

Jeans, Sir James Hopwood (1877–1946) English astrophysicist. He is most famous for his work on **stellar evolution** and the theory of the origin of the Solar System. He also developed the theory of gravitational instability, which forms the basis of most modern theories of cosmological **structure formation**. Jeans was a noted science populariser, whose books made him one of the most famous scientists in Britain during the 1920s and 1930s.

Jeans instability The gravitational process by which **structure formation** on cosmological scales is thought to occur. In the standard cosmology, the Universe on large scales and at early times is homogeneous and isotropic (see **cosmological principle**). In this sense, it has no structure at all. We know that the early Universe had to be isotropic to a high degree because of the extreme uniformity of the temperature of the **cosmic microwave background radiation** on the sky. According to the **Big Bang theory**, this radiation probes the conditions in the Universe just a few hundred thousand years from the beginning. Its isotropy on the sky means that the Universe must then have been very smooth and structure-free. On the other hand, we know that the Universe is now highly inhomogeneous, at least on relatively small scales (cosmologically speaking), because of the observed presence of stars, **galaxies** and **large-scale structure**. Structure must therefore have come into existence by some process between the recombination of the microwave background and the present epoch. The mechanism by which this happened is thought to be gravitational instability.

The basic idea behind the gravitational instability is that **gravity** is an attractive force so that, in an inhomogeneous medium, regions with above-average density tend to accrete material from their surroundings, getting progressively denser and denser, while those with below-average density become more and more depleted. In this way, small initial irregularities would have become amplified by the action of gravity and grown into large fluctuations.

However, this can happen only on scales sufficiently large for the self-gravity of an initially overdense region, which tends to make it collapse, to exceed the thermal or other pressure forces which would tend to support it. The length scale above which gravity dominates is usually called the *Jeans length*, after James **Jeans**, who worked out the theory of gravitational instability in the context of non-expanding gas clouds, way back in 1902. If they are larger than the Jeans scale, small initial irregularities grow exponentially with time, resulting in a rapid collapse into gravitationally bound, equilibrium structures. This exponential instability is so rapid that infinitesimal initial fluctuations are enough to start things off. Since no medium is completely smooth, the formation of clumps by this process is to some extent inevitable in any gas cloud. The restoring effect of pressure, however, means that fluctuations on length scales smaller than the Jeans length oscillate like acoustic waves rather than collapsing. This provides a characteristic size for the objects that form: the Jeans length is the minimum size for structures that can be produced in this way. Jeans developed the idea in the first place to explain the formation of stars within our Galaxy, and it is still the standard framework for understanding how this happens. Because of the complexities introduced into the problem by **magnetic fields**, the effect of dust and the complicated radiative effects involved when the first generation of stars form, the theory of star formation is still far from fully developed. This is one of the barriers to a full understanding of how galaxies formed, because galaxies contain vast numbers of stars as well as gas, dust and **dark matter**.

The big difference that emerges when we apply these considerations to the formation of cosmological structures is that these structures have to form within an expanding medium. This makes the cosmological version of this instability a much slower phenomenon – a power law in time, rather than an exponential function. This is essentially because, even if pressure forces can be neglected, self-gravity still has to fight against the cosmological expansion. This has several important consequences for models of structure formation in cosmology. One is that, since the instability is quite slow and the Universe has a finite age, there must have been initial irregularities with a finite amplitude: spontaneous fluctuations are not large enough to act as 'seeds', as is the case in star formation. The cosmic microwave background radiation should carry the imprint of these initial fluctuations. Another important point is that the distribution of matter on large scales should also, in a sense, 'remember' these initial fluctuations. The easiest way to understand this is to note that the typical **peculiar motions** of galaxies are around 600 km/s. The **age of the Universe** is roughly given by the inverse of the **Hubble constant**. It is easy to see, therefore, that the typical distance a galaxy can have moved during the entire evolution of the Universe is in the range 6 to 12 Mpc, for values of the Hubble constant between 50 and 100 km/s/Mpc.

But we know that there are structures in the Universe on scales as large as 100 Mpc (see **large-scale structure**), so these cannot have been formed by moving material around from a uniform initial distribution, and must be related to some kind of protostructure which existed in the initial conditions. A complete model of structure formation must include a prescription for the physical nature and statistical distribution of these **primordial density fluctuations**.

The rate of growth of fluctuations in a **cosmological model** depends on exactly how much mass there is in total, because it is the total mass that drives the instability. In fact, the growth rate is closely related to the expansion rate which, in turn, is related to the cosmological **density parameter** Ω and, possibly, to the value of the **cosmological constant** Λ; these parameters are quite uncertain at the present time.

Other factors also influence the growth rate of fluctuations on a given scale. Pressure will prevent the collapse of short-wave perturbations in a static background. In cosmology, the medium we are dealing with has several components which interact with one another in complicated ways, and there are additional effects over and above the simple pressure support mentioned above. These can damp down the small-scale modes, and can even modify the growth rate of large-scale perturbations. These effects depend on the detailed composition of the matter in the Universe, particularly any dominant component of non-baryonic dark matter, but for the **weakly interacting massive particles** (WIMPs) the detailed behaviour can be calculated. Having this non-baryonic component actually makes the problem a little easier, because WIMPs are collisionless particles, unlike ordinary gas particles.

These considerations apply to the growth rate in the so-called *linear regime*, when the density fluctuations involved are smaller than the mean density. In this regime, perturbative methods can be used to calculate the evolution of inhomogeneities until they become comparable to the mean density. But galaxies correspond to density fluctuations at least one hundred thousand times larger than the mean density: the density of material inside a galaxy is much higher than that in the surrounding **intergalactic medium**. Calculating the evolution of fluctuations into the strongly nonlinear regime is much more difficult, and generally involves the use of ***N*-body simulations** on powerful computers. In addition, the nonlinear gravitational regime will probably also involve complex hydrodynamical and radiative effects, resulting from the heating and cooling of gas and the formation of stars, as described above.

The difficulty of understanding how galaxies form, even in a model in which all the components are specified in advance, means that testing theories of galaxy formation using galaxy clustering observations is not straightforward. Galaxies need not trace the distribution of dark matter in any reliable way: in the current jargon, this means that galaxy formation may be *biased* to occur preferentially in certain regions, perhaps where the dark matter density is particularly high.

Although the Jeans instability is quite simple, there are thus many problems to be solved before we can construct a complete theory of cosmological structure formation based on it.

FURTHER READING: Narlikar, J. V., *Introduction to Cosmology*, 2nd edition (Cambridge University Press, Cambridge, 1993), Chapter 5; Lifshitz, E. M., 'On the gravitational instability of the expanding Universe', *Soviet Physics JETP*, 1946, **10**, 116; Bonnor, W. B., 'Jeans' formula for gravitational instability', *Monthly Notices of the Royal Astronomical Society*, 1957, **117**, 104.

JET see **active galaxy, quasar**.

***K* (CURVATURE CONSTANT)** see **curvature of spacetime**, **Robertson–Walker metric**.

KALUZA–KLEIN THEORY Many unified theories of the **fundamental interactions** can be made mathematically consistent only if **spacetime** has many more dimensions that the four (three of space and one of **time**) we are used to. For example, **string theories** are possible only if there are either 10 or 26 dimensions. At first sight one would be tempted to reject these theories as being inconsistent with even the most basic observations of our actual **Universe**. The idea behind *Kaluza–Klein theories* (named after Theodore Franz Eduard Kaluza (1885–1954) and Oskar Benjamin Klein (1894–1977)) is to circumvent this problem by assuming that the extra dimensions, over and above the four of everyday experience, are wrapped up so small that we cannot perceive them. Formally, this is called *compactification* or *dimensional reduction*.

Suppose, for example, that a small ball is free to move anywhere on a two-dimensional sheet. All experiments that could be done by the ball would indicate that there were two spatial dimensions. But if we were to roll up the sheet into a very narrow tube, just wide enough for the ball to fit inside, then it would be constrained to move only in one dimension. It would not be at all aware of the existence of the second dimension, and would think that space was actually one-dimensional.

Kaluza–Klein theories usually assume that the extra dimensions must somehow have been compactified to a very small scale indeed, of the order of the **Planck length**. This means that not only can we not perceive them now, but also that we are unlikely to be able to perform any experiment using **elementary particles** that would indicate their presence. Higher-dimensional theories can thus be made quite compatible with everyday experience.

It is interesting also that the contraction of extra spatial dimensions by the enormous factor required to bundle them up on the scale of the Planck length can lead to an even more enormous expansion of the three that must remain. Kaluza–Klein theories therefore lead naturally to a particular model of the **inflationary Universe**.

KANT, IMMANUEL (1724–1804) German philosopher. In 1755 he proposed a cosmogony in which the collapse of a cloud of primordial material formed the Sun and the Solar System – the essence of modern models. He correctly interpreted the Milky Way as the view we have of the Galaxy from our location within it, and (much ahead of his time) suggested that the nebulae were similar but much more distant systems, which he termed 'island universes'.

K-CORRECTION A correction that arises when the total luminosity of a source (such as a **galaxy**) at a large distance is measured. When astronomers do this they generally measure the amount of light received in only a certain part of the spectrum of **electromagnetic radiation**. The general name for this kind of measurement is *photometry*, and the part of the spectrum used is determined by the choice of *filter* (see **optical astronomy**).

When the light from a galaxy reaches the observer, however, it has suffered a **redshift** as a consequence of the **expansion of the Universe**. Any light emitted at wavelength λ_e will consequently be received by the observer at some different wavelength, λ_o, given by $(1 + z)\lambda_e$, where z is the redshift. The entire spectrum of light received will be shifted in this way, which is why **spectroscopy** can be used to measure the redshift of the source by looking at the shifted positions of **emission lines**. Photometric methods, however, do not reveal the presence of emission or **absorption lines** and are merely concerned with the total amount of light received over a relatively broad range of wavelengths determined by the filter used.

If the spectrum of radiation produced by galaxies were completely flat (apart from the emission and/or absorption features), the redshift would have no effect on photometric studies, but this is not the case. For example, galaxies do not emit very much radiation at all in the extreme ultraviolet region of the spectrum because this is beyond the limit of the Lyman series of energy transitions in the hydrogen atom. A galaxy observed at such a high redshift that this Lyman limit appears in an optical band would look very dim indeed to optical observers.

In order to reconstruct the intrinsic brightnesses of galaxies it is necessary to correct for the fact that all the galaxies in a sample are being observed in the same wavelength band, but this is not the same emitted waveband for galaxies at different redshifts. The correction of this effect using the known emission properties of galaxies is known as the *K-correction*. It plays an important role in studies of the **evolution of galaxies**, **classical cosmology** and attempts to calibrate the **extragalactic distance scale**.

KEPLER, JOHANNES (1571–1630) German astronomer. He worked out the laws of planetary motion that Isaac **Newton** later explained with his law of **gravity**. Kepler also speculated about **cosmology**, arguing, for example, that the **Universe** is finite.

L

LAMBDA (Λ) see **cosmological constant**.

LARGE-SCALE STRUCTURE The basic building blocks of the Universe are **galaxies**, but they are not the largest structures we can see. Galaxies tend not to be isolated, but to band together; the term used to describe how they are distributed on cosmological scales is *large-scale structure*.

The distribution of matter on large scales is usually determined by means of **redshift surveys**, using **Hubble's law** to estimate the distances to galaxies from their **redshifts**. The existence of structure was known for many years before redshift surveys became practicable. The distribution of galaxies on the sky is highly nonuniform, as can be seen from the first large systematic survey of galaxy positions which resulted in the famous *Lick Map*. But impressive though this map undoubtedly is, we cannot be sure whether the structures seen in it are real, physical structures or just chance projection effects. After all, we all recognise the constellations, but these are not physical associations because the stars in any one constellation lie at very different distances from us. For this reason, the principle tool of cosmography has become the **redshift survey** (for more details of the techniques used and results obtained, see under that heading).

The term used to describe a physical aggregation of many galaxies is a *cluster of galaxies* or *galaxy cluster*. Clusters vary greatly in size and richness. For example, our Galaxy, the Milky Way, is a member of the *Local Group* of galaxies, a rather small cluster, the only other large member of which is the Andromeda Galaxy. At the other extreme there are the so-called *rich clusters of galaxies*, also known as *Abell clusters*, which contain many hundreds or even thousands of galaxies in a region just few million light years across: prominent nearby examples are the Virgo and Coma Clusters. In between these two extremes, galaxies appear to be distributed in systems of varying density in a roughly **fractal** (or hierarchical) manner. The densest Abell clusters are clearly collapsed objects held together in equilibrium by their own self-gravity (see e.g. **virial theorem**). The less rich and more spatially extended systems may not be bound in this way, but may simply reflect a general statistical tendency of galaxies to clump together.

Individual galaxy clusters are still not the largest structures to be seen. The distribution of galaxies on scales larger than about 30 million light years also reveals a wealth of complexity. Recent observational surveys have shown that galaxies are not simply distributed in quasi-spherical 'blobs', like the Abell clusters, but also sometimes lie in extended quasi-linear structures called *filaments*, such as the *Perseus–Pisces chain*, or flattened, sheet-like structures such as the *Great Wall*. The latter object is a roughly two-dimensional concentration of galaxies, discovered in 1988 in the course of a galaxy survey carried out by Margaret **Geller**, John **Huchra** and other astronomers at the Harvard-Smithsonian Center for Astrophysics. This structure is at least 200 million light years by 600 million light years in size, but less than 20 million light years thick. It contains many thousands of galaxies and has a mass of at least 10^{16} solar masses. The rich clusters themselves are clustered into enormous, loosely bound agglomerations called *superclusters*. Many are known, containing anything from around ten rich clusters to more than fifty. The most prominent known supercluster is called the *Shapley Concentration*, while the nearest is the Local Supercluster, centred on the Virgo Cluster mentioned above, a flattened structure in the plane of which the Local Group is moving. Superclustering is known to exist on scales of up to 300 million light years, and possibly more, and superclusters may contain 10^{17} or more solar masses of material.

These overdense structures are complemented by vast underdense regions known as *voids*, many of which appear to be roughly spherical. These regions contain very many fewer galaxies than average, or even no galaxies at all. Voids with a density of less than 10% of the average density on scales of up to 200 million light years have been detected in large-scale redshift surveys. The existence of large voids is not surprising, given the existence of clusters of galaxies and superclusters on very large scales, because it is necessary to create regions of less-than-average density in order to create regions of greater-than-average density.

Until recently, progressively deeper redshift surveys had revealed structures of larger and larger sizes, indicating that the scale had yet to be reached where the homogeneity and isotropy required by the **cosmological principle** would become apparent. But the 1996 *Las Campanas Survey* – which contains about 25,000 galaxy positions having recessional velocities out to more than 60,000 km/s – does seem to indicate that the scale of homogeneity is at last being reached. But there is still structure on even larger scales. The temperature variations of the **cosmic microwave background radiation** on the sky (see Essay 5) correspond to structures much larger than this, extending all the way to the scale of our present **horizon**. But these structures, though large in spatial extent, are very weak in terms of the density fluctuation they represent: only one part in a hundred thousand (see also **power spectrum**). It seems, then, that the cosmological principle does indeed hold sway on the largest scales amenable to observation.

The properties of the relic radiation suggest that all the structure we see grew by a process of gravitational **Jeans instability** from small **primordial density fluctuations**. The complex network structure observed in reality is reproduced, at least qualitatively, by ***N*-body simulations** which seek to verify current models of **structure formation**. But such a qualitative comparison between the observed

and predicted properties of large-scale structure is not sufficient. In order to test theories of structure formation against such observations, we have to use objective statistical methods such as the **correlation function** or power spectrum.
SEE ALSO: **peculiar motions**.

FURTHER READING: Peebles, P. J. E., *The Large-Scale Structure of the Universe* (Princeton University Press, Princeton, 1980); Coles, P., 1996 'The large-scale structure of the Universe', *Contemporary Physics*, 1996, **37**, 429.

LAST SCATTERING SURFACE According to the standard **Big Bang theory**, the early Universe was sufficiently hot for all the matter in it to be fully ionised. Under these conditions, **electromagnetic radiation** was scattered very efficiently by matter, and this scattering kept the Universe in a state of **thermal equilibrium**. Eventually the Universe cooled to a temperature at which electrons could begin to recombine into atoms, and this had the effect of lowering the rate of scattering. This happened at what is called the *recombination era* of the **thermal history of the Universe**. At some point, when recombination was virtually complete, photons ceased to scatter at all and began to propagate freely through the Universe, suffering only the effects of the cosmological **redshift**. These photons reach present-day observers as the **cosmic microwave background radiation** (CMB). This radiation appears to come from a spherical surface around the observer such that the radius of the shell is the distance each photon has travelled since it was last scattered at the epoch of recombination. This surface is what is called the *last scattering surface*.

To visualise how this effect arises, imagine that you are in a large field filled with people screaming. You are screaming too. At some time $t = 0$ everyone stops screaming simultaneously. What will you hear? After 1 second you will still be able to hear the distant screaming of people more than 330 metres away (the speed of sound in air, v_s, is about 330 m/s). After 3 seconds you will be able to hear distant screams from people more than 1 kilometre away (even though those distant people stopped screaming when you did). At any time t, assuming a suitably heightened sense of hearing, you will hear some faint screams, but the closest and loudest will be coming from people a distance $v_s t$ away. This distance defines a 'surface of last screaming', and this surface is receding from you at the speed of sound. Similarly, in a non-expanding universe, the surface of last scattering would recede from us at the speed of light. Since our Universe is expanding, the surface of last scattering is actually receding at about twice the speed of light. This leads to the paradoxical result that, on their way to us, photons are actually moving away from us until they reach regions of space that are receding at less than the speed of light. From then on they get closer to us. None of this violates any **laws of physics** because all material objects are locally at rest.

When something is hot and cools down it can undergo a **phase transition**. For example, hot steam cools down to become water, and when cooled further it becomes ice. The

Universe went through similar phase transitions as it expanded and cooled. One such phase transition, the process of recombination discussed above, produced the last scattering surface. When the Universe was cool enough to allow the electrons and protons to fall together, they 'recombined' to form neutral hydrogen. CMB photons do not interact with neutral hydrogen, so they were free to travel through the Universe without being scattered. They *decoupled* from matter. The opaque Universe then became transparent.

Imagine you are living 15 billion years ago. You would be surrounded by a very hot opaque plasma of electrons and protons. The Universe is expanding and cooling. When the Universe cools down below a critical temperature, the fog clears instantaneously everywhere. But you would not be able to see that it has cleared everywhere because, as you look into the far distance, you would be seeing into the opaque past of distant parts of the Universe. As the Universe continues to expand and cool you would be able to see farther, but you would always see the bright opaque fog in the distance, in the past. That bright fog is the surface of last scattering. It is the boundary between a transparent and an opaque universe and you can still see it today, 15 billion years later.

Although the surface of last scattering has a temperature of 3000 K, the cosmic microwave background photons now have a temperature of about 3 K. This factor-of-1000 reduction in temperature is the result of the factor-of-1000 expansion between the time the photons were emitted and now. The photons have cooled and become redshifted as a result of the **expansion of the Universe**. For example, when the Universe is three times bigger than it is now, the CMB will have a temperature of about 1 K.

The last scattering surface is sometimes called the *cosmic photosphere*, by analogy with the visible 'surface' of the Sun where radiation produced by nuclear reactions is last scattered by the solar material. The energy source for the Sun's photons is not in the photosphere: it comes from nuclear fusion at the centre of the Sun. Similarly, the CMB photons were not created at the surface of last scattering: they were produced at a much earlier epoch in the evolution of the Universe. A tiny fraction (about one in a billion) of these photons, however, *were* created by recombination transitions at the last scattering surface. There should therefore be very weak **emission lines** in the **black-body** spectrum of the CMB radiation, but none has yet been detected.

Some interesting properties of the last scattering surface are illustrated in the Figure overleaf. Here, space is represented as two-dimensional. The time t since the Big Bang is the vertical axis; T is the temperature of the CMB and z is the **redshift** (for simplicity, the expansion of the Universe is ignored). The plane at the top corresponds to the Universe now. As stationary observers we move through time (but not space), and we are now at the apex of the cone in the NOW plane. When we look around us into the past, we can see only photons on our past **light cone**. CMB photons travel from the wavy circle in the last scattering surface along the surface of

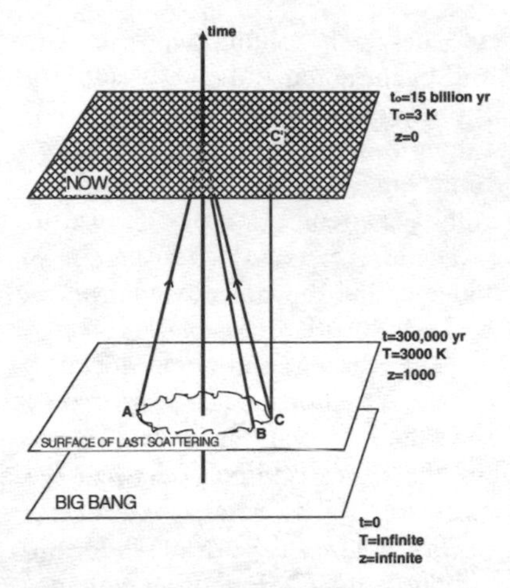

Last scattering surface The last scattering surface represented as a flat sheet. The expansion of the Universe is ignored in this diagram.

the light cone to us. The unevenness of the circle represents temperature fluctuations at the last scattering surface. The bottom two planes are at fixed times, while the NOW plane moves upwards. As it does, the size of the observable Universe (the diameter of the wavy circle) increases. The object which emitted light at C is currently at C′, and the light emitted at C is currently entering our telescopes at the apex of the cone. Points A and C are on opposite sides of the sky. If the angle between B and C is greater than a few degrees, then B and C have never been able to exchange photons with each other, so they cannot even know about each other at the time of last scattering. How can it be that their temperatures are the same? This is the essence of the cosmological **horizon problem**, and was one of the motivations for the **inflationary Universe** theory.

SEE ALSO: Essay 5.

LAWS OF PHYSICS The basic tools of physical science, sometimes called the *laws of nature*, comprising mathematical equations that govern the behaviour of matter (in the form of **elementary particles**) and energy according to various **fundamental interactions**. Experimental results obtained in the laboratory or through observations of natural physical processes can be used to infer mathematical rules which describe these data. Alternatively, a theory may be created first as the result of a hypothesis or physical principle, which receives experimental confirmation only at a later stage.

As our understanding evolves, seemingly disparate physical laws become unified in a single overarching theory. The tendency of apples to fall to the ground and the tendency of the Moon to orbit the Earth were thought to be different things before the emergence of Isaac **Newton**'s laws of motion and his theory of **gravity**. This theory was thought to be complete until the work of Albert **Einstein**, who showed that it was lacking in many aspects. A more complete (and much more mathematically intricate) theory of **general relativity** took the place of Newton's theory in 1915. In modern times, physicists are trying to unify general relativity with the rest of the theory of fundamental interactions into a **theory of everything**, a single mathematical formula from which all of physics can be derived (see also **grand unified theory**, **string theory**, **supersymmetry**).

Although this ambitious programme is far from complete, similar developments have occurred throughout the history of science, to the extent that the exact form of the laws of physics available to working scientists changes significantly with time. Nevertheless, the task of a physical cosmologist remains the same: to take whatever laws are known (or whichever hypotheses one is prepared to accept) and work out their consequences for the evolution of the **Universe** at large. This is what cosmologists have done all down the ages, from Aristotle to the present generation of early-Universe cosmologists.

But there are deep philosophical questions below the surface of all this activity. For example, what if the laws of physics were different in the early Universe – could we still carry out meaningful research? The answer to this is that modern physical theories actually predict that the laws of physics do change, because of the effects of **spontaneous symmetry-breaking**. At earlier and earlier stages in the **Big Bang theory**, for example, the nature of the electromagnetic and weak interactions changes so that they become indistinguishable at sufficiently high energies. But this change in the law is itself described by another law: the so-called *electroweak theory*. Perhaps this law itself is modified at scales on which grand unified theories take precedence, and so on right back to the very beginning of the Universe.

Whatever the fundamental rules may be, however, physicists have to assume that they apply for all times since the Big Bang. It is merely the low-energy outcomes of these fundamental rules that change with time. By making this assumption they are able to build a coherent picture of the **thermal history of the Universe** which does not seem to be in major conflict with the observations. This makes the assumption reasonable, but does not prove it to be correct.

Another set of important questions revolves around the role of mathematics in physical theory. Is nature really mathematical, or are the rules we devise merely a kind of shorthand to enable us to describe the Universe on as few pieces of paper as possible? Do we discover laws of physics, or do we invent them? Is physics simply a map, or is it the territory itself?

There is also another deep issue connected with the laws of physics pertaining to the very beginning of space and time. In some versions of **quantum cosmology**, for example, we have to posit the existence of physical laws in advance of the physical universe they are supposed to describe. This has led many early-Universe physicists to embrace a neo-Platonist philosophy in which what really exists is the mathematical equations of the (as yet unknown) theory of everything, rather than the physical world of matter and energy. But not all cosmologists get carried away in this manner. To those of a more pragmatic disposition the laws of physics are simply a useful description of our Universe, whose significance lies simply in their very usefulness.

FURTHER READING: Barrow, J. D., *The World Within the World* (Oxford University Press, Oxford, 1988); Barrow, J. D., *Pi in the Sky* (Oxford University Press, Oxford, 1992).

LEMAÎTRE, GEORGES EDOUARD (1894–1966) Belgian astronomer, cosmologist and Roman Catholic priest. In 1927 he found solutions to the equations of general relativity that corresponded to an expanding Universe, and was the first to wind the clock backwards in time to infer the presence of an initial creation event. (Alexander **Friedmann** had obtained essentially the same solution earlier, but had not considered the early stages of such a model.) Lemaître's work was publicised in Britain by Arthur **Eddington** in the early 1930s.

Lepton see **elementary particles**.

Lepton era see **thermal history of the Universe**.

Life in the Universe It is difficult to define precisely what is meant by 'life', because life exists in such a variety of forms. At the very least we can say that life, as we know it, is sustained by a complicated set of chemical reactions between (among other things) carbon, nitrogen and oxygen. These reactions often result in the appearance of very large organic molecules of various kinds. For example, the fundamental ability of life forms to reproduce stems from the replicating ability of DNA, itself one of the most complicated naturally occurring organic materials known. The complex chemistry on which biology is based can likewise be said to depend on the behaviour of the **fundamental interactions** of physics, and their action on the various **elementary particles**. In fact, chemistry is based largely on the properties of the electromagnetic interaction and on the essentials of **quantum physics**. These **laws of physics** have special properties without which complex chemistry would be impossible.

The origin of life in the Universe therefore poses an important question for cosmologists. Is the structure of the physical world around us accidental, or is there some deep reason for the emergence of the complexity associated with the development and evolution of life? Cosmologists' reactions to this question vary greatly. There are those who are so impressed by the many apparently unexplained coincidences within the laws of physics that they see in them the hand of a Creator. Their answer to these questions is therefore essentially to invoke an argument from design: the Universe had to be made the way it did in order that life should evolve within it. In philosophical terms, this is a *teleological* argument. To others, the fact that the laws of physics represent a Universe capable of sustaining life is not something to be surprised at because those laws describe *our* Universe and we already know that there is life in it.

Modern cosmological theories based on the idea of the **inflationary Universe** seem to be moving the community of cosmologists towards the second of these stances. The reason for this is that in versions of inflation called *chaotic inflation*, the laws of physics at low energies can turn out to be very different in disconnected regions of **spacetime**. It is as if the Universe consisted of a collection of bubbles, within each of which the laws of physics are different. Some of these bubbles would be conducive to chemistry and biology, while others would not. We could exist only in one of the favourable parts of this 'multiverse', so it should not surprise us that the laws of physics are such as to support our existence. This latter argument is an example of the weak version of the **anthropic principle**, whereas the teleological argument is called the strong anthropic principle.

The anthropic principle also allows us to use the existence of life as an observation about the Universe which can shed light on **cosmological models**. For example, consider the question of why the Universe is as big as it is. The scale of the observable Universe (see **horizon**) is roughly given by c/H_0, where c is the speed of light and H_0 is the **Hubble constant**. But $1/H_0$ also roughly defines the **age of the Universe**. No complex chemistry can evolve until elements like carbon and nitrogen are made, and this takes about 10 billion years of stellar evolution. (Nothing heavier than lithium is made in the Big Bang – see **nucleosynthesis**, **light element abundances**.) But the value of $1/H_0$ comes out at just this order of magnitude. Is this a coincidence? No. The Universe has to be as big as it is, because it has to be as old as it is in order for life to have evolved.

These considerations apply to the existence of any form of life based on the complex chemistry we are familiar with: they apply just as much to simple viruses as they do to more complicated species like humans. But the questions most people ask are about the existence of intelligent life, or whether there is anything resembling human life, elsewhere in the Universe. Assuming for the sake of argument that we consider human life to be intelligent, are we the only form of intelligence? Are we alone, or are there civilisations out there with technologies as advanced as our own?

If our understanding of physics is correct, it is unlikely that we will ever be able to communicate with life forms in galaxies other than our own. The fundamental barrier to the transfer of information is that nothing can travel faster than light. Given the distances between the galaxies and the speed of light, communication between life forms would simply take too long to be practicable. It will take millions of years for a signal to reach the Andromeda Galaxy from here, and millions of years longer to receive a reply. So it makes sense to focus on the question of whether there can be life within our own Galaxy, the Milky Way.

A measure of the possibility of intelligent life existing within our Galaxy was provided by Frank Drake in the form of an equation he put forward in 1961. The so-called *Drake equation* is an expression for the number N of advanced technological civilisations:

$$N = N_* \times f_p \times n_E \times f_l \times f_i \times f_c \times f_L$$

In this equation N_* is the number of stars in our Galaxy, f_p is the fraction of these stars that possess planets, n_E is the average number of Earth-like planets in each planetary system, f_l is the fraction of these that evolve life at some stage, f_i is the fraction of these systems on which life becomes intelligent, f_c is the fraction of these that develop advanced technological civilisations, and f_L is the fraction of the lifetime of the planetary system for which such civilisations survive. The final number N of civilisations is therefore given by the product of a number of quantities whose values we can only guess. Recent observational results from the **Hubble Space Telescope** suggest that planetary systems appear to be fairly common around nearby stars, but what fraction of them contain planets which are sufficiently Earth-like to develop complex chemistry is not clear. On the other hand, the last factor in the Drake equation may well be very small indeed: life on Earth has developed advanced technology only in the past century, and there is no particular reason to believe that it will last indefinitely. Putting in our best guesses for the relevant numbers leads to the conclusion that there should be only a very few advanced civilisations in the Milky Way – and we are perhaps the only one.

Another line of argument leading to the same, lonely conclusion is called the *Fermi paradox*. If there are many intelligent civilisations in the Galaxy, why have they not been here yet? Many, of course, claim that they have, and that they regularly abduct alcoholic farmers from Iowa. But we might also ask whether any intelligent civilisation would go to the trouble of launching expensive and environmentally damaging rockets to every visible star just on the off-chance that they might find some form of life there. Would it not be more sensible for them to turn their energies towards maintaining the health of their own planetary ecosystem?

It appears unlikely that life is unique in the Universe, particularly if we live in an infinite, **open universe**. But it also appears unlikely on the basis of the Drake equation that the part of our Universe within our observable horizon contains no other intelligences. Unless we can learn to live much longer than we do now, however, we will not find it easy to communicate with these distant aliens.

FURTHER READING: Shklovskii, I. S. and Sagan, C., *Intelligent Life in the Universe* (Holden-Day, New York, 1966); Newman, W. I. and Sagan, C., 'Galactic civilizations: Population dynamics and interstellar diffusion', *Icarus*, 1981, **46**, 293; Barrow, J. D. and Tipler, F. J., *The Anthropic Cosmological Principle* (Oxford University Press, Oxford, 1986); McDonough, T. R., *The Search for Extraterrestrial Intelligence* (John Wiley, Chichester, 1987); Goldsmith, D. and Owen, T., *The Search for Life in the Universe*, revised 2nd edition (Addison-Wesley, Reading, MA, 1993); Dick, S. J., *The Biological Universe* (Cambridge University Press, Cambridge, 1996); Kauffman, S., *At Home in the Universe* (Penguin, London, 1996).

LIGHT CONE The velocity of light, which is constant for all observers regardless of their state of motion, plays a fundamental role in both **special relativity** and **general relativity**. In particular, it represents the maximum speed at which particles or light signals can travel without violating our common-sense notions of cause and effect. *Light cones* are used to visualise the boundaries of regions of **spacetime** that can be connected by causal processes.

To see what a light cone is, imagine first a simple, two-dimensional, flat spacetime consisting of one spatial dimension and one time dimension. This can be represented as a graph with space in the x-direction and time in the y-direction. An observer sitting at the origin of this coordinate system travels up the y-axis. Now suppose that at time $t = 0$ this observer sends out a flash of light in the positive x-direction. This flash travels along a locus represented by a line with constant positive slope. If the observer had transmitted the flash in the negative x-direction, the locus of points describing the trajectory would be the mirror-image (about the time axis) of the previous locus. The V-shaped region between these two lines represents that part of the future spacetime that can be reached by a signal produced by the observer that travels no faster than the speed of light. If we imagine that spacetime consists of two space and one time dimension, then the future region that can be reached is a conical space formed by rotating the sloping line around the y-axis. For this reason the boundary of the future region accessible to causal influence is called the *future light cone*. (Of course, reality has three spatial dimensions, but this two-dimensional analogue has provided the name, since a three-dimensional conic surface is hard to visualise.)

We can also ask what is the region of past spacetime that can influence an observer at the origin. The boundary of this region is marked by an incoming light cone represented by an inverted V with its apex at the origin. For obvious reasons, this construction is called the *past light cone*. An illustration of the past light cone of a cosmological observer is shown in the Figure under **last scattering surface**.

These concepts apply directly in special relativity. In the general theory, spacetime need not be everywhere flat (see **curvature of spacetime**), so the light cones (as they are still called) need not be cones at all, but can be complicated curved surfaces. Locally, however, the spacetime of general relativity is always flat so that light cones are conical near their apex. The **expansion of the Universe** introduces a further complication, in that the spacetime is expanding everywhere. This further distorts the shape of both the past and future light cones.

Note that the light cone always refers to the past or future of a given event in spacetime, as defined by a particular position and a particular time. We might be interested in a more general issue, such as finding the region of past spacetime that can have influenced an observer at any time up to and including the present. This requires a generalisation of the concept of a light cone called the **horizon**.

LIGHT ELEMENT ABUNDANCES The chemical composition of the Universe is basically very simple. Setting aside any **weakly interacting massive particles** (which cannot have any chemistry), the bulk of cosmic material is in the form of hydrogen, the simplest of all chemical elements, consisting of a single proton as the nucleus (plus one electron). More than 75% of the matter in the Universe is in this simple form. Back in primordial times things were even simpler, since there had been no time that early on for stars to have burned hydrogen into helium or helium into carbon or, indeed, to have burned anything at all. According to the theory of cosmological **nucleosynthesis**, only very small amounts of very simple matter can have been made early in the Big Bang.

Apart from the hydrogen, about 25% of the material constituents (by mass) of the Universe produced by nucleosynthesis is expected to have been in the form of helium-4, the stable isotope of helium which has two protons and two neutrons in its nucleus. About a hundred thousand times rarer than this were two slightly more exotic elements. *Deuterium*, or heavy hydrogen as it is sometimes called, has a nucleus consisting of one proton and one neutron. The lighter isotope of helium, helium-3, is short of one neutron compared with its heavier version. And finally there is lithium-7, produced as a tiny trace element with an abundance of one part in ten billion of the abundance of hydrogen. Since nothing much heavier than lithium-7 is made primordially, cosmologists usually refer to these elements as the *light elements*. Roughly speaking, then, the proportions given above are the *light element abundances* predicted by the **Big Bang theory**.

So how do these computations compare with observations? Before answering this question it is important to understand why appropriate measurements are so difficult to make. Firstly, apart from hydrogen and helium, the amounts made in the Big Bang were very small, so sensitive observations are required. Secondly, and probably even more importantly, the proportions given above refer to the abundances of the light elements that emerge at the end of the nuclear fusion processes that operated in the primordial fireball. This is just a few minutes after the Big Bang itself. About 15 billion years have elapsed since then, and all kinds of processes have had time to contaminate this initial chemical mix. In particular, virtually any material we can observe astronomically will have been processed through stars in some way.

According to the theory of **stellar evolution**, the burning of hydrogen into helium is the main source of energy for most stars. We would therefore expect the fraction of matter in the **Universe** that is now in the form of helium to be significantly higher than the primordial value of about 25%. Although the expected abundance is large, therefore, the interpretation of measured values of the helium abundance is uncertain. On the other hand, deuterium can be very easily destroyed in stars (but it cannot be made in stars). The other isotopes, helium-3 and lithium-7, can be either made or destroyed in stars, depending on the temperature. The processing of material by stars is called *astration*, and because of astration uncertain corrections have to be introduced to derive estimated primordial abundances from present-day observations. We should also be aware that chemical abundances are not usually uniform throughout stars or other objects because of physical or chemical *fractionation*. An abundance measured in one part of an astronomical object (say in Jupiter's atmosphere) might not be typical of the object as a whole; such effects are known to be important, for example, in determining the abundance of deuterium in the Earth's oceans.

Despite these considerable difficulties, much effort is devoted to comparing observed abundances with these theoretical predictions. Relevant data can be obtained from studies of stellar atmospheres, interstellar and intergalactic **emission lines** and **absorption lines**, planetary atmospheres and meteorites, as well as from terrestrial measurements. Abundances of elements other than helium determined by these different methods differ by a factor of five or more, presumably because of astration and/or fractionation. The interpretation of these data is therefore rather complicated and the source of much controversy. Nevertheless, it is well established that the abundance of helium-4 is everywhere close to 25%, and this in itself is good evidence that the basic model is correct. We usually correct for the stellar burning of hydrogen into helium by looking at stars of different ages. According to the theory of stellar evolution, stars that have burnt more hydrogen into helium should also have burnt more helium into heavier elements like carbon and nitrogen. Such heavier elements are usually called *metals* by astronomers, although they are not strictly speaking metals in the true sense of the word. **Spectroscopy** can be used to estimate the helium abundance (Y) and the amount of metals (called the metallicity, Z); the best objects for such a study are clouds of ionised hydrogen, called HII regions. Stars with higher metallicity Z also have higher values of Y. We can extrapolate the known behaviour back to $Z = 0$ in order to estimate the primordial abundance of helium, usually denoted by Y_p, which cannot be higher than 24% by these measurements.

The best estimate of the helium-3 abundance comes from the *solar wind* (the constant outward stream of atomic particles from the Sun), measurements of which suggest a firm upper limit on the combined abundance of deuterium and helium-3 of about 10^{-4}. Measurements of lithium are much less certain, but it appears that the ratio of lithium to hydrogen is about 10^{-10}. Until recently, the situation with respect to deuterium was much less clear because of the effects mentioned above, and it was argued that the primordial value may have been as high as 10^{-4}. Recently, however, a much cleaner estimate of the deuterium abundance has been obtained by detecting absorption lines in **quasar** spectra that resemble weak Lyman alpha absorption. These lines are slightly shifted relative to an actual Lyman alpha absorption line because of the presence of the extra neutron in the nucleus of deuterium compared with hydrogen. Measuring a pair of lines can allow us to determine the abundance of deuterium relative to hydrogen. Recent results suggest that the primordial deuterium abundance is a few times 10^{-5}.

The recent deuterium measurements have clarified the comparison between theory and observation because the deuterium abundance depends quite sensitively on the important cosmological parameters; the helium abundance, which is a lot easier to measure, is quite insensitive to these parameters. In particular, the amount of deuterium produced is a sensitive probe of Ω_b, the **density parameter** in baryonic matter (matter that takes part in nuclear reactions: see **elementary particles, fundamental interactions**). In fact, there is also a dependence on the **Hubble**

constant, which we can represent in the dimensionless form $h = H_0/100$ kilometres per second per megaparsec. The relevant combination of these two quantities is $\Omega_b h^2$. It appears that all the measured light elements fit with the theory of nucleosynthesis provided this number is small: $\Omega_b h^2$ must lie between about 0.01 and 0.015 (or possibly a little larger). For reasonable values of the Hubble parameter, this means that no more than 10% or so of the density required to make a **flat universe** can be in the form of baryons. Since there is evidence from other arguments that the total density parameter is actually at least 0.2, this strong constraint on $\Omega_b h^2$ is the main argument for the existence of non-baryonic dark matter in the form of weakly interacting massive particles.

FURTHER READING: Kolb, E. W. and Turner, M. S., *The Early Universe* (Addison-Wesley, Redwood City, CA, 1990); Coles, P. and Ellis, G. F. R., *Is the Universe Open or Closed?* (Cambridge University Press, Cambridge, 1997), Chapter 4.

LOOKBACK TIME see **evolution of galaxies**.

LORENTZ–FITZGERALD CONTRACTION see **special relativity**.

LUMINOSITY DISTANCE Suppose we have a source of light of known brightness, say a 100-watt bulb. If we see this source at an unknown distance from us in a dark room, we can calculate how far away it is by using a simple light meter, as is used in photography. The intensity of light falls off with the square of the distance from the source, so all we need to do is measure the intensity, and then straightforwardly deduce the distance. In mathematical terms, if the flux received is denoted by l and the intrinsic power output of the source by L, then the distance is $d = \sqrt{(L/4\pi l)}$.

Now suppose we know that certain galaxies have a particular intrinsic luminosity. Can we use the same argument to calculate their distance from us? The answer is not so straightforward because of several effects. First there is the cosmological **redshift**. Light from a distance source becomes redshifted to lower energies by the **expansion of the Universe**, so that distant sources become dimmer than they would without expansion (like the bulb in the dark room). Not only does each photon carry less energy, but there are also less photons per unit time because of time-dilation effects (see **special relativity**, **general relativity**). The second effect is caused by the finite velocity of light. If we are observing an object at a sufficiently large distance for cosmological effects to come into play, then we are also observing it as it was in the past. In particular, owing to the expansion of the Universe, the object would have been nearer to us when its light was emitted than it is now, when its light is being received. Finally, the **curvature of spacetime** causes light rays to travel in paths other than the straight lines which they would be following in a Euclidean geometry. When we add these effects together, we can show that the actual distance of a source whose absolute luminosity (power output) is known and whose apparent luminosity can be measured depends sensitively upon the **cosmological model** assumed. Looking at this in a different way, we can see that the measured brightness of a source depends on its distance (or redshift) in a way that probes the geometry and deceleration rate of the Universe. This is the basis of one of the tests in **classical cosmology**.

Correcting for these complications to obtain the **proper distance** is not straightforward unless we assume a particular cosmological model. Astronomers therefore usually define the luminosity distance of an object to be the distance the object would be from the observer in a non-expanding, Euclidean universe if it produced the same measured flux. This distance will not be equal to the proper distance in general, and will also differ from the **angular-diameter distance**, but it is a useful quantity in many applications.

FURTHER READING: Berry, M. V., *Principles of Cosmology and Gravitation* (Adam Hilger, Bristol, 1989); Narlikar, J. V., *Introduction to Cosmology*, 2nd edition (Cambridge University Press, * Cambridge, 1993).

LUMINOSITY FUNCTION Not all **galaxies** have the same luminosity, and it is to take account of this that astronomers use the *luminosity function*, which is a measure of the relative abundances of galaxies of different luminosities. Mathematically, the luminosity function $\Phi(L)$ plays a role similar to a probability distribution function, and is defined in such a way that the number of galaxies per unit volume with a luminosity between L and $L + \mathrm{d}L$ is given by $\Phi(L)\,\mathrm{d}L$.

The form of the galaxy luminosity function is not predicted by theory, but it can be determined from observations if we can measure the distances to a sufficiently large sample of galaxies of different types. The form of the function that emerges from such studies is called the *Schechter function*:

$$\Phi(L) = (\Phi_*/L_*)(L/L_*)^{-\alpha}\exp(-L/L_*)$$

in which Φ_*, L_* and α are constants to be determined from observations. It is interesting above all that the value of α appears to be about 1. Mathematically, this means that if we calculate the total number of galaxies (by integrating the Schechter function over all values of L), the result diverges. There are therefore, according to this function, an infinite number of infinitely faint galaxies. This is suggestive of the existence of **dark matter**. However, the total amount of light produced, obtained by integrating the Schechter function times L over all values of L, is finite. It is by determining the total amount of light from all the galaxies in the Universe in this way that we can determine the mass-to-light ratio needed for dark matter to be able to obtain a critical-density **flat universe**. The result is that, for every star like the Sun, there has to be around 1500 times as much mass in dark form if the density parameter Ω_0 equals 1.

Note that there is no reason at all to expect there to be a universal luminosity function *a priori*. It is a requirement of $\Phi(L)$ that it should be independent of position: that the luminosities of galaxies do not depend on their position or orientation in space or on their proximity to other galaxies. This also seems to be true for galaxies of different type, say ellipticals and spirals. It would be quite easy to imagine ways in which galaxies could interact with one another over quite large scales, and thus influence one another's luminosity, but that does not seem to be the case. The luminosity function does indeed appear to be a universal function.

LYMAN-ALPHA CLOUDS see **absorption lines**, **intergalactic medium**.

Mach's principle The principle that the masses of objects are somehow determined by the gravitational effect of all the other matter in the Universe. More precisely, the inertial mass m of an object (as defined by Newton's second Law of motion, $F = ma$, as the 'reluctance' of the object to be accelerated) is asserted to be not a property intrinsic to the object, but a consequence of the net effect of all other objects. A corollary of this principle is that the concept of mass is entirely meaningless in an empty universe.

Mach's principle is a very deep physical idea of great historical and philosophical importance, but the essence of it goes back much further than Ernst Mach (1838–1916) himself. In 1686, Isaac **Newton** discussed a similar idea in the *Principia*. Newton was concerned with what happens when bodies undergo rotation. He knew that a rotating body underwent acceleration towards the centre of rotation, and he was interested in what happened, for example, when a bucket full of water was spun around a vertical axis. What we see if we do this experiment (as Newton himself did) is that the surface of the water, which is flat when the bucket is not rotating, becomes curved when it begins to rotate. This curvature is caused by the centrifugal forces experienced in the rotating frame of the bucket pushing the water outwards from the centre, making the surface of the water concave. This shape remains if we suddenly stop rotating the bucket, which shows that relative motion between the bucket and the water has nothing to do with this effect. In some sense, the acceleration is absolute.

Newton had no problem with this because his laws of motion were formulated in terms of absolute time and space, but it is at odds with the principle of relativity. What should count is the relative motion of the water. But what is it relative to? One of the first suggestions was made by Bishop Berkeley (1685–1753). He had been impressed by Galileo's principle of relativity (see **special relativity**). He claimed, as later did Mach, that the acceleration was relative, but relative to the fixed stars (or, as we would now put it, to the large-scale distribution of matter in the Universe). Because masses are measurable only in terms of forces and accelerations, Berkeley was essentially arguing that the inertia of the bucket of water is determined by cosmological considerations. The surface of the bucket would look the same if the bucket were at rest but the entire **Universe** were rotating around it.

Albert **Einstein** was profoundly influenced by Mach's version of this argument, and he sought to incorporate it explicitly in his theory of gravitation, **general relativity**; but he was not successful. Many gravitation theorists have sought to remedy this failing in alternative theories of gravity. For example, in the **Brans–Dicke theory** of gravity there is an additional **scalar field** over and above the usual matter terms in Einstein's theory. The role of this field is to ensure that the strength of gravity described by the Newtonian gravitational constant *G* is coupled to the **expansion of the Universe**; *G* therefore changes with time in this theory. This is an essentially Machian idea because the effect of changing *G* can be seen, in some senses, as changing the inertial masses of particles as the Universe expands.

FURTHER READING: Brans, C. and Dicke, R. H., 'Mach's principle and a relativistic theory of gravitation', *Physical Review Letters*, 1961, **124**, 125; Narlikar, J. V., *Introduction to Cosmology*, 2nd edition (Cambridge University Press, Cambridge, 1993), Chapter 8.

MAGNETIC FIELDS Fields associated with the electromagnetic interaction. Although the strongest **fundamental interaction** on the large scales relevant to cosmology is **gravity**, there are situations in which magnetic fields play an important role. The Earth has a significant magnetic field, as does the Sun. **Galaxies** like the Milky Way also possess a large-scale magnetic field (with a strength of a few microgauss) and there is evidence for magnetic fields in the **intergalactic medium**, particularly inside rich clusters of galaxies (see **large-scale structure**). **Active galaxies** also show evidence of strong magnetic effects: radio galaxies, for example, produce synchrotron radiation as electrons spiral around the magnetic field lines.

The magnetic fields in galaxies are thought to arise from a dynamo effect: a small initial field, generated perhaps by turbulence, becomes amplified and ordered as it is wound up by the rotation of the galactic disk. Although this is a plausible model for the generation of the fields observed, there are some problems with it. For example, it is not clear whether typical spiral galaxies have experienced enough rotation in the age of the Universe for the fields to have been sufficiently amplified. Moreover, some objects at very high **redshifts**, such as the damped Lyman-alpha systems seen in **quasar** spectra, appear also to possess magnetic fields strong enough to produce a characteristic *Faraday rotation* of the polarisation of electromagnetic radiation passing through them. The detailed mechanism by which these astrophysical magnetic fields may have been generated has yet to be elucidated in a completely satisfactory fashion.

It has also been speculated that there might be a cosmological magnetic field pervading the Universe that could have been generated early on in the **thermal history of the Universe** as a result of primordial **phase transitions**. If such a field exists, it must be very weak. Since magnetic fields are vector fields, they possess direction as well as strength. The resulting pressure forces would have caused the Universe to expand more quickly in some directions than in others, so a large-scale cosmological field would produce an anisotropic **cosmological model**. This is one of the few situations where physically realistic exact solutions of the **Einstein equations** of general relativity can be obtained that do not invoke the **cosmological principle**. However, the observed near-isotropy of the **cosmic microwave background radiation** means that a large-scale coherent magnetic field would have to be very weak.

A cosmological magnetic field need not, however, be uniform: it might be tangled up on a relatively small scale so that the effects of its anisotropy are not evident on large scales. Even this kind of field would have to be very weak. A tangled web of field lines acts as a source of pressure which behaves in a very similar way to the pressure of **radiation**. Any such field present at the time of **nucleosynthesis**, for example, would alter the rate of **expansion of the Universe** at that epoch, and the observed **light element abundances** would no longer agree with our theoretical calculations. We can argue, though, that the observed agreement requires the cosmological magnetic field to contribute no more than one part in a million to the total energy density of the Universe.

Although there are thus tight limits on the strengths of galactic or cosmic magnetic fields, they could, in principle at least, oblige us to modify any models based on the assumption that gravity alone is relevant. For example, we usually estimate the amount of **dark matter** in galaxies by using the **virial theorem**. If galactic magnetic fields were sufficiently strong, they could significantly alter the form of the equilibrium configuration of galaxies by introducing a pressure force independent of the gas pressure and **gravity**.

FURTHER READING: Parker, E. N., *Cosmical Magnetic Fields* (Clarendon Press, Oxford, 1979).

MAGNITUDES Astronomers normally quantify the amount of light received from or emitted by celestial objects in terms of a rather archaic system called the *magnitude scale*. This scale is constructed around the ancient system of ranking stars according to whether they were 1st magnitude, 2nd magnitude, and so on down to 6th magnitude, the faintest stars visible to the naked eye. The eye's response to the flux of incoming light is essentially logarithmic, so this division basically depends on the logarithm of the apparent brightness of the star. In the 19th century, when alternative non-visual photometric methods became available (first using photographic emulsions and, more recently, bolometric detectors that measure directly the amount of energy received) it became clear that a more rigorously defined system of magnitudes was required.

The *apparent magnitude*, m, of an object is defined in such a way that m is larger the fainter the object. If two objects at the same distance have intrinsic luminosities L_1 and L_2 that differ by a factor of 100, then this is defined to correspond to a difference of 5 in their apparent magnitudes m_1 and m_2. Since the scale is logarithmic, this means that

$$m_2 - m_1 = 2.5 \log_{10}(L_1/L_2)$$

(note that the brighter object has the smaller value of m). This is fine for relative measures, but some absolute point has to be fixed. This is done by defining a zero point in such a way that the apparent magnitude of the Sun is $m = -26.85$. Cosmological sources are, of course, usually very much fainter than the faintest stars. The first large-scale surveys of galaxies went down to a limit of 14.5 in apparent magnitude. This has gradually crept down over the years, to the point where the **Hubble Space Telescope** now holds the record: it has surveyed galaxies in the so-called Hubble Deep Field with apparent magnitudes as faint as 28.5. The brightest and faintest astronomical objects we can detect therefore differ by over 55 magnitudes – a brightness factor of 10^{22}.

The *absolute magnitude*, *M*, of a source is defined by notionally placing the object at some standard distance. For historical reasons, a distance of 10 parsecs is chosen: the absolute magnitude is therefore equal to the apparent magnitude the source would have if it were at 10 parsecs. Since brightness falls off with distance as an inverse-square law (at least when the distances are small compared with cosmological scales – see **luminosity distance**), this means that

$$m - M = 5 \log(D/10)$$

where D is the distance in parsecs. The absolute magnitude of the Sun turns out to be 4.72, making it a fairly ordinary star. Note that the convention of placing objects at 10 parsecs is used even for sources like galaxies, which are far more distant than 10 parsecs. The absolute magnitude of typical bright galaxies is around -22 or so. On cosmological scales the distance D in the above expression is the luminosity distance, which has to take into account the redshifting of light and the effects of the **curvature of spacetime**.

The difference between apparent and absolute magnitudes for an object is called its *distance modulus*, μ, which is therefore a logarithmic measure of the distance to the object. If we know the absolute magnitude and can measure the apparent magnitude, then the distance modulus follows immediately. The distance modulus for galaxies must be calibrated in order to determine the **extragalactic distance scale**.

MALMQUIST BIAS Attempts to use properties of astronomical objects as probes of the geometry of the **Universe** are prone to numerous systematic biases and selection effects. One of the most important such effects is called the *Malmquist bias*, named after Gunnar Malmquist (1893–1982), and it arises in any astronomical situation where a statistical approach is used to investigate the properties of populations of objects.

To understand the problem, imagine that we have a sample of galaxies selected (as most such samples are) by a limiting apparent **magnitude**. If all galaxies had the same absolute magnitude then, in a Euclidean geometry, the apparent magnitude of a galaxy would be related to its distance in a straightforward way: the fainter the magnitude, the more distant the galaxy. If we had available an independent measure of the distance to the galaxy, then any deviation from the distance inferred from its luminosity or apparent magnitude (see **luminosity**

distance) would be attributable to the **expansion of the Universe** or the **curvature of spacetime** (see **classical cosmology**).

But galaxies are not standard objects – they display a large range of luminosities, as described by the **luminosity function**. This is not in itself much of a problem, because we can imagine taking averages of the luminosities of different galaxies rather than individual measurements, and using these averages to perform tests of spatial geometry or to construct the **extragalactic distance scale**. If there is a universal luminosity function, then the average luminosity is a kind of standard property. Why can we not, therefore, simply use averages?

The problem is that if we take a sample of galaxies as described above, it will have a cut-off in apparent magnitude. This is problematic because it means that, at large distances from the observer, only very bright galaxies appear in the sample. To put it another way, only the very bright end of the luminosity function is sampled at large distances or, in statistical jargon, the distribution of luminosities is said to be *censored*. Averages over censored distributions are very difficult to handle because we are effectively dealing with a separate class of objects. It is like trying to learn about the distribution of heights in a human population by doing statistics with a sample of basketball players. Censoring of the data resulting from the exclusion of low-luminosity objects has the effect of introducing a progressively upward bias into the average luminosity of the remaining objects as deeper samples are taken.

The effect is exacerbated if galaxies are clustered (which they are: see **large-scale structure**), because we then have to take into account the fact that galaxies are not spread out uniformly in distance from the observer, but tend to occur in clumps. This particularly nasty version of the effect is called *inhomogeneous Malmquist bias*, and it particularly plagues studies of **peculiar motions**.

MASSIVE COMPACT HALO OBJECT (MACHO) A comparison between the predictions of primordial **nucleosynthesis** with the observed **light element abundances** places a firm upper limit on the amount of matter than can exist within the **Universe** in the form of *baryons* (see **elementary particles**). In many respects the upper limit on the cosmological baryon density emerging from these considerations is very important. Observations of the amount of **dark matter** in **galaxies** and **large-scale structure** suggest that the value of the **density parameter** Ω is around 0.2 at least. This value is much larger than the baryon density permitted by nucleosynthesis, and has therefore led to the inferred presence of large quantities of dark matter in the form of non-baryonic particles: the **weakly interacting massive particles** or WIMPs.

But **nucleosynthesis** constrains the baryon density from below as well as from above. Direct calculations of the amount of visible matter (obtained by adding up all contributions from the **luminosity function** of galaxies) show that the luminous baryons fall well short of the lower limit imposed by the light element abundances. The inescapable conclusion, therefore, is that most of the baryons in the Universe are themselves dark.

The dark baryons assumed to be present in galaxies and other systems are generally thought to be in the form of *massive compact halo objects*, or MACHOs for short. This hypothetical component of the mass distribution is usually assumed to be in the form of low-luminosity stars (*brown dwarfs*) or massive Jupiter-like planets of about one-tenth of a solar mass. Stars any more massive than this would produce enough light to be seen directly because nuclear reactions would have initiated in their central regions. This does not happen for objects in the MACHO mass range. Such objects are therefore in principle difficult to see directly, but they are in principle detectable by their very weak **gravitational lensing** effect. The lensing effects of MACHOs (which are events of very low probability) are sought by continuously monitoring fields containing millions of stars in the Large Magellanic Cloud, a small satellite galaxy of the Milky Way. Lensing occurs as light travelling from one of these stars to us encounters a dark object in the halo of our Galaxy. Because the intervening objects have small masses and are moving across the line of sight to the star, they do not produce strong focusing effects as in other lensing systems, but merely cause a short-lived amplification of the light from the star. Dozens of *microlensing* events thought to be produced by MACHOs have been detected since the first (in 1993), though the interpretation of these data remains controversial. It is still unclear what the masses of the MACHOs should be or how they are distributed in the galactic halo.

SEE ALSO: Essay 6, **intergalactic medium**.

FURTHER READING: Riordan, M. and Schramm, D., *The Shadows of Creation: Dark Matter and the Structure of the Universe* (Oxford University Press, Oxford, 1993); Alcock, C. *et al.*, 'Possible gravitational microlensing of a star in the Large Magellanic Cloud', *Nature*, 1993, **365**, 621.

MATHER, JOHN CROMWELL (1946–) US space scientist. He led the team at the NASA Goddard Space Flight Center which planned the **Cosmic Background Explorer** (COBE) mission and, with George **Smoot**, he saw the project through to completion. He was responsible in particular for the experiments on board COBE which accurately measured the **black-body** spectrum of the **cosmic microwave background radiation**, and studied the **infrared background**.

MATTER ERA see **thermal history of the Universe**.

MAXWELL, JAMES CLERK (1831–79) Scottish physicist. He constructed the theory of electromagnetism (see **fundamental interactions**), and showed that light is a form of **electromagnetic radiation**. He also did fundamental work on statistical mechanics and **thermodynamics**.

MESON see **elementary particles**.

MESZAROS, PETER (1943–) Hungarian-born astrophysicist, based in the USA. Besides the eponymous **Meszaros effect**, he is known for his work on gamma-ray bursters (see **gamma-ray astronomy**).

MESZAROS EFFECT According to the standard theory of the gravitational **Jeans instability**, massive structures will eventually collapse once their self-gravity exceeds the restoring force provided by their internal pressure. When this happens to less massive structures, they do not collapse but instead perform acoustic oscillations. The dividing line between the two is called the *Jeans length*, which determines a characteristic size for structures that can form by condensing out of a cloud of gas or dust by gravitational processes.

The Jeans instability is the idea behind most theories of **structure formation** in cosmology, but there are several important additional physical processes which make it necessary to modify the standard theory. One such process, which is important in certain models, is called the *Meszaros effect*; sometimes the horrendous alternative term 'stagspansion' is used. This effect arises during the early stages of the evolution of density fluctuations during the radiation era (see **thermal history of the Universe**).

When the Universe was dominated by radiation, its expansion rate was different to the rate during the matter era. For example, in the simplest case of a **flat universe** model the scale factor increases with time to the power of two-thirds when it is matter-dominated, but only to the power of one-half when it is radiation-dominated. We can study the growth of fluctuations in the matter distribution in both these phases by using **perturbation theory**. It is found that fluctuations grow in simple proportion to the scale factor during matter domination, but only logarithmically with the scale factor during radiation domination. Thus, for most of the history of the Universe, fluctuations on small length scales grow much more slowly than we would expect from the standard Jeans instability calculation.

In practical terms this means that the shape of the **power spectrum** of **primordial density fluctuations** changes: small-scale fluctuations are suppressed, but larger ones are not. An initial power spectrum of a pure power-law shape, as is predicted in many models of the **inflationary Universe**, thus develops a bend in it. The wavelength at which this bend appears corresponds roughly to the scale of the cosmological horizon at the time when the Universe ceased to be radiation-dominated. The Meszaros effect is the main physical process that alters the shape of the initial power spectrum in the cold-dark-matter theory of cosmological structure formation.

FURTHER READING: Coles, P. and Lucchin, F., *Cosmology: The Origin and Evolution of Cosmic Structure* (John Wiley, Chichester, 1995), Chapter 11; Meszaros, P., 'The behaviour of point masses in an expanding cosmological substratum', *Astronomy and Astrophysics*, 1974, **37**, 225.

****METRIC** In the theory of **general relativity**, the **tensor** g_{ij} describing the geometry of four-dimensional **spacetime**, which is related to the properties of matter described by the energy–momentum tensor. The four-dimensional interval between two infinitesimally separated events can be written in the general form

$$ds^2(x) = g_{ij}(x)\,dx^i\,dx^j$$

where the repeated indices are summed using the summation convention for tensors. The coordinates x^j are written in this general way so as to include space and time in the same format: the index j runs from 0 to 3, so that 0 is the **time** coordinate and (1, 2, 3) are spatial coordinates. The metric contains all the information required to specify the geometrical properties of spacetime: ds^2 represents the spacetime interval between two points x^j and $x^j + dx^j$. The mathematical form of the metric depends on the choice of coordinates used to describe the space, but the geometrical structure is the same regardless of the coordinate system used: Cartesian, polar, cylindrical or whatever. This is demonstrated by the fact that the intervals are invariant: changing coordinates changes the labels on individual points in spacetime, but does not change the intervals ds between them. It may help to think of the equation defining the metric as a kind of tensor generalisation of Pythagoras's theorem in which the sum contains terms not just in the coordinates taken one at a time (i.e. x^2, y^2, z^2) but also terms like xy and yz.

If $ds^2 > 0$, the interval is said to be *time-like*; ds/c would then be the time interval measured by a clock moving freely between x^j and $x^j + dx^j$. On the other hand, if $ds^2 < 0$ the interval is *space-like*; the modulus of ds then represents the length of a ruler with ends at x^j and $x^j + dx^j$ as measured by an observer at rest with respect to the ruler. If $ds^2 = 0$, the interval is said to be *light-like* or *null*. This last type of interval is important because it means that the two points x^j and $x^j + dx^j$ can be connected by a light ray.

The metric can be very complicated in very inhomogeneous situations where general relativity is employed, and the choice of an appropriate coordinate system may become highly problematic. In cosmology, however, there is a relatively simple special form of the metric, compatible with the **cosmological principle**, where a natural choice of coordinates presents itself without difficulty: the **Robertson–Walker metric**.

SEE ALSO: **curvature of spacetime**, **tensor**.

FURTHER READING: Berry, M. V., *Principles of Cosmology and Gravitation* (Adam Hilger, Bristol, 1989); Schutz, B. F., *A First Course in General Relativity* (Cambridge University Press, Cambridge, 1985); Rindler, W., *Essential Relativity; Special, General, and Cosmological*, revised 2nd edition (Springer-Verlag, New York, 1979); Misner, C. W., Thorne, K. S. and Wheeler, J. A., *Gravitation* (W. H. Freeman, San Francisco, 1972); Weinberg, S., *Gravitation and Cosmology: Principles and Applications of the General Theory of Relativity* (John Wiley, New York, 1972).

MICHELSON–MORLEY EXPERIMENT see **aether**.

MICROLENSING see **gravitational lensing, massive compact halo object**.

MILNE, EDWARD ARTHUR (1896–1950) English astrophysicist and mathematician. He was one of the originators of the **cosmological principle**, and also worked on **stellar evolution**. He constructed an interesting

cosmological model based entirely on **special relativity**.

MINKOWSKI, HERMANN (1864–1909) German mathematician. A former teacher of Albert **Einstein**, he proposed the introduction of four-dimensional **spacetime** into the formalism of **special relativity**. Einstein resisted the idea at first, but it later provided him with the methods required to build his theory of **general relativity**.

MISSING MASS see **dark matter**.

N

* *N*-BODY SIMULATIONS Standard theories of cosmological **structure formation** embody the idea of **gravitational instability** – the simple fact that gravity, being an attractive force, tends to make fluctuations in density grow. The theory of this **Jeans instability** for small fluctuations is well-established, but it is not possible to obtain exact mathematical solutions of this theory except in very simple cases (such as, for example, a perfectly spherical fluctuation). When the fluctuations are small, however, a general solution of the problem can be found using techniques based on **perturbation theory**. The early stages of the evolution of structure, the so-called *linear regime*, can be solved quite accurately by these methods. However, the complexity of the physical behaviour of fluctuations in the *nonlinear regime* makes it impossible to study the details of the later stages exactly using exact analytical methods. For this task we turn to numerical simulation methods. The situation here is rather similar to that of weather forecasting, in which, though the basic physical principles governing the behaviour of the atmosphere seem to be reasonably well known, it is impossible to predict the weather at all except by using a powerful computer to simulate what is going on. Experience shows, however, that even with the most powerful computers in the world, the results are not all that reliable.

In cosmology, the use of numerical simulations is widespread, even forming part of the international 'grand challenge' laid down to computer experts as a test of their abilities. The idea behind these methods is to represent part of the expanding Universe as a box containing a large number *N* of point masses interacting through their mutual **gravity**. This box, typically a cube, must be at least as large as the scale at which the Universe becomes homogeneous if it is to provide a 'fair sample' which is representative of the Universe as a whole (see **large-scale structure**, **cosmological principle**). A common trick is to assume that the cube has periodic boundary conditions in all directions; in practice, this has the effect that a particle moving out through one face of the box reappears through the opposite face. This assists in some of the computational techniques by allowing very efficient methods based on the *Fourier transform* to be employed in summing the *N*-body forces. A number of numerical techniques are used, differing for the most part only in the way the forces on each particle are calculated.

The simplest way to compute the nonlinear evolution of a cosmological fluid is to represent it as a discrete set of particles, and then sum the (pairwise) interactions between them directly to calculate the Newtonian forces of gravity between every pair. Such calculations are often called *particle–particle* or PP computations or, more usually, *direct summation*. With the adoption of a (small) time-step, the net acceleration resulting from the force on each particle can be used to update its velocity. The particle can then be moved to a new position in the box. New positions are then calculated for all the particles, and used to recalculate the inter-particle forces, and so on. (These techniques are not intended to represent the motion of a discrete set of particles: the particle configuration is an approximation to a continuous fluid.)

There is a numerical problem with the summation of the gravitational forces that are acting between particles: the Newtonian gravitational force between two particles increases as the particles approach each other according to the inverse-square law of Newtonian gravity. It is therefore necessary to choose an extremely small timestep to resolve the large velocity changes this induces. A very small timestep would require the consumption of enormous amounts of computer time and, in any case, computers cannot handle the divergent force terms when the particles are arbitrarily close to each other. These problems are usually avoided by treating each particle, not as a point mass, but as an extended body. The effect of this is to modify the Newtonian force between particles by replacing Newton's $1/r^2$ law by a law in which the dependence on r is of the form $1/(\epsilon^2 + r^2)$, where ϵ is called the *softening length*. The form of this modified force law avoids producing infinite forces at zero separations. This is equivalent to replacing point masses by extended bodies with a size of the order of ϵ. Since we are not supposed to be dealing with the behaviour of a set of point masses anyway, the introduction of a softening length is quite reasonable, but it does mean that we cannot trust results obtained for scales of the order of ϵ or less.

Our simulation, then, contains N particles. At each timestep every particle is acted on by all other $N - 1$ particles in the box. This requires a total of $\frac{1}{2}N(N - 1)$ evaluations of the Newtonian gravitational force at each timestep. (This is like the problem of having N people in a room, each shaking hands once with everyone else: the total number of handshakes is $\frac{1}{2}N(N - 1)$.) This is the crucial limitation of these methods: they tend to be very slow for large numbers of particles, with the required computational time scaling roughly as N^2. The maximum number of particles for which it is practical to use direct summation is of the order of 10^5, which is not sufficient for realistic simulations of large-scale structure formation.

The usual method for improving on direct N-body summation for computing interparticle forces is some form of *particle–mesh* scheme, usually abbreviated to PM. In such a scheme the forces are solved by assigning mass points to a regular grid and then solving *Poisson's equation* on it (see **gravity**). The use of a regular grid with periodic boundary conditions (as mentioned above) makes it possible to use *fast Fourier transform* (FFT) methods to solve for the gravitational potential, which speeds up the process considerably. The FFT algorithm, a mainstay of most signal processing methods, is much faster than direct summation: if there are N particles, the time taken is $N \log N$ – a substantial improvement for large N over the direct summation technique. The price to be paid for this reduction in computation time is that the Fourier summation method implicitly requires the simulation box to have periodic boundary conditions. However, since this is probably the most reasonable way of simulating a 'representative' part of the Universe, it does not seem too high a price to pay for the increased speed.

The potential weakness with this method is the comparatively poor force resolution on small scales because of the finite spatial size of the computational mesh. A substantial increase in spatial resolution can be achieved by using instead a hybrid *particle–particle–particle–mesh* method, which solves the short-range forces directly (PP) but uses the mesh to compute those of longer range (PM); hence such codes are usually known by the initials P3M (PP + PM). Here, the short-range resolution of the algorithm is improved upon by adding a correction to the mesh force. This contribution is obtained by summing directly all the forces from neighbours within some fixed distance R of each particle. Alternatively, we can use a modified force law on these small scales to assign any particular density profile to the particles, similar to the softening procedure mentioned previously. This part of the force calculation may well be quite slow, so it is advantageous merely to calculate the short-range force at the start for a large number of points spaced linearly in radius, and then find the actual force by simple interpolation. The long-range part of the force calculation is done by a variant of the PM method described earlier.

Variants of the PM and P3M techniques are now the standard workhorses for cosmological clustering studies. Different research groups have slightly different interpolation schemes and choices of softening length. Which of the two techniques should be used depends in general on the degree of clustering it is wished to probe. Strongly nonlinear clustering in dense environments probably requires the force resolution of P3M. For analysing structure on a larger scale, where no attempt is made to probe the inner structure of highly condensed objects, PM is probably good enough. It should, however, be recognised that the short-range forces are not computed exactly, even in P3M, so the apparent extra resolution may not necessarily be imparting anything of physical significance.

An alternative procedure for enhancing the force resolution of a particle code while keeping the demand on computational time within reasonable limits is to adopt a hierarchical subdivision procedure. The generic name given to this kind of technique is *tree code*. The basic idea is to treat distant clumps of particles as single massive 'pseudo-particles'. The usual algorithm employs a mesh which is divided into cells hierarchically in such a way that every cell which contains more than one particle is divided into eight sub-cells: its dimensions are halved in every direction. If any of the resulting sub-cells contains more than one particle, that cell is sub-divided again. There are some subtleties involved with communicating particle positions up and down the resulting 'tree', but it is basically quite straightforward to treat the distant forces using the coarsely grained distribution contained in the high level of the tree, while short-range forces use the finer grid. The greatest problem with such codes is that, though they run quite quickly in comparison with particle–mesh methods with the same resolution, they do require considerable memory resources. Their use in cosmological contexts has so far therefore been quite limited, one of the problems being the difficulty of implementing periodic boundary conditions in such algorithms.

These codes deal only with the gravitational instability of matter. The full problem of structure formation also requires the much more complex behaviour of gas and radiation to be addressed. The incorporation of hydrodynamical processes into cosmological simulations is one area of intense research interest. This has led, for example, to refinements of the particle codes mentioned above into hydrodynamical methods called *smoothed particle hydrodynamics* (SPH), and so-called *Eulerian schemes* which are based on techniques borrowed from aeronautical engineering. These methods are the state-of-the-art prototypes in cosmological numerical research, and are therefore not fully tested as yet.

Whichever code is used, we need to be able to set up the initial conditions for a numerical simulation in a manner appropriate to the cosmological scenario under consideration. For most models this requires a random-phase realisation of the **power spectrum** (see primordial **density fluctuations**). This is usually achieved by setting up particles in such a way that they lie exactly on the grid positions, then moving them slightly so as to create a density field with the required statistical properties. The initial particle velocities are set up in a similar manner.

The field of numerical simulations is one in which there is enormous activity, with the development of international teams, such as the UK-led Virgo consortium, using the world's most powerful supercomputers. An example of a cosmological simulation performed by a group at Los Alamos is shown in the Figure overleaf, in which the gradual development of structure according to the theory of gravitational instability can be clearly seen. Cosmologists are continually running such simulations for different models of structure formation and comparing them with observations of the large-scale structure of the galaxy distribution in order to test models.

FURTHER READING: Hockney, R. W. and Eastwood, J. W., *Computer Simulation Using Particles* (Adam Hilger, Bristol, 1988); Makino, J. and Taiji, M., *Scientific Simulations with Special-Purpose Computers* (John Wiley, Chichester, 1998); Bertschinger, E. and Gelb, J. M., 'Cosmological *N*-body simulations', *Computers in Physics*, 1991, **5**, 164.

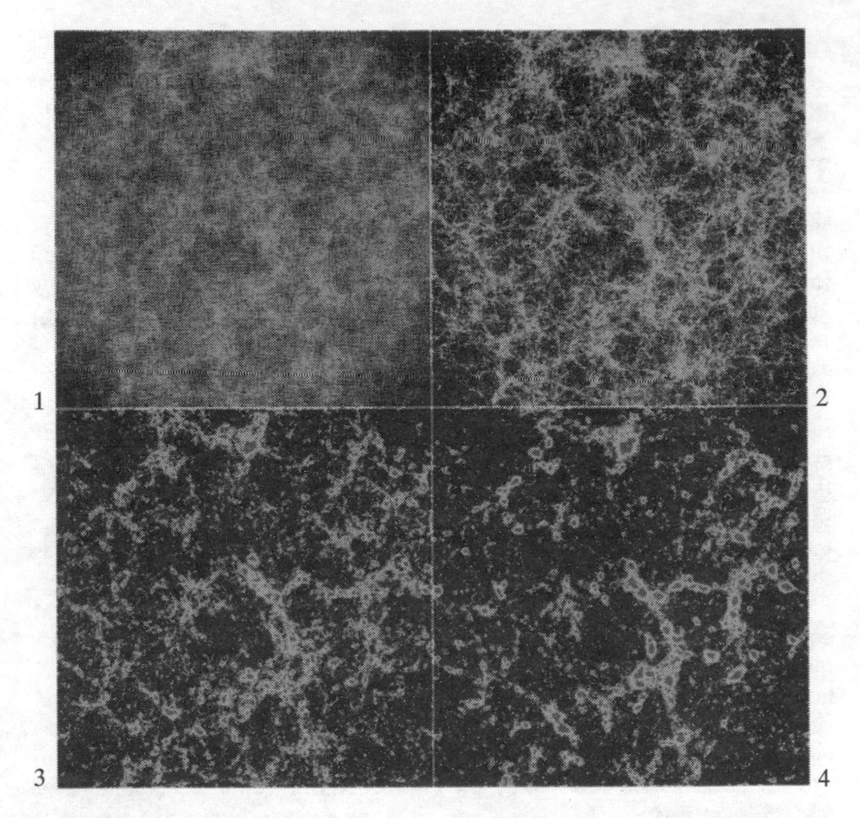

***N*-body simulations** Example of an *N*-body computation of clustering in an expanding Universe: the sequence 1–4 shows how matter gets more and more clumpy as time goes on.

Neutrino see **elementary particles**.

Neutron see **elementary particles**.

Newton, Sir Isaac (1642–1727) English physicist and mathematician. He introduced into science the modern concept of the quantitative law of physics. His monumental book *Principia*, first published in 1687, dominated the subject of natural philosophy (as physics was then called) for two centuries. In it, Newton laid down the principles of the science of mechanics and the law of universal gravitation, setting out his laws of motion and deriving Johannes **Kepler**'s laws of planetary motion.

No-boundary hypothesis see **quantum cosmology**.

No-hair theorems see **black hole**.

Nucleosynthesis The process by which complex atomic nuclei are made from **elementary particles**. In cosmology, it refers specifically to the process by which light atomic nuclei are created from protons and neutrons (baryons) in the first few minutes of the **thermal history of the Universe** according to the **Big Bang theory**. By 'light' is meant no heavier or more complex than lithium, for reasons explained below.

The observed abundances of the relevant elements are discussed in detail elsewhere (see **light element abundances**); here it is helpful to start with some ball-park figures for the most important nuclei. Roughly speaking, the abundance by mass of helium-4 (which is made of two protons and two neutrons), usually denoted by the symbol Y, is about 25%. This corresponds to about 6% of all atomic nuclei in the Universe, the helium nucleus having about four times the mass of the hydrogen nucleus. The abundance of the light isotope helium-3 (two protons, but only one neutron) is a few times 10^{5}. The heavy isotope of hydrogen, usually called *deuterium* (one proton and one neutron) is of the same order as that of helium-3. The abundance of lithium-7 is even smaller: around one part in ten billion. The rest of the material in the Universe is basically hydrogen (one proton and no neutrons), not counting the potentially large amounts of non-baryonic **dark matter** (which do not play any role in nucleosynthesis and which are therefore ignored in this discussion).

In the present matter era, nucleosynthesis occurs predominantly in stellar interiors during the course of **stellar evolution**. Stellar processes, however, generally destroy deuterium more quickly than it is produced, because the strength of the **electromagnetic radiation** present in stars causes deuterium to photodissociate into its component protons and neutrons. Nuclei heavier than lithium-7, on the other hand, are essentially made only in stars. In fact there are no stable nuclei with atomic weight 5 or 8, so it is difficult to construct elements heavier than helium by adding helium nuclei (also known as alpha-particles) to other helium nuclei or to protons. In stars, however, collisions between alpha-particles do produce small amounts of unstable beryllium-8 from which carbon-12 can be made by adding another alpha-particle; a chain of synthesis reactions can therefore develop, leading to elements heavier than carbon. In the cosmological context, at the temperature of a billion degrees characteristic of the onset of nucleosynthesis, the density of the Universe is too low to permit the synthesis of significant amounts of carbon-12 in this way (the density of matter at this time being roughly that of water). It is clear, therefore, that the elements heavier than helium-4 are largely made in stellar interiors. On the other hand, the percentage of helium observed is too high to be explained by the usual predictions of stellar evolution. For example, if our Galaxy maintained a constant luminosity (through continual nucleosynthesis in stars) for about 10 billion years, the total energy radiated would correspond to the fusion of 1% of the original hydrogen into helium, in contrast to the 6% that is observed.

It is interesting that the difficulty of explaining the nucleosynthesis of helium by stellar processes alone was recognised as early as the 1940s by Ralph **Alpher**, Hans Bethe and George **Gamow**, who themselves proposed a model of cosmological nucleosynthesis. Difficulties with this model, in particular an excessive production of helium, persuaded Alpher and Robert **Herman** in 1948 to consider the idea that there might have been a significant radiation background at the epoch of nucleosynthesis. They estimated that this background should have a present temperature of around 5 K, not far from the value it is now known to have (2.73 K).

The calculation of the proportions of light nuclei produced in the primordial fireball requires a few assumptions to be made about some of the Universe's properties at the relevant stage of its evolution. In addition to the normal assumptions going into the **Friedmann models**, it is necessary to restrict the number of possible species of neutrino (see **elementary particles**) to no more than three, and to assume that there are no additional sources of pressure (such as **magnetic fields**). Most important, however, is the assumption that the Universe was in a state of **thermal equilibrium** early on, at temperatures of more than a billion degrees.

Before nucleosynthesis began, protons and neutrons would have been continually interconverting via weak nuclear interactions (see **fundamental interactions**). The relative numbers of protons and neutrons can be calculated on the assumption that they were in thermal equilibrium and, while the weak interactions were fast enough to maintain equilibrium, the neutron/proton ratio would have been continually adjusting itself to the cooling surroundings. At some critical point, however, the weak nuclear reactions became inefficient and the ratio could no longer adjust. What happened then was that the neutron/proton ratio became 'frozen out' at a particular value (about one neutron for every six protons). This ratio is fundamental in determining the eventual abundance of helium-4. But there was an obstacle to this, which meant that nucleosynthesis had to pause for several seconds.

To make helium by adding protons and neutrons together, it is first necessary to make deuterium. But deuterium is easily disrupted by radiation: if it gets hit by a photon, it splits into a proton and a neutron. As soon as any deuterium is made, it is destroyed again. This delay is called the *deuterium bottleneck*. While this nuclear traffic-jam persisted, no helium could be made. Moreover, the neutrons that froze out before this happened could spontaneously decay (with a half-life of about 10 minutes) into a proton, an electron and a neutrino. The result of the delay was thus that slightly fewer neutrons were available for the subsequent cooking of helium.

When the temperature of the radiation bath fell below a billion degrees, the radiation was no longer strong enough to dissociate deuterium, which lingered long enough for further reactions to occur. Two deuterium nuclei can weld together to make helium-3, with the ejection of a neutron. Helium-3 can capture a deuterium nucleus, making helium-4 and ejecting a proton. These two reactions happened very quickly, with the result that virtually all neutrons ended up in helium-4, and only traces of the intermediate deuterium and helium-3 survived. The abundance by mass of helium-4 that comes out naturally is about 25%, just as required. Likewise, the amounts of intermediate nuclei are also close to the observations. The Figure shows an example of a detailed computation of these abundances.

The apparent agreement between theory and observation is not, however, the whole story. While the rough figures match very well, the exact abundances of the light elements produced depend in a complicated way on the total amount of baryonic matter. This is usually expressed as the corresponding value of the **density parameter** in baryons, Ω_b. (Non-baryonic dark matter does not participate in nuclear reactions; the total value of Ω can therefore be larger than Ω_b.) Increasing Ω_b tends to increase the amount of helium-4,

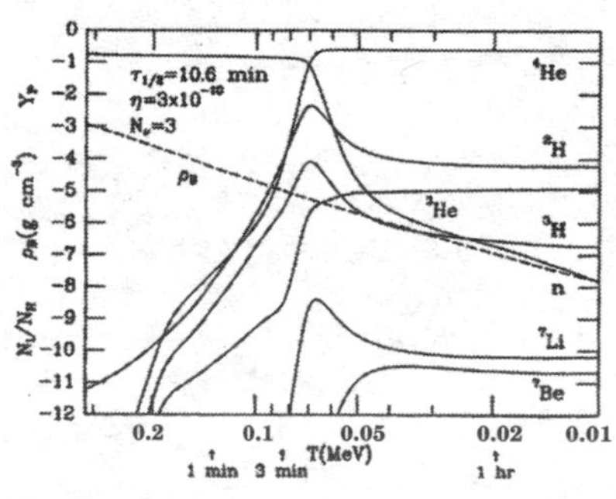

Nucleosynthesis During the first few minutes of the Big Bang, the abundances of various light isotopes built up through a series of nuclear reactions to values that match present-day observations fairly well.

but only slightly. Increasing Ω_b, however, also tends to reduce drastically the amount of deuterium and helium-3. The variation of lithium-7 with the baryon density is more complicated, principally because it has two possible formation mechanisms: direct formation via the fusion of helium-3 and helium-4 if the density is low, and electron capture by beryllium-7 if the density is high. At intermediate densities the production of lithium-7 is slightly suppressed. The fact that the abundances depend on Ω_b in different ways suggests that measuring two of them independently should provide a strong test of the theory: observations of both must match the predictions of the theory for one particular value of the baryon density. Recent results indicate that this is difficult to accomplish, but that the favoured value for Ω_b is probably no more than 10%. (For more details of this comparison, see **light element abundances**.)

While these theoretical calculations do seem to account reasonably well for the observations, this fact does not in itself rule out possible alternative theories of nucleosynthesis. For example, **cosmological models** which are based on non-standard theories of **gravity** predict different expansion rates and therefore different freeze-out values of the neutron/proton ratio. Likewise, the presence of primordial magnetic fields or degenerate neutrinos (see **elementary particles**) might alter the predictions. The attitude of most cosmologists, however, is that if it works it must be right, and the agreement between theory and observations within the standard model is usually taken to rule out novel physics and non-standard theories. However, we should always keep an open mind about alternative theories, particularly since they are by no means completely excluded by observations.

FURTHER READING: Weinberg, S., *The First Three Minutes: A Modern View of the Origin of the Universe* (Fontana, London 1983).

NUMBER COUNTS see **source counts**.

O

Observable Universe see **horizon, Universe**.

Olbers' paradox One of the most basic astronomical observations we can make, just with the unaided eye, is to note that the night sky is dark. This fact is so familiar to us that we do not stop to think that it might be difficult to explain, or that anything important can be deduced from it. But quite the reverse is true. The observed darkness of the sky at night was long regarded by many outstanding intellects as a paradox that defied explanation – the so-called *Olbers' paradox*.

The starting point from which this paradox is developed is the assumption that our Universe is static, infinite, homogeneous and Euclidean. Before 20th-century developments in observation (**Hubble's law**) and theory (**cosmological models** based on **general relativity**), these assumptions would all have appeared quite reasonable to most scientists. In such a universe, the intensity of light received by an observer from a source falls off as the inverse square of the distance between the two. Consequently, more distant stars or galaxies appear fainter than nearby ones. A star infinitely far away would appear infinitely faint, which suggests that Olbers' paradox is resolved by the fact that distant stars (or galaxies) are simply too faint to be seen. But we have to be careful here. Imagine, for argument's sake, that all stars shine with the same brightness. Now divide the Universe into a series of narrow concentric spherical shells, something like an onion. The light from each source within a shell of radius r falls off as the square of r, but the number of sources increases in the same manner. The observer therefore receives the same amount of light from each shell, regardless of the value of r. An infinite universe contains an infinite number of shells and therefore produces an infinite answer. The brightness is not going to be infinite in practice because nearby stars will block out some of the light from stars beyond them. But in any case the sky should be as bright as the surface of a star like the Sun. This is emphatically not what is observed. It might help to think of this in another way, by imagining yourself in a very large forest. You may be able to see some way through the gaps between nearby trees, but if the forest is infinite every possible line of sight will end with a tree.

As is the case with many other famous eponyms, this puzzle was not actually first discussed by the man whose name is now attached to it: Heinrich Wilhelm Matthäus Olbers (1758–1840). His discussion was published in 1826, but Thomas **Digges** struggled with this problem as early as 1576. At that time, however, the mathematical technique of adding up the light from an infinite set of narrow shells, which relies on the differential calculus, was not known. Digges therefore simply concluded that distant sources must just be too faint to be seen, and did not worry about the problem of the number of sources. Johannes **Kepler** was also interested in this problem, and in 1610 he suggested that the Universe must be finite in spatial extent. Edmond Halley (of cometary fame) also addressed this issue about a century later, in 1720, but did not make significant progress. The first discussion that would nowadays be regarded as a correct formulation of the problem was published in 1744, by Philippe de Chéseaux. Unfortunately, his solution was not correct: he imagined that the intervening space somehow absorbed the energy carried by light on its path from source to observer. Olbers himself came to a similar conclusion.

Later students of this conundrum included Lord Kelvin, who speculated that the extra light is absorbed by dust. This is no solution to the problem either because, while dust may initially simply absorb optical light, it would soon heat up and re-radiate the energy at infrared wavelengths, and there would still be a problem with the total amount of **electromagnetic radiation** reaching an observer. To be fair to Kelvin, however, at the time of his speculation it was not known that heat and light were both forms of the same kind of energy, and neither was it obvious that they could be interconverted in this way. To show how widely Olbers' paradox was known in the 19th century, it is worth mentioning that Friedrich Engels, Manchester factory owner and co-author with Karl Marx of the *Communist Manifesto*, also considered it in his book *The Dialectics of Nature*. In this discussion he singles out Kelvin for particular criticism, mainly for the reason that Kelvin was a member of the aristocracy.

Probably the first inklings of a correct resolution of Olbers' paradox were contained not in a dry scientific paper, but in a prose poem entitled *Eureka* published in 1848 by Edgar Allan Poe. Poe's astonishingly prescient argument is based on the realisation that light travels at a finite speed. This in itself was not a new idea, the first calculation of c having been made by Ole Römer almost two centuries earlier. But Poe appreciated that light just arriving from distant sources must have set out a very long time in the past. In order to receive light from them now, therefore, they had to have been burning in the distant past. If the Universe has only lasted for a finite time, then shells cannot continue to be added out to infinite distances, but only as far as the distance given by the speed of light multiplied by the **age of the Universe** (ct). In the days before scientific **cosmology**, many believed that the Universe had to be very young: the Biblical account of the Creation made it only a few thousand years old, so the problem simply did not arise.

Of course, we are now familiar with the ideas that the Universe is expanding (and that light is consequently redshifted), that it may not be infinite and that space may not be Euclidean.

All these factors have to be taken into account when the brightness of the sky is calculated in different cosmological models. The fundamental reason why Olbers' paradox is not a paradox is the finite lifetime, not necessarily of the Universe, but of the structures that can produce light. According to **special relativity**, mass and energy are equivalent. If the density of matter is finite, then so is the amount of energy it can produce by nuclear reactions. Any object that burns matter to produce light can therefore only burn for a finite time before it fizzles out. Moreover, according to the **Big Bang theory** all matter was created at a finite time in the past anyway, so Olbers' paradox thus receives a decisive knockout combination.

Although Olbers' paradox no longer stands as a paradox, the ideas behind it still form the basis of important cosmological tests. The brightness of the night sky may no longer be feared infinite, but there is still expected to be a measurable glow of background light produced by distant sources too faint to be seen individually. In principle, in a given cosmological model and given certain assumptions about how **structure formation** proceeded, we can calculate the integrated flux of light from all the sources that can be observed at the present time, taking into account the effects of redshift, spatial geometry, and what we know of the formation and evolution of various sources. Once this is done, we can compare predicted light levels with observational limits on the background glow, which are now quite well-known in certain wavebands.

FURTHER READING: Harrison, E., *Darkness at Night: A Riddle of the Universe* (Harvard University Press, Cambridge, MA, 1987).

OMEGA (Ω) see **density parameter**.

OPEN UNIVERSE Any **cosmological model** in which the **curvature of spacetime** is negative. In such a universe the normal rules of Euclidean geometry do not necessarily hold. For example, the sum of the interior angles of a triangle is less than 180°, and parallel lines actually diverge from each other. The **Friedmann models** that describe open universes are those in which the **density parameter** $\Omega < 1$ and the **deceleration parameter** $q < \frac{1}{2}$. These models are also infinite in spatial extent. They also never recollapse in the future, for the deceleration generated by **gravity** is not sufficient to cause the **expansion of the Universe** to cease; they therefore expand for ever, with ever-decreasing density. Broadly speaking, present determinations of the density parameter suggest that the Friedmann models which best fit the data best correspond to an **open universe**.

FURTHER READING: Coles, P. and Ellis, G. F. R., 'The case for an open universe', *Nature*, 1994, **370**, 609.

OPTICAL ASTRONOMY The branch of astronomy that concerns itself with observations made in the *optical* (or *visible*) part of the spectrum of **electromagnetic radiation** at wavelengths in the very narrow wavelength region between about 400 nm (violet) and 800 nm (red). Until the mid-20th century, all astronomical observations were performed in the optical waveband. Despite this waveband being quite narrow, optical observations still play a major role in astronomy in general, and cosmology in particular.

In the early 17th century the invention of the optical telescope led to a huge increase in the number of objects that could be observed and in the resolution with which they could be seen. Since **Galileo** first pointed his telescope at the planets, telescopes have gradually grown in size and sophistication. Nevertheless, for over two centuries the kinds of observation that were possible were relatively crude. Positions on the sky could be recorded, and visual appearances sketched by hand. The development of photographic plates in the mid-19th century allowed images to be taken over long exposures that could reveal details invisible to the human eye, even with the aid of the largest telescopes.

The character of astronomy changed dramatically in the 19th century with the advent of **spectroscopy**. Optical spectroscopy, first of the Sun by Joseph von Fraunhofer, and then of more distant stars, allowed astronomers to study the physics and chemistry of astronomical bodies, and gave birth to the science of astrophysics.

Observational cosmology began as a branch of optical astronomy in the early decades of the 20th century, when Vesto **Slipher** and Edwin **Hubble** showed that the spiral 'nebulae' were extragalactic. Hubble himself then made spectroscopic observations of these nebulae which showed them to be other galaxies receding from us. This established the **expansion of the Universe** and prepared the ground for the emergence of the **Big Bang theory** and all the cosmological developments that have followed from it.

Since the Second World War, astronomers have been able to explore a much larger part of the spectrum of electromagnetic radiation (see **radio astronomy**, **infrared astronomy**, **ultraviolet astronomy, X-ray astronomy, gamma-ray astronomy**), but most astronomical observations are still made in the optical part of the spectrum. As these new wavebands have begun to be explored, new techniques have been devised for optical observations (including, for example, optical interferometry). Optical telescopes have also been getting bigger and bigger, with 10-metre reflecting telescopes now in operation. The **Hubble Space Telescope** also does much of its work in the optical waveband.

The reason for this continued attention to quite a narrow part of the electromagnetic spectrum is that there is so much interesting physics connected with the behaviour of stars, and their stellar evolution can be probed using observations in visible light. This is the part of the spectrum our eyes are adapted to see, because the Sun is responsible for our very existence.

SEE ALSO: Essay 4.

FURTHER READING: Florence, R., *The Perfect Machine: Building the Palomar Telescope* (HarperCollins, New York, 1994); Graham-Smith, F. and Lovell, B., *Pathways to the Universe* (Cambridge University Press, Cambridge, 1988); Hubble, E., *The Realm of the Nebulae* (Yale University Press, Newhaven, CT,

1936); Preston, R., *First Light: The Search for the Edge of the Universe* (Random House, New York, 1996); Tucker, W. and Tucker, K., *The Cosmic Inquirers: Modern Telescopes and Their Makers* (Harvard University Press, Harvard, 1986).

Oscillating universe see **closed universe**.

P

Pauli, Wolfgang (1900–1958) Austrian–Swiss–US physicist. He discovered the exclusion principle of **quantum mechanics**, for which he was awarded the 1945 Nobel Prize for Physics, and which is important in providing the pressure forces that maintain equilibrium in degenerate stars (see **stellar evolution**).

Parallel universes see **quantum physics**.

Payne-Gaposchkin, Cecilia Helena (1900–1979) English astronomer, who from 1922 worked at Harvard University. In her thesis of 1925 she examined the **light element abundances** of elements in stars of various ages and throughout the Universe, and established that hydrogen is the major constituent of stars.

Peculiar motions The large-scale motions of **galaxies** are described in broad terms by **Hubble's law**. The pattern of motions resulting from this law is often called the *Hubble flow*: all observers see themselves as the centre of expansion, so that all motions are radially outward, and galaxies are moving away at a speed proportional to their distance. This behaviour is readily explained in the standard **cosmological models** as being a consequence of the homogeneity and isotropy of the **Universe** as embodied in the **cosmological principle** (see also **expansion of the Universe**).

However, the real pattern of galaxy motions is not exactly of this form. This is because, while the cosmological principle might hold in a broad sense on large scales, the Universe is not exactly homogenous: it contains galaxies distributed in a complicated hierarchy of **large-scale structure**. Local fluctuations in the density from the uniformity required of a completely homogeneous Universe give rise to fluctuations in the local gravitational field from place to place. These gravitational fluctuations tend to deflect galaxies from the paths they would follow in a pure Hubble flow.

These departures from the pure Hubble expansion are usually called *peculiar motions*, though there is nothing particularly peculiar about them: their occurrence is entirely expected given the observed inhomogeneous nature of the real Universe. Peculiar motions modify the form of Hubble's law by adding a term to the right-hand side:

$$v = H_0 d + v_p$$

where v_p is that component of the peculiar motion that lies in the line-of-sight direction from the galaxy to the observer. Unlike the pure Hubble expansion, which is radial, the peculiar motions are generally randomly directed in space. Since the total velocity v is inferred from the **redshift**, however, only the part that lies in the radial direction can be detected directly from the spectrum. The typical peculiar motions of galaxies are several hundred kilometres per second, so we can see from the above equation that peculiar motions can swamp the Hubble flow entirely for nearby objects. Some nearby galaxies, such as the Andromeda Galaxy, are even moving *towards* the Milky Way. At large distances, however, the discrepancies are very small compared with the Hubble flow.

Peculiar motions are one cause of the observed scatter in the Hubble diagram of velocity against distance. This scatter might suggest that their presence is merely an irritation, but peculiar motions are very important because they raise the possibility of measuring the amount of matter responsible for their generation (see **dark matter**). This is possible because the size of the peculiar motion depends on the density field, and is not simply a random error. Basically, the more matter there is in the Universe – in other words, the higher the value of the **density parameter** Ω – the larger should be the size of any peculiar motions.

There are two basic ways to study peculiar motions. In the first, and simplest, they are measured not directly but by using a **redshift survey** of a large number of galaxies to make a map in 'redshift space'. In other words, we measure the redshift and hence the total velocity of a sample of galaxies, and then assume that Hubble's law holds exactly. The presence of peculiar motions will distort the map, because velocity is not simply proportional to distance, but also depends to some extent on the matter density. This manifests itself in two ways. In very dense regions of strong gravitational forces, the peculiar velocities are large and random (see **virial theorem**). All the galaxies are in a small volume in real space, but because of their huge peculiar motions they are spread out in redshift space. What we see in a redshift survey are therefore not near-spherical blobs (which is what clusters really are), but 'fingers' stretched along the line-of-sight to the observer. These features are called, somewhat irreverently, the *fingers of God*. In more extended systems such as superclusters, the peculiar motions are not so large but they are discernible because they are coherent. Imagine a spherical supercluster which is gradually collapsing. Consider what happens to objects on the edge of the structure: galaxies on the far side of the structure will be falling towards the observer, while those on the near side will be falling away from the observer. This squashes the structure in redshift space compared with what it is like in real space. We can use statistical arguments to quantify the distortions present in redshift-space maps, and hence attempt to work out how much mass there is causing them.

The second way of studying peculiar motions is to attempt to measure them directly. This requires us to measure the distance to the galaxy directly (thus introducing all the difficulties inherent in the **extragalactic distance scale**), and then to subtract the Hubble flow to obtain the peculiar velocity v_p. This technique has been used for many years to map the flow of relatively nearby galaxies, and some interesting features have emerged. The Local Group of galaxies, for example, is moving towards the centre of the Virgo Cluster of galaxies at about 200 km/s; this is essentially the kind of infall motion discussed above. On larger scales, coherent peculiar motions are expected to be smaller than this, but there is evidence of flows on quite large scales. For example, a study of the motions of a sample of spiral galaxies by Vera **Rubin** and co-workers in 1976 revealed an apparent anisotropy in the expansion of the Universe on a scale of around 100 million light years. This has become known as the *Rubin–Ford effect*.

The fact that we do not see a purely isotropic expansion reflects the fact that the Universe is not homogeneous on these scales. It has been claimed that there is motion on scales much larger than this. Astronomers have postulated a *Great Attractor* – a hypothetical concentration of matter with a mass of more than 10^{16} solar masses, located about 150 million light years from our Galaxy in the direction of the borders of the constellations Hydra and Centaurus – which may be pulling surrounding galaxies, including the Milky Way, into itself. Although there is clearly a concentration of galaxies at the place where the Great Attractor has to be located to account for these large-scale galaxy motions, more recent studies indicate that no single object is responsible, and that the observed bulk flows of galaxies are probably caused by the concerted gravitational effect of several distinct clusters of galaxies.

An alternative approach to the study of peculiar motions is to look at the properties of the dipole anisotropy of the **cosmic microwave background radiation**. Because of our motion with respect to the **frame of reference** in which this radiation is isotropic, we see a characteristic cosine variation in the temperature on the sky. The size of this variation is roughly v/c, because of the **Doppler effect**, and the direction of maximum temperature gives the direction of our motion. The microwave background therefore supplies us with the best measured peculiar motion in cosmology: our own!

SEE ALSO: Essays 2 and 5.

FURTHER READING: Burstein, D., 'Large-scale motions in the Universe: A review', *Reports on Progress in Physics*, 1990, **53**, 421.

PEEBLES, PHILLIP JAMES EDWIN (1935–) Canadian-born cosmologist, who was a student of Robert **Dicke** at Princeton during the 1960s and has remained there ever since. He worked on the interpretation of the **cosmic microwave background radiation** discovered by Arno **Penzias** and Robert **Wilson**, and went on to pioneer the theoretical study of cosmological **structure formation**.

PENROSE, ROGER SIR (1931–) English mathematician. He has a broad span of interests, but is best known for his work on the **singularity** theorems for **black holes** and, later with Stephen **Hawking**, the singularity theorems in cosmology.

PENZIAS, ARNO ALLAN (1933–) German-born US physicist. With

Robert **Wilson** he accidentally discovered the **cosmic microwave background radiation** while working with an antenna designed for use with communications satellites. The two men shared the 1978 Nobel Prize for Physics.

* **PERTURBATION THEORY** Often in physics we have to apply relatively simple physical laws to extremely complicated situations. Although it is straightforward to write down the equations necessary to calculate what a given physical system will do, it may be difficult to solve these equations unless the situation has a particular symmetry, or if some aspects of the problem can be neglected. A good example lies in the construction of **cosmological models** using **general relativity**. The **Einstein equations** are extremely complicated, and no general solutions are available. However, if we assume that the Universe is completely homogeneous and isotropic (i.e. if we invoke the **cosmological principle**), then the special symmetry implied by the **Robertson–Walker metric** drastically simplifies the problem. We ends up with the **Friedmann models**, and only the relatively simple *Friedman equation* to solve.

But the Universe is not exactly homogeneous and isotropic now, even if the **cosmic microwave background radiation** suggests that it must have been so earlier on (see **large-scale structure**). One of the most important questions asked by cosmologists is how this structure came about. In order to make a theory of **structure formation**, surely we need to solve the Einstein equations for the general case of an inhomogeneous and anisotropic universe? To be precise, the answer to this question is 'yes', but to be reasonably accurate the answer is 'no'. Even though our **Universe** is not exactly homogeneous, it is almost so. If we calculate the expected departures from the Robertson–Walker metric for all the mass concentrations we know about, we find them to be small – about one part in a hundred thousand. So we need solutions of the Einstein equations that describe an almost but not quite homogeneous universe. For this we need a model which is almost a Friedmann model, but not quite.

The mathematical technique for generating solutions of equations that are *almost* the same as solutions you already know is called *perturbation theory*, and it is used in many branches of physics other than cosmology. The basic idea can be illustrated as follows. Suppose we have to calculate $(1.0001)^9$ without using a calculator. This problem can be thought of as being almost like calculating 1^9, because the quantity in brackets is not far from 1; and 1^9 is just 1. Suppose that we represent the extra 0.0001 we have to deal with by the symbol ϵ. The problem now is to calculate the product of nine terms $1 + \epsilon$: $(1 + \epsilon)(1 + \epsilon) \ldots (1 + \epsilon)$. Now imagine multiplying out this expression term by term. Since there are nine brackets each with two terms (a 1 and an ϵ), there are $2^9 = 512$ combinations altogether – quite a task. The first term would be a 1, which is obtained by multiplying all the 1's in all the brackets. This would be the biggest term, because there are no other terms bigger than 1 and all the terms containing ϵ are much smaller. If we multiplied the ϵ in the first bracket by the 1's in all the others, we would get ϵ. By taking one ϵ and eight 1's in every possible way from the nine brackets we would get nine terms altogether, all of which are ϵ. Now, any other terms made in more complicated ways that this, like five 1's and four ϵ's, would result in powers of ϵ (in this case ϵ^4). But because ϵ is smaller than 1, all these terms are much smaller than ϵ itself, and very much smaller than 1. It should therefore be a good approximation just to keep the nine terms in which ϵ appears on its own, and ignore terms that contain ϵ^2 or ϵ^3 or higher powers of ϵ. This suggests that we can write, approximately,

$$(1 + \epsilon)^9 \approx 1 + 9\epsilon$$

Going back to our original problem, we can put $\epsilon = 0.0001$, from which we find that the approximate answer to be 1.0009. In fact, the right answer is 1.000 900 36. So our approximation of taking only the lowest-order correction (ϵ) to a known solution ($1^9 = 1$) works very well in this case.

The way to exploit this idea in cosmology is to begin with the equations that describe a Friedmann model for which the Robertson–Walker metric (which we denote here by g) holds. We know how to handle these equations, and can solve them exactly. The we write the equations again, not in terms of g but in terms of some other quantity $g' = g + h$, where h is a small correction like ϵ in the above example – in other words, a *perturbation*. If h is small, we can neglect all the terms of order higher than h and obtain a relatively simple equation for how h evolves. This is the approach used to study the growth of small **primordial density fluctuations** in the expanding Universe.

Of course, the approach breaks down when the small correction becomes not so small. The method used above does not work at all well for $(1.1)^9$, for example. In the study of structure formation by means of the **Jeans instability**, the fluctuations gradually grow with time until they become large. We then have to abandon perturbation methods and resort to another approach. In the example above, we have to reach for a calculator. In cosmology, the final nonlinear stages have to be handled in a similar brute-force way, by running ***N*-body simulations**.

FURTHER READING: Coles, P. and Lucchin, F., *Cosmology: The Origin and Evolution of Cosmic Structure* (John Wiley, Chichester, 1995), Part 3.

* **PHASE TRANSITION** A change from a *disordered* high-energy state into an *ordered* low-energy state, undergone by matter as it cools down. For example, a liquid is quite disordered, while a crystal is highly regular and ordered. In the very early stages of the Big Bang, it is thought that, as matter cooled, it underwent many such changes, during the course of which the Universe's state of symmetry (roughly speaking, the properties of **elementary particles** and the apparent form of the **fundamental interactions**) was altered. This happened in such a way that the present-day low-energy, low-temperature Universe does not appear to possess the symmetries that it should have in unified theories of the fundamental interactions, such as **grand unified theories**.

To visualise how this might happen, imagine standing on the ridge of a roof. On either side of you, the roof slopes away in a completely symmetrical fashion. If you fall, there is nothing about the roof that would make you fall down the left or the right side. But if you do fall, perhaps blown over by a random gust of wind, you can fall only one way. As you descend to your low-energy state, you have broken the symmetry of the situation (and possibly a few limbs). This analogy emphasises the point that the outcomes of given physical laws do not necessarily have the same symmetry as the laws themselves. Likewise, as the Universe was cooling it need not necessarily have respected the symmetry of laws that govern its behaviour. This can happen in particular when *phase transitions* are involved. (For a detailed description of the circumstances in which such effects can occur, and of the different kinds of phase transition possible, see **spontaneous symmetry-breaking**.)

The model of spontaneous symmetry-breaking has been widely used to study the behaviour of particle interactions in theories of the fundamental interactions. Because phase transitions of the required type are expected to appear in the early Universe according to standard particle physics models, the initial stages of the Big Bang are often described as the *era of phase transitions*. One important idea, which we shall refer to later, is that we can identify an order parameter Φ, which is small in the high-temperature disordered state and grows during the phase transition, with the value of a particular quantum **scalar field**, such as the so-called *Higgs field* that occurs in some **quantum field theories**. The free energy related to the scalar field can be related to the effective potential of the field $V(\Phi)$ (which describes its interactions).

In the standard **Big Bang theory**, the **thermal history of the Universe** during the period lasting from the **Planck time** or thereabouts (where the temperature was some 10^{32} K) until the moment when quarks combined into hadrons, which happened when the temperature was about 10^{12} K, is punctuated by a series of phase transitions (not all of which are well understood). Some of these transitions and their consequences are described below. It is convenient to express the temperature in terms of the equivalent energy in electronvolts (eV), the unit favoured by experimental particle physicists. The Planck temperature, 10^{32} K, corresponds to about 10^{19} GeV (where 1 GeV is 10^9 eV) – far higher than the energies that can be reached in any terrestrial accelerator experiment. The quark–hadron transition took place at around 300 MeV, (where 1 MeV is 10^6 eV), which is well within the reach of large accelerators such as that at CERN.

Starting at 10^{19} GeV, the first major landmark reached by the cooling Universe was the energy scale at which grand unified theories (GUTs) are thought to have begun to break symmetry, at around 10^{15} GeV. In the period before this, quantum gravitational effects are thought to have become negligible, and the particles to have been held in thermal equilibrium by means of interactions described by a GUT. At the GUT temperature, which was reached about 10^{-37} seconds after the initial bang, the GUT symmetry began to break. If we assume for the sake of argument that the gauge group describing the GUT physics is the simplest possible, SU(5) (see **gauge theory**), then at 10^{15} GeV the relevant symmetry group would have changed for a period into a different one, SU(3)⊗SU(2)⊗U(1) for example, or perhaps some other symmetry. The GUT transition is expected to have resulted in the formation of magnetic monopoles, and perhaps to have given rise to other **topological defects**.

A GUT, which unifies the electroweak interactions with the strong interactions, puts leptons and hadrons on the same footing and thus allows processes which do not conserve baryon number *B* (violation of baryon number conservation is not allowed in either quantum chromodynamics or electroweak theory). It is thought, therefore, that processes could occur at the GUT temperature which might create the baryon–antibaryon asymmetry that is observed now (see **baryogenesis**). Baryons themselves did not form until much later, when the quark–hadron transition had taken place, but at energies where the GUT symmetry is not broken, quarks and antiquarks are equivalent, so a baryon–antibaryon asymmetry cannot have resulted from this phase. The conditions necessary for the eventual creation of a baryon–antibaryon asymmetry were stated in 1967 by Andrei **Sakharov**. It seems that these conditions prevailed at the GUT scale or slightly lower, depending on the particular version of GUT or other theory adopted; baryogenesis can even occur at much lower energies, around the electroweak scale. Even though this problem is complicated and therefore rather controversial, with reasonable hypotheses we can arrive at a baryon–antibaryon asymmetry of the same order as that observed: about one in a thousand billion (see also **antimatter**). It is worth noting also that, if the Universe is initially lepton-symmetric, the reactions that violate conservation of baryon number can also produce an excess of leptons over antileptons which is equal, in the case of SU(5) GUTs, to that of the baryons over the antibaryons. This is simply because the GUTs unify quarks and leptons. In a GUT the value of the baryon asymmetry actually produced depends only on fundamental parameters of the theory. This means that, even if the Universe is inhomogeneous, the value of the asymmetry should be the same in any region.

When the temperature fell below 10^{15} GeV, the strong and electroweak interactions decoupled. The superheavy bosons that mediate the GUT interaction now rapidly disappeared through annihilation or decay processes. In the moment of symmetry-breaking the order parameter Φ, whose appearance signalled the phase transition proper, could assume a different 'sign' or 'direction' in adjoining spatial regions. It is thus possible that Φ could have changed rapidly with spatial position, between one region and another. (This is similar to the 'Bloch walls' which, in a ferromagnet, separate the different magnetised domains of magnetisation.) The 'singular' regions in which Φ was discontinuous would have had a structure that depended critically on the symmetry that had been broken, giving rise to different possible types of topological defect.

Although the strong interaction separated from electroweak interactions at about 10^{-37} seconds, the electroweak theory kept its form until much later, about 10^{-11} seconds after the beginning. It is probable that phase transitions occurred in this period which are not yet well understood. The particles we are considering here range in energy from 10^2 to 10^{15} GeV; within the framework of the SU(5) model discussed above there are no particles predicted to have masses in this range of energies, which is consequently known as the *grand desert*. Nevertheless, there remain many unresolved questions about this epoch. In any case, of the end of this period we can safely say that, to a good approximation, the Universe would have become filled with an ideal gas containing leptons and anti-leptons, the four vector bosons, quarks and antiquarks and gluons (see **elementary particles**). At a temperature of a few hundred GeV there would have been a spontaneous breaking of the SU(2)⊗U(1) symmetry that describes the electroweak theory, through a phase transition which was probably of first order. All the leptons acquired masses (with the probable exception of the neutrinos), while the intermediate vector bosons gave rise to the massive bosons now known as the W and Z particles, and to photons. The massive bosons disappeared rapidly by decay and annihilation processes when the temperature fell below about 90 GeV.

The last phase transition in this sequence occurred when the temperature had fallen still further, to around 300 MeV. In the framework of QCD theory, the strong interactions then became very strong indeed and led to the confinement of quarks into hadrons, signalling the start of the quark–hadron phase transition, after which came the (very short-lived) hadron era. The remaining events from that time onward are described in the entry on the **thermal history of the Universe**.

SEE ALSO: Essay 4.

FURTHER READING: Kolb, E. W. and Turner, M. S., *The Early Universe* (Addison-Wesley, Redwood City, CA, 1990).

PHOTON see **electromagnetic radiation**, **elementary particles**.

PLANCK, MAX KARL ERNST LUDWIG (1858–1947) German physicist. He was the originator of **quantum theory** with his ideas on the origin of the spectrum of **black-body** radiation. A planned satellite mission by the European Space Agency (ESA) to map the **cosmic microwave background radiation** is named Planck Surveyor in his honour. He also has a fundamental constant named after him: the Planck constant, h.

* **PLANCK ERA** see **thermal history of the Universe**.

PLANCK LENGTH, PLANCK TIME The theory of **general relativity** has to be modified if it is to be applied to situations where the matter density is extremely high, in order to take account of the effects of to **quantum physics** (see also **quantum gravity**). In cosmology, this basically means that the theory cannot be used in its basic form when quantum effects manifest themselves on the scale of the **horizon**.

When do we expect quantum corrections to become significant? Of course, in the absence of a complete theory (or even *any* theory at all) of quantum gravity, it is not possible to give a precise answer to this question. On the other hand, we can make fairly convincing general arguments that yield estimates of the timescales and energy scales where we expect quantum gravitational effects to be large, and for which therefore we should distrust calculations based only on the classical (non-quantum) theory of general relativity. It turns out that the limit of validity of general relativity in the **Friedmann models** is fixed by the *Planck time*, which is of the order of 10^{-43} seconds after the Big Bang.

The Planck time t_P is the time for which quantum fluctuations governed by the Heisenberg uncertainty principle exist on the scale of the *Planck length*, $l_P = ct_P$, where c is the speed of light. From these two scales we can construct other Planck quantities such as the *Planck mass*, m_P, the *Planck energy*, E_P, and so on. Starting from the Heisenberg uncertainty principle in the form

$$\Delta E\, \Delta t \approx h/2\pi$$

and ignoring any factors of 2π from now on, we can see that on dimensional grounds alone we can identify the energy term with some mass m_P through the relation $E = m_P c^2$. Assuming that Δt can be represented as the Planck time t_P, we have

$$m_P c^2 t_P \approx h$$

We can express the Planck mass as a *Planck density* ρ_P times a *Planck volume* (or rather the cube of a Planck length). We want to bring **gravity** into these considerations, in the shape of the Newtonian gravitational constant G. We can do this by noting that the free-fall collapse time for a self-gravitating body of density ρ is given by $t^2 = 1/G\rho$. Replacing m_P by $\rho_P(ct_P)^3$ and then ρ_P by $1/Gt_P{}^2$ in the above expression leads to

$$c^5 t_P^2/G \approx h$$

which finally leads us to an expression for the Planck time in terms of fundamental constants only:

$$t_P \approx \sqrt{(hG/c^5)}$$

which is around 10^{-43} seconds. The Planck length is simply this multiplied by the speed of light, c, and is consequently around 10^{-33} cm. This, for example, is about the size of the cosmological **horizon** at the Planck time (assuming that the concept of a horizon is meaningful at such an early time). The *Planck density* is phenomenally high: about 10^{96} grams per cubic centimetre. Interestingly, however, the Planck mass itself is not an outrageous number: $m_P = \sqrt{(hc/G)} \approx 10^{-5}$ g. We can carry on with this approach to calculate the *Planck energy* (about 10^{19} GeV) and the *Planck temperature* (the Planck energy divided by the Boltzmann constant, which gives about 10^{32} K).

In order to understand the physical significance of the Planck time and all the quantities derived from it, it is useful to think of it in the following manner, which ultimately coincides with the derivation given above. We can define the *Compton time* for a particle of mass m to be $t_C = h/mc^2$; this represents the time for which it is permissible to violate the conservation of energy by an amount equal to the mass of the particle, as deduced from the uncertainty principle. For example, a pair of virtual particles of mass m can exist for a time of about t_C. We can also defined the *Compton radius* of a body of mass m to be equal to the Compton time times the velocity of light: $l_C = ct_C = h/mc$. Obviously both these quantities decrease as m increases. These scales indicate when phenomena which are associated with quantum physics are important for an object of a given mass.

Now, the *Schwarzschild radius* of a body of mass m is given by $l_S = 2Gm/c^2$. This represents, to within an order of magnitude, the radius that a body of mass m must have for its rest-mass energy mc^2 to equal to its internal gravitational potential energy $U \approx Gm^2/l_S$. General relativity leads us to the conclusion that no particle (not even a photon) can escape from a region of radius l_S around a body of mass m; in other words, speaking purely in terms of classical mechanics, the escape velocity from a body of mass m and radius l_S is equal to the velocity of light. We can similarly define a *Schwarzschild time* to be the quantity $t_S = l_S/c = 2Gm/c^3$; this is simply the time taken by light to travel a **proper distance** l_S. A body of mass m and radius l_S has a free-fall collapse of the order of t_S. Note that both t_S and l_S increase as m increases.

We can easily verify that, for a mass equal to the Planck mass, the Compton and Schwarzschild times are equal to each other and to the Planck time. Likewise, the relevant length scales are all equal. For a mass greater than the Planck mass, that is to say for a *macroscopic* body, $t_C < t_S$ and $l_C < l_S$, and quantum corrections are expected to be negligible in the description of the gravitational interactions between different parts of the body. Here we can describe the self-gravity of the body using general relativity or even, to a good approximation, Newtonian theory. On the other hand, for bodies of the order of the Planck mass, that is to say for *microscopic* entities such as elementary particles, $t_C > t_S$ and $l_C > l_S$, and quantum corrections will be important in a description of their self-gravity. In the latter case we must use a theory of quantum gravity in place of general relativity or Newtonian gravity.

At the cosmological level, the Planck time represents the moment before which the characteristic timescale of the expansion is such that the cosmological horizon, given roughly by l_P, contains only one particle (with mass equal to the Planck mass) for which $l_C \geq l_S$. On the same grounds as above, we therefore have to take into account quantum effects on the scale of the cosmological horizon.

It is interesting to note the relationship between the Planck quantities and the properties of **black holes**. According to theory, a black hole of mass M, because of quantum effects, emits **Hawking radiation** like a **black body**. The typical energy of photons emitted by the black hole is kT, where the temperature T is given by

$$T \approx hc^3/4\pi kGM$$

The time needed for such a black hole to evaporate completely (i.e. to lose all its rest-mass energy Mc^2 via Hawking radiation) is given by

$$t \approx G^2M^3/hc^4$$

By taking these two equations and inserting $M = m_P$, we arrive at the interesting conclusion that a Planck-mass black hole evaporates on a timescale of the order of the Planck time.

These considerations show that quantum gravitational effects are expected to be important not only at a cosmological level at the Planck time, but also continuously on a microscopic scale for processes operating over distances of about l_P and times of about t_P. In particular, the components of the **metric** describing **space-time** geometry will suffer fluctuations of the order of l_P/l on a length scale l and of the order of t_P/t on a timescale t. At the Planck time, the fluctuations are 100% on the spatial scale l_P of the horizon and on the timescale t_P of the expansion. We might imagine the Universe at very early times as behaving like a collection of Planck-mass black holes, continually evaporating and recollapsing in a Planck time. This picture is very different from the idealised, perfect-fluid universe described by the Friedmann models, and it would not be surprising if deductions from these equations, such as the existence of a **singularity**, were found to be invalid in a full quantum description.

FURTHER READING: Kolb, E. W. and Turner, M. S., *The Early Universe* (Addison-Wesley, Redwood City, CA, 1990).

****POWER SPECTRUM** $P(k)$ While the **Universe** may be roughly homogeneous and isotropic on the scale of our **horizon**, as required by the **cosmological principle**, the distribution of galaxies in space is decidedly inhomogeneous on scales smaller than this (see **large-scale structure**). In order to quantify the lumpiness of the matter distribution revealed by **redshift surveys** and to relate this to models of **structure formation**, cosmologists employ a variety of statistical tools, the most common of which is called the *power spectrum* and which is usually given the symbol $P(k)$.

The power spectrum is defined in the mathematical language of Fourier series. The simplest way to define it is to define a *fluctuation field* $\delta(\boldsymbol{x})$ at different spatial locations $\boldsymbol{x}$ in terms of the actual density of matter $\rho(\boldsymbol{x})$ at the position $\boldsymbol{x}$, and subtract the mean density of matter ρ_0:

$$\delta(\boldsymbol{x}) = (\rho(\boldsymbol{x}) - \rho_0)/\rho_0$$

Because this departure from the mean density is divided by the mean density, the resulting δ is dimensionless; it is usually called the *density contrast*.

Suppose we consider a part of the Universe contained within a cubic volume V of side L ($V = L^3$). Assuming (for mathematical purposes only) that the Universe outside this cube consists of periodic replications of what is inside, we can expand δ in a Fourier series representing a superposition of plane waves with different amplitudes:

$$\delta(\boldsymbol{x}) = \Sigma\, \delta(\boldsymbol{k}) \exp(i\boldsymbol{k}\cdot\boldsymbol{x})$$

The sum is taken over all the waves that fit into the periodic box. These waves have wave-vectors $\boldsymbol{k}$ of the form $(k_x, k_y, k_z) = (n_x, n_y, n_z)2\pi/L$, where the n_x etc. are integers. The

Fourier coefficients $\delta(\boldsymbol{k})$ for each mode are complex numbers having both an amplitude and a phase. If instead of the volume V we had chosen a different volume V', we would have found that the same series expansion would be possible but that the coefficients were a different set of complex numbers. We can imagine repeating this box-shifting idea for a huge number of boxes all around the Universe. It would then be possible to calculate statistical averages over these Fourier coefficients, such as the average squared modulus of $\delta(\boldsymbol{k})$, which we can write as $\langle\delta(\boldsymbol{k})\delta^*(\boldsymbol{k})\rangle$, where the * denotes a complex conjugate. We can then define the quantity $\delta_{\boldsymbol{k}}$ as

$$\Sigma \langle\delta(\boldsymbol{k})\delta^*(\boldsymbol{k})\rangle = (1/V) \Sigma\, \delta_{\boldsymbol{k}}^2$$

where the summation is now over all modes, and the average is over all possible boxes with the same volume V. We can interpret $\delta_{\boldsymbol{k}}$ as being a kind of average amplitude for waves with wave-vector $\boldsymbol{k}$. If the distribution of matter is statistically homogeneous and isotropic, then this average will not depend on the direction of $\boldsymbol{k}$, just its magnitude k. The final step in this definition is now to let the volume V tend to infinity; the right-hand side of the above equation is then

$$(1/V) \Sigma\, \delta_{\boldsymbol{k}}^2 = (1/2\pi^2) \int P(k)k^2\, \mathrm{d}k$$

where $P(k)$ is the power spectrum or, more precisely, the *power spectral density function*. Despite its rather involved derivation, $P(k)$ is quite simple to visualise: it represents the contribution to the fluctuation field from waves of wavenumber k (and therefore wavelengths $\lambda = 2\pi/k$). A completely flat power spectrum corresponds to white noise (i.e. equal power at all frequencies), while a power spectrum that is sharply peaked at some particular value of k has a characteristic length scale. A single plane wave has a power spectrum consisting of a single spike.

In most theories of cosmological structure formation there are initial **primordial density fluctuations** specified by a very simple power-law form: $P(k) \propto k^n$. In most versions of the **inflationary Universe** the power spectrum has an index n very close to 1, which is called the *scale-invariant* or *Harrison–Zel'dovich spectrum*. The situation is a little more complicated than this in realistic models of structure formation, because this initial spectral shape can be modified by non-gravitational processes in the period before recombination: **Silk damping**, **free streaming** and the **Meszaros effect** can all alter the shape of $P(k)$. Given any particular detailed model, however, these effects can be modelled and they are usually represented in terms of a transfer function $T(k)$ for the model such that

$$\text{Output } P(k) = \text{Input } P(k) \times T(k)^2$$

Knowing $T(k)$ then allows us to define the initial conditions for a detailed computation of the evolution of density fluctuations.

The advantages of the power spectrum as a way of characterising density fields are many. Firstly, in the gravitational **Jeans instability**, as long as the fluctuations are small (δ much less than 1) we can solve the equations describing their evolution using **perturbation theory**. We find that the shape of the power spectrum is not changed by gravitational evolution: only the amplitude increases as a function of time. To put this another way, in perturbation theory all the Fourier modes evolve independently, so that the amplitude of each mode increases at the same rate. Assuming that the power spectrum decreases with decreasing k (increasing wavelength), which it must do if the Universe is to be smooth on large scales, this means that the power spectrum for small k should retain its primordial shape. This allows astronomers to probe the primordial density fluctuations directly by using observations of galaxy clustering.

On smaller scales where nonlinear (non-perturbative) effects are important, the power spectrum changes shape but it does not lose its usefulness. We can compute the power spectrum in an **N-body simulation** of a model, for example, and use it to test the model in question against observations, from which the power spectrum can be estimated quite easily.

Another advantage for the initial stages of clustering evolution is linked to the statistical properties of the initial fluctuations. In most theories, including the **inflationary Universe** models, the initial seed irregularities are Gaussian. This means that the power spectrum alone furnishes a complete statistical description of the fluctuations. Know the power spectrum, and you know it all.

There is also an important connection between $P(k)$ and the method that has been historically important for quantifying the distribution of galaxies in space: the two-point **correlation function**, $\xi(r)$. Together these two functions form a Fourier transform pair, so they provide completely equivalent information about the fluctuation field. A very similar concept to $P(k)$ can be used to describe the fluctuations in temperature on the (two-dimensional) celestial sphere, rather than in three-dimensional space (see Essay 5 for more details).

FURTHER READING: Harrison, E. R., 'Fluctuations at the threshold of classical cosmology', *Physical Review* D, 1970, **1**, 2726; Zel'dovich, Ya. B., 'A hypothesis unifying the structure and entropy of the Universe', *Monthly Notices of the Royal Astronomical Society*, 1972, **160**, 1P; Coles, P., 'The large-scale structure of the Universe', *Contemporary Physics*, 1996, **37**, 429; Coles, P. and Lucchin, F., *Cosmology: The Origin and Evolution of Cosmic Structure* (John Wiley, Chichester, 1995), Chapter 14.

PRIMORDIAL DENSITY FLUCTUATIONS The standard theory of how cosmological **structure formation** is thought to have occurred is based on the idea of gravitational instability (**Jeans instability**), according to which small initial irregularities in the distribution of matter become amplified by the attractive nature of **gravity**. This idea explains, at least qualitatively, how it is possible for the high degree of inhomogeneity we observe around us to have arisen from a much more regular initial state. But gravitational instability would not have worked unless there were small fluctuations in the density at early times, so a complete theory of structure formation must explain how these initial fluctuations got there and predict their vital characteristics.

Since the mid-1980s there have emerged two rival views of how these initial *seed* fluctuations might have arisen, and they gave rise to two distinct theories of how the **large-scale structure** of the Universe was put in place. One of these models involved the idea of **topological defects** created during a **phase transition** in the early Universe. Phase transitions can be thought of as acting like regions of trapped energy, and they do drastic things to the distribution of matter around them: cosmic strings and global textures, in particular, were thought to have affected the early Universe sufficiently to seed density fluctuations directly. The theory of these kinds of fluctuation is difficult, however, because the physics is essentially nonlinear from the start. Although many researchers worked on these *defect theories* in the 1980s and early 1990s, their efforts have produced few concrete predictions.

The alternative picture involves the **inflationary Universe**. Inflation relies on the existence of a quantum **scalar field** whose vacuum energy drives the Universe into an accelerated expansion, ironing out any wrinkles and simultaneously decreasing the **curvature of spacetime** virtually to zero (see **flatness problem**, **horizon problem**). But the scalar field also produces small fluctuations because of quantum fluctuations in it, essentially arising from Heisenberg's uncertainty principle (see **quantum mechanics**). The initial fluctuations arising from inflationary models are much simpler than in the case of topological defects: the quantum fluctuations are small, so that methods from **perturbation theory** can be used. They are also statistically simple: the density field resulting from quantum fluctuations is of the simplest form known in probability theory – the form which is known as a Gaussian random field. The properties of such a field are described completely by the **power spectrum** of the fluctuations. In inflationary models the appropriate power spectrum is of a form known as the *scale-invariant spectrum*, which was derived (for an entirely different purpose) in the 1970s independently by Edward Harrison and by Yakov **Zel'dovich**.

This scale-invariant spectrum is particularly important because we know from very simple arguments that the Universe has to possess fluctuations that are nearly scale-invariant. The term 'scale-invariant' means that fluctuations in the **metric** (the equivalent in Newtonian language to fluctuations in the gravitational potential) have the same amplitude on all scales. We know that the fluctuations in the **cosmic microwave background radiation** have an amplitude of around 10^{-5}; since these are thought to be generated by the **Sachs–Wolfe effect**, this number is of the same magnitude as that of the metric fluctuations. The scale in question here is the scale of our **horizon**: several thousand megaparsecs. But we can also look at individual galaxy clusters, which are less than 10 Mpc across. The random motions in these clusters, of around a thousand kilometres per second, can be related using the **virial theorem** to the gravitational potential energy of the cluster; we find that the gravitational potential energy is about 10^{-5} of the total rest mass. This is no coincidence if the initial fluctuations are scale-invariant, as these are two independent measurements of the metric fluctuations on two very different length scales.

The detection of temperature anisotropies in the cosmic microwave background radiation (the famous **ripples**) with properties which are consistent with an inflationary origin left most cosmologists in little doubt that the density fluctuations were indeed of this nature. Subsequent finer-scale observations and more detailed calculations of the predictions of defect theories have confirmed this preliminary view. Topological defects are now, to most cosmologists, of only abstract theoretical interest.

SEE ALSO: Essays 1, 3 and 5.

FURTHER READING: Harrison, E. R., 'Fluctuations at the threshold of classical cosmology', *Physical Review* D, 1970, **1**, 2726; Zel'dovich, Ya. B., 'A hypothesis unifying the structure and entropy of the Universe', *Monthly Notices of the Royal Astronomical Society*, 1972, **160**, 1P.

PRIMORDIAL FIREBALL see **Big Bang theory**.

** **PROPER DISTANCE, PROPER TIME** Both **special relativity** and **general relativity** are theories in which space and **time** are welded together in a single four-dimensional construction called **spacetime**. This may make the theories extremely elegant and powerful, but it does introduce some problems with the formulation of unambiguous definitions of spatial and temporal intervals.

Spacetime in special relativity, for example, is constructed in such a way that the interval between two events at times (t, x, y, z) and $(t + dt, x + dx, y + dy, z + dz)$ is given by

$$ds^2 = c^2\, dt^2 - (dx^2 + dy^2 + dz^2)$$

a mathematical form known as the *Minkowski metric*. Different observers might possess different clocks and be in different coordinate systems because of their motion, but the interval ds is always the same for all observers undergoing relative motion. Two events that appear instantaneous to one observer, but at different spatial positions, might appear to be happening at different times in the same place to a second observer moving relative to the first one. Space and time get mixed up in this way.

It is nevertheless possible to define 'standardised' measures of distance and time by introducing the concepts of *proper time* and *proper distance*. For example, intervals of proper time are those measured by a clock which is at rest in the frame of the measurement. If we try to make a measurement of the time interval between two events using a clock that is moving with respect to them, we end up with some other measurement that disagrees with proper time. Likewise, proper distances in special relativity are those measured with a ruler that is at rest in the frame of the measurement.

These concepts can be applied to cosmology, although in this case we have to use the more complicated form of the metric laid down by the cosmological principle, i.e. the **Robertson–Walker metric**:

$$ds^2 = c^2\, dt^2 - a(t)^2[dr^2/(1 - kr^2) + r^2(d\theta^2 + \sin^2\theta\, d\phi^2)]$$

If anything, this form of the metric makes it easier to see how to define proper distances and times in cosmology than in special relativity. For example, the coordinate t is itself a cosmological proper time coordinate because it is by definition the same for all observers moving with the cosmological expansion. The fact that the Universe is homogeneous and isotropic automatically synchronises clocks for these fundamental observers because they can measure time according to the local density of matter. Proper distances would be distances measured by an observer at rest with respect to the expansion, so such distances must correspond to intervals with $dt = 0$ in the metric. Proper distances from an observer to distant objects can thus be expressed as follows:

$$d = \int ds = \int a(t)\, dr/\sqrt{(1 - kr^2)}$$

There is a constant of integration because t is constant; and the integral is from the origin at 0 to the *comoving coordinate* (see **Robertson–Walker metric**) of the object in question, which we can call x. Hence the proper distance from an observer to an object is just $a(t)f(x)$, where the form of $f(x)$ depends just on the sign of k: $f(x) = x$ if $k = 0$ (**flat universe**); $f(x) = \sin^{-1}(x)$ if $k = 1$ (**closed universe**); $f(x) = \sinh^{-1}(x)$ if $k = -1$ (**open universe**). The coordinate x stays with the object for all time, so the proper separation simply scales as $a(t)$; hence the global **expansion of the Universe**. Since the proper distance increases with proper time, it makes sense to call it a *proper velocity*; this velocity obeys **Hubble's law**.

Simple though proper distances are, they are not measurable by astronomical techniques. One always makes observations, not along constant time **hypersurfaces** with $dt = 0$, but using light which travels along null paths with $ds = 0$. For this reason astronomers generally use alternative measures of distance, such as **angular-diameter distance** or **luminosity distance**.

FURTHER READING: Narlikar, J. V., *Introduction to Cosmology*, 2nd edition (Cambridge University Press, Cambridge, 1993); Rindler, W., *Essential Relativity: Special, General, and Cosmological*, revised 2nd edition (Springer-Verlag, New York, 1979).

PROTOGALAXY see **evolution of galaxies, structure formation**.

PROTON see **elementary particles**.

Q

** **QUANTUM COSMOLOGY** The field of quantum cosmology attempts to provide a consistent theory of the very earliest stages of the **Universe**. In doing so, it therefore seeks to extend the standard **Big Bang theory** into areas where it is currently incomplete, particularly to the era between the **Planck time** and the **initial singularity**. Ultimately the intention is to account in a consistent way for the birth of the Universe. Unfortunately, this task requires the existence of a satisfactory theory of **quantum gravity**, and there is no such theory at present. There is therefore no well-accepted set of tools with which to tackle quantum cosmology. On the other hand, it is a field in which there is considerable activity, and considerable controversy has been generated even by preliminary studies.

The central concept in quantum physics is that of the *wavefunction*, a complex function usually given the symbol ψ. In the simplest possible case of a single-particle system as described by **quantum mechanics**, the wavefunction is simply a function of the position $\boldsymbol{x}$ of the particle and the time t: $\psi = \psi(\boldsymbol{x}, t)$. Although the interpretation of ψ is by no means straightforward, it is generally accepted that the square of the modulus of ψ determines the probability of finding the particle at position $\boldsymbol{x}$ at time t. One popular mathematical formulation of **quantum physics** uses the concept of a *sum-over-histories*. In this formulation, the probability of the particle in question arriving at the event in **spacetime** labelled $(\boldsymbol{x}, t)$ is given by an integral over all possible paths of the particle leading to that location in spacetime. Each of these paths is weighted by a quantity called the *action* and given the symbol $S(\boldsymbol{x}', t)$. Different paths have different actions, so not all paths are counted to the same extent, as we shall see shortly. Each path, also called a *history*, is described by a curve of the form $\boldsymbol{x}'(t)$, giving its spatial position $\boldsymbol{x}'$ as a function of t, which can be thought of as the intersection of the particle's history with a 'time-like' **hypersurface** labelled by t (see **spacetime**). The required sum-over-histories is then of the form

$$\psi(\boldsymbol{x}, t) = \int \mathrm{d}x' \, \mathrm{d}t' \exp[\mathrm{i}S(\boldsymbol{x}', t')]$$

where the integration is a path integral with respect to an appropriate measure on the space of all possible histories. The upper limit of integration will be the event in spacetime given by $(\boldsymbol{x}, t)$, and the lower limit will depend on the initial state of the system. The integral is taken over all possible paths $(\boldsymbol{x}', t')$, starting at the initial state and ending at the end state. The action S describes the forces to which the particle is subjected as it moves. We cannot say that the particle takes any one definite path in quantum mechanics; somehow it seems to take all paths simultaneously in order to arrive at its destination.

This sum-over-histories formalism is often used in standard quantum mechanics, but it also seems to be the one that appears the most promising for the study of quantum gravity. To make any progress with quantum cosmology, however, we have to make some simplifying assumptions. First, we assume that the Universe is finite and closed (the relevant integrals appear to be undefined in an **open universe**). We also have to assume that the spatial topology of the Universe is fixed; the topology is not determined in **general relativity**. The relevant action for general relativity is denoted by S_E (the E stands for Einstein). This is one of the major deficiencies in quantum gravity. There is no choice for the action of spacetime coupled to matter fields that yields a mathematically satisfactory quantum theory. One problem is that theories made in this way appear not to be *renormalisable*, which means they have unwanted and uncontrollable infinities that cannot be made to disappear like they can in standard quantum field theories.

In fact, there is no reason why the Einstein action S_E should keep its form as we move to higher and higher energies. For example, it has been suggested that wherever gravitational fields are very strong, the action for general relativity might pick up terms which depend in more complicated ways on the **curvature of spacetime** than those included in Einstein's theory when the gravitational fields are very strong. Indeed, so-called *second-order gravity theories* constructed in this way have proved to be of considerable theoretical interest because they can be shown to be in some sense equivalent to general relativity with the addition of a **scalar field**. Such theories lead inevitably to a model for the inflationary Universe, but they also violate the conditions necessary for the existence of an initial **singularity**. Since, however, we have no good reason in this context to choose one action above any other, for this discussion it makes sense to assume that the Einstein action is the appropriate one to take. We shall also simplify things even further by assuming that we are dealing with an empty universe (i.e. one in which there are no matter or radiation fields).

To formulate cosmology in a quantum manner, we first have to think of an appropriate analogue for the history outlined above. It is perhaps most sensible to start by trying to determine a wavefunction for the spatial configuration of the Universe at a particular time. In general relativity the such a configuration is given simply by the three-dimensional geometry (3-geometry) of a space-like **hypersurface**. Let this geometry be described by a three-dimensional **metric** $h_{\mu\nu}(\boldsymbol{x})$ that describes only space and not time (μ and ν take only three values, corresponding to the three spatial coordinates). Just as the position of a particle can be thought of as the intersection of the history of the particle with a time-like hypersurface, the 3-geometry can be thought of as a slice through four-dimensional spacetime at a particular time. This entire spacetime has a

4-geometry described by some 4-metric g_{ij} (in this metric i and j run over four possible coordinates, three of space and one of time). General relativity is a four-dimensional theory, so the action depends explicitly on the 4-metric. If we want to make an integral that looks like the one above for the particle wavefunction, then the sum has to be taken over all the 4-metrics g_{ij} that produce the 3-metric $h_{\mu\nu}(\boldsymbol{x})$ when they are sliced at a given time. The appropriate sum-over-histories integral is therefore of the form

$$\Psi[h_{\mu\nu}(\boldsymbol{x})] = \int \mathrm{d}g_{ij} \exp(\mathrm{i}S_{\mathrm{E}}(g_{ij}))$$

The wavefunction Ψ is therefore defined over the space of all possible 3-geometries consistent with the initial assumptions (i.e. a **closed universe** with a fixed topology). Such a space is usually called a *superspace* – if you like, a space of possible spaces. The integral is taken over appropriate 4-geometries consistent with the 3-geometry $h_{\mu\nu}$. The usual quantum-mechanical wavefunction ψ evolves according to the Schrödinger equation discussed under **quantum physics**; the function Ψ evolves according to similar equation called the *Wheeler–de Witt equation*, which is of a similar basic form, but which is too complicated to write down here. The wavefunction Ψ is rather grandiosely called the *wavefunction of the Universe*. The term is something of an exaggeration because it does not take account of matter or radiation. By analogy with ordinary quantum theory, we can regard the square of the modulus of Ψ as representing the probability that the Universe will find itself with a particular spatial geometry.

It is the determination of what constitutes the appropriate set of histories over which to integrate that is the crux of the problem of quantum cosmology, even in this extremely simplified setting. The problem is analogous to the problem of not knowing the lower limit of integration for the moving particle problem. One suggestion, by James Hartle and Stephen **Hawking**, is that the sum on the right-hand side of the above equation should be taken only over compact Euclidean 4-geometries. This essentially involves changing the time coordinate to **imaginary time**, so that time appears with the same sign as the spatial coordinates in the metric (i.e. the signature of the 4-metric is changed from the usual Lorentz signature to a Euclidean signature. In this case the appropriate 4-geometries have no boundary (like the surface of the Earth), so this is often called the *no-boundary condition* or *no-boundary hypothesis*. Among other advantages, the relevant Euclidean integrals can then be made to converge in a way in which Lorentzian ones apparently cannot. Other choices of initial condition have, however, been proposed. Alexander Vilenkin, among others, has proposed a model in which the Universe undergoes a sort of quantum tunnelling from a vacuum state. This corresponds to a definite creation, whereas the Hartle–Hawking proposal has no creation in the usual sense of the word. It remains to be seen which, if any, of these formulations is correct.

FURTHER READING: Barrow, J. D., *Theories of Everything* (Oxford University Press, Oxford, 1991); Hartle, J. B. and Hawking, S. W., 'The wave function of the Universe', *Physical Review* D, 1983, **28**, 2960; Hawking, S. W., *A Brief History of Time* (Bantam, New York, 1988); Vilenkin, A., 'Boundary conditions in quantum cosmology', *Physical Review* D, 1986, **33**, 3560.

* **Quantum field theory** Modern quantum theory is more sophisticated than the simplified version discussed under **quantum mechanics**. Theories of the **fundamental interactions** and the **elementary particles** are more complicated mathematical structures called *quantum field theories*. In these theories the fundamental entities are not particles at all, but fields, like the electromagnetic field. Space is pervaded by these fields, of which there are many in complicated theories like **grand unified theories**. The equations that result are *field equations*, and there are solutions of these equations that represent quantised oscillations of these fields. These oscillations represent particles: different modes of oscillation correspond to different particle properties. For example, the oscillations in the electromagnetic field are the photons; those in the Higgs field (see below) are called Higgs bosons.

Different fields play different roles in other theories of the fundamental interactions. Some of these fields represent the known elementary particles, while other fields are brought in to enable **spontaneous symmetry-breaking** to occur. For example, in the electroweak theory a field called the *Higgs field* was introduced to explain why the low-energy state of this theory has a broken symmetry of electromagnetism on the one hand, and weak nuclear interactions on the other. Oscillations of this field would then represent a particle called the *Higgs particle* which might be detectable in particle accelerator experiments.

The Higgs field is an example of a **scalar field**, a class of quantum fields that is very important in the cosmology of the early Universe. In particular, most theories of the **inflationary Universe** rely on the properties of such fields. Scalar fields possess no inbuilt sense of direction, and the number of particle states corresponding to their quantum behaviour is restricted. Alternatives are *vector fields*; an example from classical physics is the **magnetic field**, which is depicted conventionally as lines of force in particular directions.

Quantum field theory has been very successful in helping physicists construct unified models of the fundamental interactions. Indeed, the first complete quantum field theory, *quantum electrodynamics* (QED), is one of the most accurate physical theories ever constructed: the so-called *Lamb shift* (a small difference between two energy levels in the hydrogen spectrum) predicted by this theory has been verified to great precision by experimentalists. Nevertheless, there are some considerable conceptual difficulties with these theories, including the appearance of divergent terms in the calculations. These problems led to the introduction of the technique of *renormalisation*, a device for separating out infinite terms from the theory. Attempts to extend quantum field theory to **gravity** have failed, largely because the resulting theories are not renormalisable in the same way. Moreover, attempts to calculate the energy density of the vacuum state of the Universe using these theories yield enormously large answers; this is the essence of the problem of the **cosmological constant**.

SEE ALSO: Essay 3; **gauge theory**.

FURTHER READING: Weinberg, S., *The Quantum Theory of Fields* (2 vols) (Cambridge University Press, Cambridge, 1995).

QUANTUM GRAVITY The best theory of **gravity** that seems to be available at the present time is **general relativity**. This is a classical theory, in the sense that Maxwell's equations of electromagnetism (see **fundamental interactions**) are also classical, in that they involve entities that are continuous rather than discrete and describe behaviour that is deterministic rather than probabilistic. For example, general relativity is essentially a theory of the interaction between **spacetime** and matter fields. It is a requirement of this theory that both these components be smooth (mathematically, they have to satisfy certain differentiability conditions). On the other hand, **quantum physics** describes a fundamental lumpiness: everything consists of discrete packets or *quanta*. Likewise, the equations of general relativity allow us to calculate the exact state of the Universe at a given time in the future if sufficient information is given for the present or for some time in the past. They are therefore deterministic. The quantum world, on the other hand, is subject to the uncertainty embodied in Heisenberg's uncertainty principle.

Classical electromagnetism is perfectly adequate for many purposes, even for the extreme radiation fields encountered in astrophysics, but the theory does break down sometimes. Physicists therefore sought (and eventually found) a successful theory of electromagnetism which became known as quantum electrodynamics, or QED (see **quantum field theory**). While Einstein's theory of general relativity seems quite accurate for most purposes, it was also natural to attempt to construct a quantum theory of gravity to complete the picture. Einstein himself always believed that his theory was incomplete in this sense, and would eventually need to be replaced by a more complete theory. By analogy with the breakdown of classical electromagnetism, we might expect this to be necessary for very strong gravitational fields, or extremely short length scales. Attempts to build such a theory have been largely unsuccessful, mainly for complicated technical reasons to do with the fact that the theory of general relativity is not *renormalisable*.

The lack of a consistent quantum theory of gravity leaves cosmologists with the problem of not knowing how to describe the earliest stages of the Big Bang, where the density of matter was so high that quantum corrections to gravity are expected to have been important. We can estimate the scales of length and time on which this happens: they are usually called the **Planck length** and **Planck time**. For example, our understanding of the Universe breaks down completely for times before the Planck time, which is about 10^{-43} seconds after the Big Bang itself.

Although there is nothing resembling a complete picture of what a quantum theory of gravity might involve, there are some interesting speculative ideas. For example, since

general relativity is essentially a theory of spacetime, space and time themselves must become quantised in quantum gravity theories. This suggests that, although space and time appear continuous and smooth to us, on scales of order the Planck length (around 10^{-33} cm), space is much more lumpy and complicated, perhaps consisting of a foam-like topology of bubbles connected by **wormholes** that are continually forming and closing again on a timescale of order the Planck time. It also seems to make sense to imagine that quantised **gravitational waves**, or *gravitons*, might play the role of the gauge bosons in other fundamental interactions, such as the photons in the theory of quantum electrodynamics.
SEE ALSO: **quantum cosmology**.

* **QUANTUM MECHANICS, QUANTUM PHYSICS, QUANTUM THEORY** One of the great revolutions that occurred in physics was the introduction of quantum theory, or quantum physics, during the early years of the 20th century. This changed for ever the simple, mechanistic view of the world founded upon Isaac **Newton**'s laws of motion. A Universe running according to Newtonian physics is *deterministic*, in the sense that if we knew the positions and velocities of all the particles in a system at a given time, we could predict their behaviour at all subsequent times simply by applying Newton's laws. Quantum mechanics changed all that, since one of the essential components of this theory is the principle (now known as Heisenberg's uncertainty principle) that, at a fundamental level, the behaviour of particles is unpredictable.

In the world according to quantum theory, every entity has a dual nature. In classical physics there were two distinct concepts used to describe natural phenomena: waves and particles. Quantum physics tells us that these concepts do not apply separately to the microscopic world. Things that we previously imagined to be particles (point-like objects) can sometimes behave like waves; phenomena that we previously thought of as waves can sometimes act like particles. For example, **electromagnetic radiation** can behave like a wave phenomenon: we can display interference and diffraction effects using light rays, for example. Moreover, James Clerk **Maxwell**'s theory of electromagnetism (see **fundamental interactions**) showed that this radiation was actually described mathematically by a wave equation: the wave nature of light is therefore predicted by this theory. On the other hand, Max **Planck**'s work on the **black-body** spectrum showed that light could also behave as if it came in discrete packets which he called *quanta*. Planck hesitated to claim that these quanta could be identified with particles, and it was Albert **Einstein**, in his work on the photoelectric effect for which he was later awarded the 1921 Nobel Prize for Physics, who made the step of saying that light was actually made of particles. These particles later became known as *photons*. So how can something be both a wave and a particle? The answer is that real entities cannot be described exactly by either concept, but behave sometimes as if they were waves, and sometimes as if they were particles.

Imagine a medieval monk returning to his monastery after his first trip to Africa. During his travels he chanced upon a rhinoceros, and is now faced with the task of describing it to his incredulous brothers. Since none of them has ever seen anything resembling a rhino in the flesh, he has to proceed by analogy. The rhinoceros, he says, is in some respects like a dragon and in others like a unicorn. The brothers then have a reasonable picture of what the beast looks like. But neither dragons nor unicorns exist in nature, while the rhinoceros does. It is the same with our quantum world: reality is described neither by idealised waves nor by idealised particles, but these concepts can give some impression of certain aspects of the way things really are.

The idea that energy comes in discrete packets (quanta) was also successfully applied to the theory of the hydrogen atom, by Niels **Bohr** in 1913, and to other aspects of atomic and nuclear physics. The existence of discrete energy levels in atoms and molecules is fundamental to the field of **spectroscopy**, which nowadays is a major area of observational astrophysics.

But the acceptance of the quantised nature of energy (and light) was only the start of the revolution that founded modern quantum mechanics. It was not until the 1920s and the work by Erwin **Schrödinger** and Werner **Heisenberg** that the dual nature of light as both particle and wave was finally elucidated. For while the existence of photons had become accepted, there had been no way to reconcile the idea with the well-known wave behaviour of light. What emerged in the 1920s was a theory of quantum physics built upon *wave mechanics*. In Schrödinger's version of quantum theory, the behaviour of all systems is described in terms of a *wavefunction* (usually denoted by ψ) which evolves according to an equation called the *Schrödinger equation*. In general this equation is extremely complex; for a system consisting of a single particle it reduces to

$$(ih/2\pi)\,\partial\psi/\partial t = -(h^2/8\pi^2 m)\,\nabla^2\psi + V(\boldsymbol{x}, t)\psi$$

where h is the Planck constant, m is the mass of the particle and V is the potential function that describes the forces acting on the particle. The wavefunction ψ will be a function of both space and time, and in general it will also be a complex function (with both real and imaginary parts). Since the Schrödinger equation is essentially of the same form as a wave equation, the wavefunction can undergo interference and diffraction, and all the other phenomena associated with waves. There are other formulations of quantum mechanics, including the *sum-over-histories* approach discussed under **quantum cosmology**.

So where does the particle behaviour come in? The answer is that $\psi(\boldsymbol{x}, t)$ is not like, for example, an electromagnetic wave which we think of as existing (in some sense) at the point in space and time labelled by $(\boldsymbol{x}, t)$: it is a *probability wave*. In fact, the square of the modulus of ψ represents the probability of finding the particle at place $\boldsymbol{x}$ and time t. Quantum theory asserts that this is all we can know about the system: we cannot predict with certainty exactly where the particle will be at a given time – just the probability.

An important aspect of this *wave–particle duality* is the uncertainty principle mentioned above. This has many repercussions for physics, but the simplest involves the position of a particle and its momentum. Let us suppose that the particle is moving along a one-dimensional line, so that its position can be described in terms of one coordinate x; we denote its momentum by p. *Heisenberg's uncertainty principle* states that we cannot know the position and momentum of a particle independently of each other. The better we know the position x, the less well we know the momentum p. If we could pinpoint the particle exactly, then its momentum would be completely unknown. This principle is quantitative: if the uncertainty in position (measured by the 'spread' of the wavefunction) is Δx and the uncertainty in momentum is Δp, then the product $\Delta x \, \Delta p$ can never be less than a fundamental number: $h/2\pi$, where h is the Planck constant. These considerations do not apply only to position and momentum, but also to energy and time and to other pairs of quantities that are known as *conjugate variables*. It is a particularly important consequence of the energy–time uncertainty principle that empty space can give birth to short-lived *virtual particles*: if the rest mass of the particles is m, then they can spring out of nothing for a fleeting existence given by h/mc^2.

Exactly how we are to interpret this probabilistic approach is open to considerable debate. For example, consider a system in which particles travel in a beam towards two closely separated slits. The wavefunction ψ corresponding to this system displays an interference pattern because the 'probability wave' passes through both slits. If the beam is powerful, it will consist of huge numbers of photons. Statistically, therefore, the photons should land on a screen behind the slits according to the probability dictated by the wavefunction. Since the slits set up an interference pattern, the screen will show a complicated series of bright and faint bands where the waves sometimes add up in phase and sometimes cancel one another. This seams reasonable, but suppose we turn the down the power of the beam. In principle, we could turn down the power until there is only one photon at any time travelling through the slits. The arrival of each photon can be detected on the screen. By running the experiment for a reasonably long time we could build up a pattern on the screen. Despite the fact that only one photon at a time is travelling through the apparatus, the screen still shows the pattern of fringes. In some sense each photon must turn into a wave when it leaves the source, travel through both slits, interfering with itself on the way, and then turn back into a photon in order to land in a definite position on the screen. If we were simply to block one of the slits, no pattern would be seen, just a single bright area where all the photons land.

So what is going on? Clearly each photon lands in a particular place on the screen. At this point we know its position for sure. What does the wavefunction for this particle do at this point? According to one interpretation – the so-called *Copenhagen interpretation* – the wavefunction collapses so that it is concentrated at a single point. This happens whenever an experiment is performed and a definite result is obtained. But before the outcome is settled, nature itself is indeterminate: the photon really does not go through either one of the slits: it is in a 'mixed' state. The act of measurement changes the wavefunction, and it therefore changes reality. This has led many to speculate about the interaction between consciousness and quantum 'reality': is it consciousness that causes the wavefunction to collapse?

A famous illustration of this conundrum is provided by the paradox of *Schrödinger's cat*. Imagine that there is a cat inside a sealed room, which contains a vial of poison. The vial is attached to a device which will break it and poison the cat when a quantum event occurs, for example the emission of an alpha-particle by a lump of radioactive material. If the vial breaks, death is instantaneous. Most of us would accept that the cat is either alive or dead at a given time. But if we are to take the Copenhagen interpretation seriously, then the cat is somehow both: the wavefunction for the cat comprises a superposition of the two possible states. Only when the room is opened and the state of the cat is 'measured' does it 'become' either alive or dead.

An alternative to the Copenhagen interpretation is that nothing physically changes at all when a measurement is performed. What happens is that the observer's state of knowledge changes. If we accept that the wavefunction ψ represents what is known by the observer rather than what is true in reality, then there is no problem in having it change when a particle is known to be in a definite state. This view suggests a 'hidden variable' interpretation of quantum mechanics. Perhaps at some level things are deterministic, but we simply do not know the values of the determining variables until an experiment is performed.

Yet another view is the *many worlds interpretation*. In this view, every time an experiment is performed (e.g. every time a photon passes through the slit device) the Universe, as it were, splits into two: in one half-Universe the photon goes through the left-hand slit, and in the other it goes through the right-hand slit. If this happens for every photon, we end up with an enormous number of *parallel universes*. All possible outcomes of all possible experiments occur in this ensemble. The many worlds interpretation is probably favoured by most quantum cosmologists.

SEE ALSO: **quantum cosmology, quantum field theory**.

FURTHER READING: Squires, E., *The Mystery of the Quantum World* (Adam Hilger, Bristol, 1986); Deutsch, D., *The Fabric of Reality* (Allen Lane, London, 1997).

QUARK see **elementary particles**.

QUASAR The first quasars to be found were detected by their strong radio emission, but they were found to be so small that, like stars but unlike **galaxies**, they could not be resolved with optical telescopes. For this reason they became known as *quasi-stellar radio sources*, or quasars for short. Later on, other such objects were found which did not emit radio waves at all, so the name was changed to *quasi-stellar object* (QSO), but the name quasar has stuck. It seems that only one in about two hundred quasars is actually 'radio-loud'.

Quasars have been found at very high **redshifts** indeed – as great as $z = 4.9$. Objects with such high redshifts are so far away that their light has taken more than 90% of the **age of the Universe** to reach us (assuming, of course, that the redshift of this source is caused by the **expansion of the Universe**). Since these objects were first discovered in the 1960s, there has been a considerable debate about whether they really are at the distances inferred for them from **Hubble's law**. Most astronomers accept that they are, but there are notable dissenters such as Halton Arp, who continues to produce images that appear to show quasars physically interacting with galaxies which have much lower redshift. If the quasar redshifts are cosmological in origin, then these pictures must be dismissed as chance projections: after all, we all have holiday photographs that appear to show trees growing out of our loved ones' heads.

If the quasars are at cosmological distances, they must be phenomenally luminous in virtually all regions of the spectrum of **electromagnetic radiation**. Typically, quasars radiate more than a thousand times as much energy as all the stars in the Milky Way put together. Moreover, they are variable on a timescale of a few hours: this shows that much of their radiant energy must be emitted from a region smaller than a few light hours across. Since they emit so much energy from such a small region, it is thought that quasars are powered by the accretion of matter onto a central **black hole** with a mass of perhaps 100 million solar masses. Matter falling into the black hole loses energy, which escapes as radiation, thus powering the activity. The existence of these objects at such high redshifts means that some structures must have formed very early in the evolution of the Universe. Not only must the black hole itself have been created, but there must also have been enough material in the surrounding region to feed it. This has important implications for theories of **structure formation**, especially those in which galaxies and **large-scale structure** are predicted to form only at relatively recent cosmological epochs.

Since accretion onto a black hole is the same mechanism that powers **active galaxies**, quasars may be thought of as extreme examples of this class of object. It is quite possible that most galaxies that are sufficiently massive to play host to a black hole of the required size may have done so at some stage in the past. Quasars are also used to study the properties of the **intergalactic medium**, through the pattern of **absorption lines** seen in their spectra and apparently caused by matter along the line of sight to the quasar from the observer.

FURTHER READING: Arp, H. C., *Quasars, Redshifts, and Controversies* (Interstellar Media, Berkeley, CA, 1987).

R

Radiation In astronomy the term 'radiation' is usually taken to mean **electromagnetic radiation** in any of its various forms, but in cosmology it is often used in a broader sense. Although photons of electromagnetic radiation are massless, in the theory of **general relativity** they do exert a gravitational effect by virtue of their nonzero total energy density. This is essentially because of the famous equivalence between mass and energy predicted by **special relativity** and described by the relation $E = mc^2$. The **cosmic microwave background radiation** therefore contributes to the gravitational deceleration of the Universe as measured by the **density parameter**. However, because the **black-body** temperature is so small, the total energy density is tiny compared with the energy stored in the rest-mass energy of all the matter: in fact, it is about one part in a hundred thousand. The present Universe is therefore *matter-dominated*.

The effect of radiation on the evolution of the Universe is rather different from that of ordinary matter (or pressureless matter, sometimes referred to as *dust*). In the **Friedmann models**, for example, the **expansion of the Universe** by a factor of a (the *scale factor*) results in a lowering of the matter density by a factor of a^3. The total number of particles in the Universe does not change, but the number per unit volume goes down as they are diluted. Since the mass of each one stays the same, the total density of matter must decrease.

But radiation behaves differently. As the Universe expands, the number of photons is conserved – just as the number of material particles is – but because of the cosmological **redshift**, the energy carried by each photon goes down by a factor of $1/a$. The total energy density of the radiation therefore falls as $1/a^4$. Looking at this the other way around, we can conclude that, though the gravitational effect of the cosmic background is negligible now, at early times its effect would have been significant. Since the relative importance now is around 10^{-5} and scales as a^3/a^4, there must have been a time when matter and radiation contributed equally, corresponding to a scale factor about 10^{-5} times smaller than the present value. At times earlier than this the Universe would have been *radiation-dominated*.

A radiation-dominated universe behaves differently from a matter-dominated one. For example, a flat matter-dominated Friedmann model evolves according to $a(t) \propto t^{2/3}$, while for a flat radiation-dominated model, $a(t) \propto t^{1/2}$. In fact, things are more complicated than this because the behaviour which has been discussed so far in the context of electromagnetic radiation also occurs for any other relativistic particles, whether or not they are massless. Another radiation background predicted by the standard **Big Bang theory** is the *cosmic neutrino background*, produced at the start of the lepton era of the **thermal history of the Universe** (see **elementary particles**). These neutrinos, which may well be massless, contribute to the expansion rate to a similar extent to the photons of the microwave background. In the standard Big Bang theory, it is possible to turn back the clock to reveal the thermal history of the Universe at earlier and earlier times. Working back in this way, we find that the temperature of the black-body radiation rises to such an extent that it is possible for massive particles to be created directly from radiation in a process called *pair creation*. These particles would typically have been created with velocities close to the velocity of light, and their behaviour would therefore no longer be dominated by their rest mass. To all intents and purposes these ultra-relativistic particles also behave just like electromagnetic radiation. For this reason it is unavoidable that the behaviour of the early Universe, at least while it is controlled by physics below the scale of **grand unified theories**, is dominated not by matter but by radiation.

RADIATION ERA see **thermal history of the Universe**.

RADIO ASTRONOMY The branch of astronomy that concerns itself with observations made in the radio part of the spectrum of **electromagnetic radiation**, which covers a huge range of wavelengths from about 1 mm to 30 m. Radio broadcasting equipment makes use of longer wavelengths than this, but these cannot penetrate the atmosphere and are therefore inaccessible, at least from the surface of the Earth. Many different kinds of astronomical object emit in the radio region of the spectrum by thermal processes (as with **black-body** radiation or bremsstrahlung) or non-thermal processes (as with synchrotron radiation, produced by electrons spiralling in **magnetic fields**). It is possible also to perform spectroscopy in the radio region, particularly with the prominent 21-cm hydrogen absorption line that allows radio astronomers to study the properties of relatively cold gas in our Galaxy.

Radio astronomy began in the 1930s with the work of Karl Jansky (after whom the unit of flux density in the radio region is now named), but it was only after the Second World War, following and the rapid development of radar instrumentation, that major research groups got under way. Powerful radio sources were soon identified with objects radiating at visible wavelengths, such as the Crab Nebula (a **supernova** remnant), and the pattern of radio emission of our own Galaxy was mapped out. Radio astronomy began to have an impact on cosmology as the design of radio telescopes was improved and bigger ones were built. First, radio galaxies (see **active galaxies**) and **quasars** were discovered. The first systematic counts of radio sources on the sky as a function of their brightness then showed clear evidence for the **evolution of galaxies** (see **classical cosmology**). This effectively discredited the

steady state theory even before the discovery of the **cosmic microwave background radiation**, again using radio techniques, dealt the knockout punch.

Radio telescopes come in various forms and employ different techniques to gather the very small amount of energy that comes from extragalactic radio sources. Many, such as the famous Lovell Telescope at Jodrell Bank, employ a parabolic dish to focus radio waves onto a detector. Techniques have been devised that allow different telescopes to be used together as an *interferometer*. In a method called *aperture synthesis*, the signals from a number of telescopes can be linked to recreate the observational sensitivity of a much larger instrument. Very-long-baseline interferometry (VLBI) synthesises an interferometer thousands of kilometres across from telescopes all around the world.

Despite its past and present importance to astronomy, the future of radio astronomy is threatened by the increasing encroachment of commercial digital communication systems, such as mobile phones, into regions of the radio band that were previously reserved for astronomical use.

SEE ALSO: Essays 4 and 5.

FURTHER READING: Hey, J. S., *The Radio Universe*, 3rd edition (Pergamon Press, Oxford, 1983); Rybicki, G. and Lightman, A. P., *Radiative Processes in Astrophysics* (John Wiley, New York, 1979).

RECOMBINATION see **cosmic microwave background radiation**.

RECOMBINATION ERA see **thermal history of the Universe**.

REDSHIFT (z) The change in position of **emission lines** or **absorption lines** in the spectrum of **electromagnetic radiation** produced by a **galaxy**, **quasar** or other extragalactic object. For a line emitted at wavelength λ_e and observed at wavelength λ_o, the redshift is simply given by

$$1 + z = \lambda_o/\lambda_e$$

Because the observed wavelength is greater than the emitted one for objects moving with the **expansion of the Universe**, the redshift is positive. The simplest way to see how this effect arises it is to regard the wavelength as being stretched by the cosmological expansion as it travels from source to receiver. If the *cosmic scale factor* is $a(t_e)$ at emission and $a(t_o)$ at reception, the redshift is given by

$$1 + z = a(t_o)/a(t_e)$$

(see **expansion of the Universe**). The redshift of a distant source therefore acts as a direct measure of the **time** when the light was emitted. It is possible to solve the Friedmann equation that determines the expansion rate of the standard **Friedmann models** to calculate the behaviour of $a(t)$; this can then be used in the **Robertson–Walker metric** to calculate a relation between redshift z and cosmological **proper time** t at emission. Because redshifts are directly observable quantities, many cosmologists refer to the various stages of evolution of the **thermal history of the Universe** in terms of redshift rather than of time. For example, the **cosmic microwave background radiation** was produced at an epoch corresponding to a redshift of the order of 1000, the epoch of domination by radiation at z of the order of 100,000, and so on. The Big Bang itself happened at the origin of time where the scale factor was zero, the redshift of was therefore infinite.

This interpretation of redshifts in terms of the expansion of the Universe is accepted by most cosmologists, but there was once considerable controversy over this issue. For example, quasars have been observed at such high redshifts, corresponding to lookback times greater than 90% of the **age of the Universe**, that their energy output must be phenomenal. (As of the end of 1997, the astrophysical object with the highest known redshift was a pair of gravitationally lensed galaxies with a redshift of 4.92.) This, together with apparent associations between quasars and galaxies with very different redshifts, has led some cosmologists – including Geoffrey **Burbidge** and Halton Arp (see **quasar**) – to question the cosmological interpretation.

Alternative interpretations are possible because there are other effects beside the **Doppler effect** (which is the origin of the cosmological redshift) that could in principle produce a redshift. For example, according to Einstein's theory of **general relativity**, strong gravitational fields give rise to a gravitational redshift: light loses energy has it climbs out of a gravitational potential well. Other ideas included the so-called *tired-light cosmologies* based on alternative theories of **gravity** which do not produce an expanding Universe at all. There have also been claims, hotly disputed by mainstream scientists, that quasar redshifts are quantised. If this is true, it would again be difficult to explain within the framework of standard **cosmological models** based on general relativity.

Although these alternative ideas have not been definitely excluded, the cosmological interpretation leads to a very coherent view of the distribution of matter in the Universe and its evolution with time. It is therefore quite reasonable for cosmologists to trust the standard view until definite evidence is provided to the contrary.

FURTHER READING: Arp, H. C., *Quasars, Redshifts, and Controversies* (Interstellar Media, Berkeley, CA, 1987); Franx, N., *et al.* 'A pair of lensed galaxies at $z = 4.92$ in the field of CL1358+62', *Astrophysical Journal*, 1997, **486**, L75.

REDSHIFT SURVEY A method used to map the distribution of **galaxies** in space, by using **spectroscopy** to obtain **redshifts** for a large controlled sample of galaxies selected according to some well-defined criterion. It is usual to start with a *parent catalogue*. This is a catalogue which lists all galaxies discovered on some part of the sky down to a particular apparent **magnitude** limit, and gives their angular position (and probably an estimated magnitude). A follow-up programme is then initiated to obtain redshifts for these galaxies, or a well-defined subset of them.

The need to measure redshifts in this way stems from the fact that galaxies seen on the sky do appear to be clustered. The *Shane–Wirtanen counts* that were used to build the famous *Lick Map* of a million galaxies in 1967 revealed a frothy visual distribution of galaxies and voids. But since this map contains no information about the distance to the galaxies, we cannot be sure that the pattern seen is a real pattern or simply the effect of chance projections. The most prominent constellations, for example, are recognisable entities, but the stars in them are at greatly different distances from the Sun. Could the same be true for the filaments and voids seen in the Lick Map? The answer requires three-dimensional information.

According to **Hubble's law**, the redshift of a galaxy is simply proportional to its distance from the observer, so we can use redshifts to obtain estimates of radial distance to supplement the angular coordinates of the galaxy on the sky.

A famous example of this approach is the survey made by the Harvard-Smithsonian Center for Astrophysics (CfA), which published its first results in 1986. This was a survey of the redshifts of 1061 galaxies found in a narrow strip on the sky as seen in the original *Palomar Sky Survey* published by Fritz **Zwicky** in 1961. The CfA survey has subsequently been extended to several more strips by the same team. Among its many notable discoveries was the famous *Great Wall* (see **large-scale structure**). Another important example is the more recent *QDOT Survey* (1990) of a randomly selected subset of one in six of the galaxies identified in an all-sky survey by the satellite IRAS (see **infrared astronomy**). The advantage of using a parent sky catalogue compiled by a satellite is that it is possible to cover the whole sky with such observations, whereas observations performed from the ground are inevitably restricted to only part of the celestial sphere. The QDOT survey was instrumental in ruling out the then-fashionable cold dark matter theory of **structure formation** because it revealed the presence of much larger structures than that theory predicted. So far the largest redshift survey to have been published is the *Las Campanas Redshift Survey* (see the Figure), which covers about 25,000 galaxies.

In the 1980s and early 1990s, redshift surveys were slow and laborious processes because it was necessary to point a telescope at each galaxy in turn, take a spectrum, calculate the redshift and then move to the next galaxy. To acquire several thousands of spectra would take months of telescope time which, because of the competition for resources, would usually be spread over several years. More recently the invention of *multi-fibre methods* in conjunction with wide angular fields has allowed astronomers to capture as many as 400 redshifts in one pointing of the telescope. Two new redshift surveys are already under way using these methods: the *Sloan Digital Sky Survey* run by a joint Princeton–Chicago collaboration using a dedicated telescope, and the Two-Degree Field (2dF) run by the United Kingdom and Australia using the Anglo-Australian Telescope. These surveys will map the local Universe to unprecedented depths, and will examine about a million galaxies each.

Such will be the statistical quality of the new generation of redshift surveys that it will be possible to obtain information not only about the galaxies' positions, but also about their **peculiar motions**. This will tell us more about the amount of **dark matter** distributed on cosmological scales.

SEE ALSO: Essays 2 and 4.

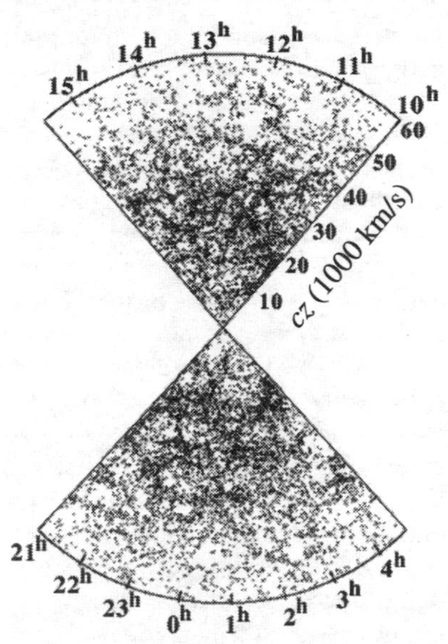

Redshift survey The largest to be completed to date: the *Las Campanas Redshift Survey*. This cone diagram shows the galaxies found in a narrow strip on the sky with redshift z used to estimate their distance. Around 25,000 galaxies are shown.

FURTHER READING: de Lapparent, V., Geller, M. J. and Huchra, J. P., 'The large-scale structure of the Universe', *Astrophysical Journal*, 1986, **302**, L1; Saunders, W. *et al.*, 'The density field of the local Universe', *Nature*, 1991, **349**, 32; Shechtman, S. *et al.*, 'The Las Campanas Redshift Survey', *Astrophysical Journal*, 1996, **470**, 172.

Relativity see **general relativity**, **special relativity**.

Riemann, (Georg Friedrich) Bernhard (1826–66) German mathematician. Although he was not the first to consider them, he contributed greatly to the study of the geometry of curved spaces, and his name is associated with a number of related mathematical constructs. Although he died long before Albert **Einstein** began to work on his theory of general relativity, Einstein was profoundly influenced by Riemann's ideas and much of the theory is based on concepts originated by him.

Ripples Small variations of the apparent sky temperature of the **cosmic microwave background radiation** detected by the **Cosmic Background Explorer** (COBE) satellite in 1992. The discovery received world-wide media coverage, and radio, TV and newspapers cast around for a suitable description of what had been found. The name that stuck was *ripples*, which is actually a fairly good description of the very low-amplitude, large-scale wavelike fluctuations displayed in the COBE map. The term *wrinkles* is also sometimes used.

The significance of these ripples is described more fully in Essay 5. In a nutshell, the temperature variations observed are generally thought to have been produced by small **primordial density fluctuations** via a process called the **Sachs–Wolfe effect**. These ripples have immense implications for theories of **structure formation**, and appear to favour models of the **inflationary Universe**.

COBE had a rather limited angular resolution of about 10°, which is why only the long-wavelength fluctuations (large-scale **anisotropy**) could be seen. A new generation of experiments will detect finer-scale ripples, which should have different properties and physical origin. In particular, experiments are being devised to search for **Sakharov oscillations** on an angular scale of about 1°.

SEE ALSO: **gravitational waves**.

FURTHER READING: Smoot, G. and Davidson, K., *Wrinkles in Time* (William Morrow, New York, 1993); Smoot, G. F. *et al.*, 'Structure in the COBE Differential Microwave Radiometer first-year maps', *Astrophysical Journal Letters*, 1992, **396**, L1.

** **ROBERTSON–WALKER METRIC** The most general form of a Riemannian **metric** that is compatible with the global homogeneity and isotropy required by the **cosmological principle**. It is named after Howard Percy Robinson (1903–61) and Arthur Geoffrey Walker (1909–).

In general, a metric relates physical distances or time intervals between events separated in **spacetime** to the coordinates used by observers to describe the locations of such events. The metric is therefore a mathematical function of the coordinates chosen, but it expresses something which does not depend on coordinates: the geometrical configuration of spacetime itself. General relativity deals with a four-dimensional spacetime in which the separation between the space and **time** coordinates is not obvious. In a homogeneous and isotropic cosmology, however, it is possible to define a unique time coordinate, called cosmological **proper time**, and three related spatial coordinates. In general, as we shall see, the spatial part of the metric describes a curved space which is either expanding or contracting with cosmological proper time.

Because general relativity is a geometrical theory, any model based on the cosmological principle must display the appropriate geometrical properties of general homogeneous and isotropic spaces. Let us suppose that we can regard the matter in the Universe as a continuous fluid, and assign to each element of this fluid a set of three spatial coordinates x^α (where $\alpha = 1, 2$ or 3). Thus, any point in spacetime can be labelled by the coordinates x^α, corresponding to the fluid element which is passing through the point, and a time parameter which we take to be the proper time t measured by a clock moving with the fluid element. The coordinates x^α are called *comoving coordinates*. We can show from simple geometrical considerations only (i.e. without making use of any of the **Einstein field equations** of general relativity) that the most general spacetime metric describing a universe in which the cosmological principle is obeyed is of the form

$$ds^2 = c^2\,dt^2 - a(t)^2[dr^2/(1 - kr^2) + r^2(d\theta^2 + \sin^2\theta\,d\phi^2)]$$

where we have used spherical polar coordinates because of the assumption of isotropy around every point. The coordinates (r, θ, ϕ) are comoving coordinates which remain fixed for any object moving with the **expansion of the Universe** (r is, by convention, dimensionless); t is cosmological proper time. The quantity $a(t)$, which has yet to be determined, is a function which has dimensions of length and which is called the *cosmic scale factor*, or sometimes the *expansion parameter*. The constant k is called the *curvature constant*, and it is scaled to take the values 1, −1 or 0 only.

This expression can be obtained from the general form of a spacetime metric,

$$ds^2 = g_{ij}\,dx^i\,dx^j$$

if the distribution of matter is

uniform. The space is then homogeneous, so the proper time measured by an observer is directly related to the density measured at the observer's location. This immediately means that we can pull out the first term in the Robertson–Walker metric and write the desired form as

$$ds^2 = c^2 dt^2 - g_{ab}\, dx^a dx^b = c^2 dt^2 - dl^2$$

where the interval dl represents the **proper distance** between two objects at a fixed time t. This coordinate system is called the *synchronous gauge*, and is the most commonly used way of defining time in cosmology. Other ways are, however, possible and indeed useful in other circumstances.

Before we tackle the job of finding the three-dimensional (spatial) metric tensor g_{ab}, we shall cut our teeth on the simpler case of an isotropic and homogeneous space of only two dimensions. Such a space can be either the usual Cartesian plane (flat Euclidean space with infinite radius of curvature), or a spherical surface of radius R (a curved space with positive Gaussian curvature $1/R^2$) or the surface of a hyperboloid (a curved space with negative Gaussian curvature).

In the first case the metric, in polar coordinates (ρ, ϕ), is of the form

$$dl^2 = a^2(dr^2 + r^2\, d\phi^2)$$

where we have introduced a dimensionless coordinate ($r = \rho/a$); the arbitrary constant a has the dimension of length. For the surface of a sphere of radius R the metric can be written as

$$ds^2 = a^2(d\theta^2 + \sin^2\theta\, d\phi^2)$$

(as when using latitude and longitude on the surface of the Earth; $a = R$, the radius of the sphere); this can be rewritten as

$$dl^2 = a^2\,[dr^2/(1 - r^2) + r^2\, d\phi^2]$$

by using the dimensionless variable $r = \sin\theta$. The hyperboloidal case is entirely analogous, except that we use $r = \sinh\theta$ and the resulting expression has a changed sign:

$$dl^2 = a^2\,[dr^2/(1 + r^2) + r^2\, d\phi^2]$$

The Robertson–Walker metric is obtained by exactly similar reasoning, except that we have to use spherical polar coordinates instead of plane polar coordinates. The upshot of this is that nothing changes except that the terms in $r^2 d\phi^2$ are replaced by terms in $r^2(d\theta^2 + \sin^2\theta\, d\phi^2)$, as in the form given above for the Robertson–Walker metric itself. The values of $k = 1$, 0, and -1 in the final form correspond respectively to the *hypersphere* (3-sphere), Euclidean three-dimensional space and a space of constant negative curvature (see **curvature of spacetime**).

The geometrical properties of Euclidean space ($k = 0$) are well known. The properties of the hypersphere ($k = 1$) are quite complicated. This space is closed (i.e. it has finite volume) but it has no boundaries. This property is clear by analogy with the two-dimensional sphere: this is a closed surface, with finite area and no boundary. In the three-dimensional case the volume of the space is finite.

The properties of a space of constant negative curvature ($k = -1$) are more similar to those of Euclidean space: the hyperbolic space is open (i.e. infinite). All the relevant formulae for this space can be obtained from those describing the hypersphere by replacing trigonometric functions by hyperbolic functions.

When $k \neq 0$ the parameter a is related to the curvature of space. In fact, the Gaussian curvature is given by $C_G = k/a^2$; as expected, it is positive for the closed space and negative for the open space. The Gaussian curvature radius $R_G = \sqrt{C_G} = \sqrt{k}/a$ is respectively positive or imaginary in these two cases. In cosmology, the term 'radius of curvature' is used to describe the modulus of R_G; with this convention a always represents the radius of spatial curvature. Of course, in a flat universe the parameter a has no geometrical significance and it can be scaled arbitrarily.

Whether the space is closed, open or flat depends on whether the **density parameter** Ω is greater than 1, equal to 1 or less than 1.

SEE ALSO: **closed universe**, **flat universe**, **open universe**.

FURTHER READING: Narlikar, J. V., *Introduction to Cosmology*, 2nd edition (Cambridge University Press, Cambridge, 1993); Rindler, W., *Essential Relativity: Special, General, and Cosmological*, revised 2nd edition (Springer-Verlag, New York, 1979).

ROTATION CURVES The material in spiral galaxies rotates about the centre at high speed, up to hundreds of kilometres per second. We can measure this speed quite easily using the **Doppler effect**. Imagine a rotating disk galaxy oriented in such a way that its plane is presented to us edge-on. Material on one side will be approaching, and that on the other will be receding. Consequently one side will have a spectrum that is shifted towards blue colours, while the other side will be shifted to the red. We can therefore use **spectroscopy** to plot a graph of the velocity of material as a function of its distance from the centre of rotation. Such a curve is called a *rotation curve*.

In this respect a spiral galaxy is similar our Solar System, in which the planets are in orbit around the Sun. The difference is that most of the mass of the Solar System resides in the Sun, but most of the mass of a galaxy does not lie in near the centre of its rotation. We can gauge this from the fact that the speeds of the planets in their orbits decrease with increasing distance from the Sun, while rotation curves show that the matter in spiral galaxies has a roughly constant velocity out to tens of thousands of light years from the centre. There must be material distributed throughout the galaxy to generate this constant velocity, while in the Solar System there is no such material and the rotation curve therefore drops off with distance. Moreover, it is known that the amount of starlight coming from a galaxy falls off very rapidly with distance from the centre.

Putting the flat rotation curve together with the falling light curve leads us to conclude that material is being pulled around by matter not associated with the galactic stars: in other words, there must be **dark matter**. The detailed shape of rotation curves can be fitted by adding a dark-matter halo to the disk component (see the Figure). The usual way of quantifying this dark matter is to calculate the *mass-to-light ratio*, which is the ratio of the total mass (inferred from dynamics) to the total

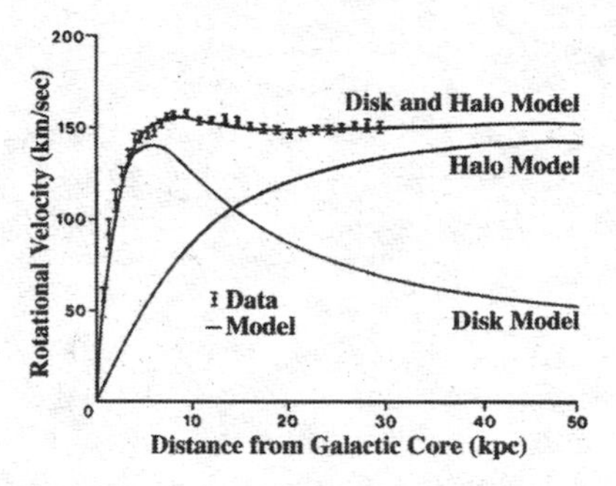

Rotation curves An example of an observed rotation curve for a spiral galaxy (the data points with error bars) together with the predictions of various dynamical models. The material contained in the galactic disk cannot by itself explain the flat rotation curve, but the addition of a 'halo' component produces a model that matches the observations quite well.

luminosity (obtained by adding up all the starlight). It is convenient to give the result in terms of the mass and luminosity of the Sun, which therefore has a mass-to-light ratio of 1. A typical galaxy has a mass-to-light ratio in the range 10 to 30, implying that it contains at least 10 times as much matter as is present in the form of stars like the Sun.

RUBIN, VERA COOPER (1928–) US astronomer, who has spent most of her working life at the Carnegie Institution, Washington, DC. Since 1950 she has studied the rotation of galaxies, discovering that their outer regions rotate more rapidly than expected (see **rotation curves**), suggesting the presence of **dark matter**. Her work has also revealed the Rubin–Ford effect (see **peculiar motions**).

RUBIN–FORD EFFECT see **peculiar motions**.

RYLE, SIR MARTIN (1918–84) English radio astronomer, who worked at Cambridge. His development of **radio astronomy** instruments and techniques, including the radio interferometer and aperture synthesis, made possible the first extensive surveys of radio sources, including the definitive '3C' catalogue. The large number of faint sources discovered (many subsequently turned out to be **quasars**) provided evidence for the **evolution of galaxies**, which helped to discredit the **steady state theory** and supported the **Big Bang theory**.

S

* **Sachs–Wolfe effect** On large angular scales, the most important of various physical processes by which the **primordial density fluctuations** should have left their imprint on the **cosmic microwave background radiation** in the form of small variations in the temperature of this radiation in different directions on the sky. It is named after Rainer Kurt Sachs (1932–) and Arthur Michael Wolfe (1939–). The effect is essentially gravitational in origin. Photons travelling from the **last scattering surface** to an observer encounter variations in the **metric** which correspond to variations in the gravitational potential in Newtonian **gravity**. These fluctuations are caused by variations in the matter density ρ from place to place. A concentration of matter, in other words an upward fluctuation of the matter density, generates a gravitational potential well. According to **general relativity**, photons climbing out of a potential well will suffer a gravitational redshift which tends to make the region from which they come appear colder. There is another effect, however, which arises because the perturbation to the metric also induces a time-dilation effect: we see the photon as coming from a different spatial **hypersurface** (labelled by a different value of the cosmic scale factor $a(t)$ describing the **expansion of the Universe**).

For a fluctuation ϕ in the gravitational potential, the effect of gravitational redshift is to cause a fractional variation of the temperature $\Delta T/T = \phi/c^2$, where c is the speed of light. The time dilation effect contributes $\Delta T/T = -\delta a/a$ (i.e. the fractional perturbation to the scale factor). The relative contributions of these two terms depend on the behaviour of $a(t)$ for a particular cosmological model. In the simplest case of a **flat universe** described by a matter-dominated **Friedmann model**, the second effect is just $-\frac{2}{3}$ times the first one. The net effect is therefore given by $\Delta T/T = \phi/3c^2$. This relates the observed temperature anisotropy to the size of the fluctuations of the gravitational potential on the last scattering surface.

It is now generally accepted that the famous **ripples** seen by the **Cosmic Background Explorer** (COBE) satellite were caused by the Sachs–Wolfe effect. This has important consequences for theories of cosmological **structure formation**, because it fixes the amplitude of the initial **power spectrum** of the primordial density fluctuations that are needed to start off the gravitational **Jeans instability** on which these theories are based.

Any kind of fluctuation of the metric, including gravitational waves of very long wavelength, will produce a Sachs–Wolfe effect. If the primordial density fluctuations were produced in the **inflationary Universe**, we would expect at least part of the COBE signal to be due to the very-long-wavelength gravitational waves produced by quantum fluctuations in the **scalar field** driving inflation.

SEE ALSO: Essay 5.

FURTHER READING: Sachs, R. K. and Wolfe, A. M., 'Perturbations of a cosmological model and angular variations of the cosmic microwave background', *Astrophysical Journal*, 1967, **147**, 73.

SAKHAROV, ANDREI DMITRIEVICH (1921–89) Soviet physicist and noted political dissident. His early work in the field of physics was related to nuclear weapons, but during the 1960s he turned his attention to cosmology and pioneered the theory of **baryogenesis**. He also studied the properties of acoustic waves under the extreme conditions that apply in the early Universe (see **Sakharov oscillations**).

* **SAKHAROV OSCILLATIONS** The large-scale pattern of angular fluctuations (**ripples**) in the temperature of the **cosmic microwave background radiation** detected by the **Cosmic Background Explorer** (COBE) satellite is thought to have been generated by the **Sachs–Wolfe effect**. On smaller angular scales, we would expect to see a different behaviour. In fact, the level of **anisotropy** seen on angular scales of a degree or less should be much higher than that detected by the 10° resolution of the COBE satellite. The characteristic increase in the fluctuation amplitude on these smaller scales is usually called the *Doppler peak*, but this is an extremely inappropriate name for the effect. The physical origin of the enhanced temperature fluctuations was originally worked out by Andrei **Sakharov** (though in a different context) during the 1960s. A more fitting description of this phenomenon is therefore *Sakharov oscillations.*

The physics behind these oscillations is discussed in some detail in Essay 5, so only a brief description is given here. What happens is essentially that, during the *plasma era* of the **thermal history of the Universe**, fluctuations on intermediate length scales oscillated like longitudinal compression waves because they were smaller than the relevant *Jeans length* defined in the theory of gravitational **Jeans instability**. These waves were similar to sound waves in air, except that the medium in which they were oscillating was a two-component fluid of matter and radiation. The cosmological compression waves were also standing waves, acting as if they were in a cavity whose size was fixed by the scale of the cosmological **horizon** at the time.

When such a wave is oscillating there are basically two effects that can cause it to alter the temperature of the radiation as seen by an observer. If a region of such a wave is undergoing a compression, then both matter and radiation are squeezed together. Not only is the region then denser, it is also hotter. But during the oscillations, matter and radiation also move into and out of the compressed region, thus inducing a **Doppler effect** and a consequent increase (or decrease) in the observed temperature according to whether the fluid is moving towards (or away from) the observer. These two effect both contribute, but they are not generally in phase with each other: the phase of maximum compression corresponds to the phase of minimum velocity.

Calculating the net result for waves of different wavelengths is quite complicated, but it is clear that the degree to which the velocity and density effects tend to reinforce each other varies from wave to wave. Some waves would therefore have produced relatively high temperature fluctuations, and others lower fluctuations. When we look at the pattern of temperature fluctuations seen on the sky we see a series of bumps in the angular power spectrum corresponding to this complicated phase effect: the Sakharov oscillations.

SEE ALSO: Essay 5.

FURTHER READING: Sakharov, A. D., 'The initial stage of an expanding Universe and the appearance of a nonuniform distribution of matter', *Soviet Physics JETP*, 1966, **22**, 241.

SANDAGE, ALLAN REX (1926–) US astronomer and observational cosmologist. A former student of Edwin **Hubble**, he became the foremost observer of galaxies of his generation, doing important work on **quasars**, the evolution of **galaxies**, the **extragalactic distance scale** and **classical cosmology**.

** **SCALAR FIELD** In the standard **Friedmann models** on which the **Big Bang theory** is based, the material components of the Universe are generally described as if they were perfect classical fluids with a well-defined density ρ and pressure p. Models of such perfect fluids can describe most of the **thermal history of the Universe** quite adequately, but for the very early Universe they are expected to break down. At very high temperatures it is necessary to describe matter using **quantum field theory** rather than fluid mechanics, and this requires some alterations to be made to the relevant **cosmological models**.

One idea which emerges from these considerations, and which is now ubiquitous in modern cosmology, is the idea that the dynamical behaviour of the early Universe might be dominated by a variety of quantum field called a *scalar field*. A scalar field is characterised by some numerical value, which we shall call Φ. It can be a function of spatial position, but for the purposes of illustration we take it to be a constant. (A *vector field* would be characterised by a set of numbers for each spatial position, like the different components of spin, for example.) We can discuss many aspects of this kind of entity without having to use detailed **quantum theory** by introducing the concept of a *Lagrangian action* to describe its interactions. The Lagrangian for a scalar field can be written in the form

$$L(\Phi) = \tfrac{1}{2}(\mathrm{d}\Phi/\mathrm{d}t)^2 - V(\Phi)$$

The first of these terms is usually called the *kinetic term* (it looks like the square of a velocity), while the second is the *potential term* (V is a function that describes the interactions of the field). The Lagrangian action is used to derive the equations that show how Φ varies with time, but we do not need them for this discussion. The appropriate energy–momentum **tensor** to describe such a field in the framework of **general relativity** can be written in the form

$$T_{ij} = -pg_{ij} + (p + \rho)U_i U_j$$

where g_{ij} is the metric, and U_i and U_j are components of the 4-velocity. To simplify this equation, and the following ones, we have introduced a convention from particle physics in which $h/2\pi = c = 1$. The energy density ρ and pressure p in this expression are effective quantities, given by

$$\rho = \tfrac{1}{2}(\mathrm{d}\Phi/\mathrm{d}t)^2 + V(\Phi)$$

and

$$p = \tfrac{1}{2}(\mathrm{d}\Phi/\mathrm{d}t)^2 - V(\Phi)$$

If the kinetic term is negligible with respect to the potential term, the effective equation of state for the field becomes $p = -\rho$. This is what happens during the **phase transitions** that are thought the drive the **inflationary Universe** model. Under these conditions the field behaves in exactly the same way as an effective **cosmological constant** with

$$\Lambda = 8\pi G\rho/c^2$$

where we have replaced the required factor of c^2 in this expression.

Despite the fact that the scalar field can behave as a fluid in certain situations, it is important to realise that it is not like a fluid in general. If Φ is oscillating, for example, there is no definite relationship between the effective pressure and the effective energy density.

SEE ALSO: Essay 3, **singularity**, **spontaneous symmetry-breaking**.

FURTHER READING: Kolb, E. W. and Turner, M. S., *The Early Universe* (Addison-Wesley, Redwood City, CA, 1990).

SCHMIDT, MAARTEN (1929–) Dutch-born astronomer, who moved to the USA in 1959. He is known for his work on **quasars**, in particular for his discovery in 1963 of the immense redshift of the lines in the spectrum of the quasar designated 3C 273. This, and his subsequent finding that the number of quasars increases with distance, provided important support for the **Big Bang theory**.

SCHRAMM, DAVID NORMAN (1945–97) US physicist and cosmologist who from 1974 worked at the University of Chicago. He was influential in bringing together the disciplines of cosmology, particle physics and astrophysics. His work showed that the Universe is dominated by **dark matter**, and he contributed to the theory of **nucleosynthesis**.

SCHRÖDINGER, ERWIN (1887–1961) Austrian physicist. He founded the field of **quantum mechanics**, and devised the famous Schrödinger equation which describes the evolution of the wavefunction. He shared the 1933 Nobel Prize for Physics with Paul **Dirac**.

SCHWARZSCHILD, KARL (1873–1916) German astronomer. He discovered the solution to Einstein's theory of **general relativity** that represents a spherically symmetric **black hole**. At the time he was serving on the Russian front, where he was to die the same year as he found the famous solution.

SHAPLEY, HARLOW (1885–1972) US astronomer. He discovered Cepheid variables in the **globular clusters**

that surround the Galaxy, and used Henrietta Leavitt's period–luminosity law for Cepheids to estimate of the size of the Galaxy and the Sun's position within it. The early 1920s saw the so-called Great Debate between Shapley, who initially maintained that the spiral 'nebulae' were small and nearby, and Heber Curtis, who took the view that they were independent, distant galaxies. The 1932 *Shapley–Ames Catalog* of galaxies revealed that they are unevenly distributed, and grouped into clusters (see **large-scale structure**).

SILK, JOSEPH (1942–) English cosmologist, based in the USA. He has made many contributions to the theory of the **cosmic microwave background radiation** and cosmological **structure formation**, including **Silk damping**.

SILK DAMPING When the theory of the gravitational **Jeans instability** is applied in the context of the **Big Bang theory** as part of a theory of **structure formation**, it is essential to take into account a number of physical processes that modify the evolution of density fluctuations. The phenomenon of *Silk damping*, named after Joseph **Silk**, is one such effect that applies when we are dealing with adiabatic fluctuations in a medium comprising of baryonic matter and **radiation**.

During the *plasma era* of the **thermal history of the Universe**, the baryonic matter was fully ionised. Under these conditions, free electrons were tightly coupled to the cosmic background radiation by a process known as *Thomson scattering*. Although this coupling was very tight because the collisions between electrons and photons were very rapid, it was not perfect. Photons were not scattered infinitely quickly, but could travel a certain distance between successive encounters with electrons. This distance is called the *mean free path*.

According to the classic theory of Jeans instability, fluctuations would have oscillated like acoustic waves when they are shorter than the so-called *Jeans length*. These waves were longitudinal, and corresponded to a sequence of compressions and rarefactions of the medium through which the waves were travelling. Acoustic waves persisted because there was a restoring force caused by pressure in the regions of compression, which eventually turned them into rarefied regions. In the plasma era most of this pressure was supplied by the photons.

Consider what would have happened to a wave whose wavelength was smaller than the mean free path of the photons. The photons would have leaked out of a compressed region before they had a chance to collide with the electrons and produce a restoring force. This is called *photon diffusion*. The photons moved out of the compression region and into the neighbouring regions of rarefaction, thus smoothing out the wave. Rather than oscillating as acoustic waves, small-scale fluctuations therefore became smoothed out, and this is what is termed Silk damping.

A similar phenomenon occurs with sound waves in air. Here the restoring force is caused by the air pressure, but because air is not a continuous medium, but is made of molecules with a finite mean free path, any wave which is too short (i.e. of too high a frequency) will not be able to sustain oscillations. High-frequency oscillations in air therefore get attenuated, just as acoustic waves do.

Silk damping causes the smoothing of **primordial density fluctuations** on length scales smaller than those of clusters of galaxies. The implication for a theory of structure formation based on this idea is therefore that individual **galaxies** must have formed in a 'top-down' manner by the fragmentation of larger objects. Modern theories of structure formation which include non-baryonic dark matter do not suffer greatly from this effect, so that galaxies can form from smaller objects in a 'bottom-up' fashion.

FURTHER READING: Silk, J., 'Fluctuations in the primordial fireball', *Nature*, 1967, **215**, 1155; Coles, P. and Lucchin, F., *Cosmology: The Origin and Evolution of Cosmic Structure* (John Wiley, Chichester, 1995), Chapter 12.

SINGULARITY A 'pathological' mathematical behaviour in which the value of a particular variable becomes infinite. To give a very simplified example, consider the calculation of the Newtonian force due to gravity exerted by a massive body on a test particle at a distance *r*. This force is proportional to $1/r^2$, so if we tried to calculate the force for objects at zero separation, the result would be infinite. Singularities are not always signs of serious mathematical problems: sometimes they are caused simply by an inappropriate choice of coordinates. For example, something strange and akin to a singularity happens on some of the maps of the world in a standard atlas. These maps look quite sensible until we look at regions very near to the poles. In the standard Mercator projection, the north and south poles appear not as points, as they should, but are spread out into a straight line along the top and bottom of the map. If you were to travel to one of the poles you would not find anything catastrophic happening. The singularity that causes these lines to appear is an example of a coordinate singularity, and it can be transformed away simply by using a different kind of projection.

Singularities occur with depressing regularity in solutions of the equations of **general relativity**. Some of these are coordinate singularities like the one discussed above, and are not particularly serious. However, Einstein's theory does predict the existence of real singularities where real physical quantities (such as the matter density) become infinite. The **curvature of spacetime** can also become infinite in certain situations.

Probably the most famous example of a singularity is that which lies at the centre of a **black hole**. This appears in the original Schwarzschild solution corresponding to a hole with perfect spherical symmetry. For many years, physicists thought that the existence of a singularity of this kind was merely a consequence of the rather artificial nature of this spherical solution. However, a series of mathematical investigations, culminating in the singularity theorems devised by Roger **Penrose**, showed that no special symmetry is required, and also that singularities can arise as a result of gravitational collapse.

As if to apologise for predicting these singularities in the first place, general relativity does its best to hide them from us. A *Schwarzschild black hole* is surrounded by an *event horizon* that effectively protects outside observers from the singularity itself. It seems likely that all singularities in general relativity are protected in this way, and so-called *naked singularities* are not thought to be physically realistic.

There is also a singularity at the very beginning in the standard **Big Bang theory**. This again is expected to be a real singularity, where the temperature and density both become infinite. In this respect the Big Bang can be thought of as a kind of time-reversal of the gravitational collapse that forms a black hole. As with the Schwarzschild solution, many physicists thought that the initial cosmological singularity could be a consequence of the special symmetry required by the **cosmological principle**. But this is now known not to be the case. Stephen **Hawking** and Roger Penrose have generalised Penrose's original black hole theorems to show that a singularity invariably exists in the past of an expanding Universe in which certain very general conditions apply. So is it possible to avoid this singularity, and if so, how?

It is clear that the initial cosmological singularity might be nothing more than a consequence of extrapolating deductions based on the classical theory of general relativity into a situation where this theory is no longer valid. Indeed, as Einstein himself wrote in 1950:

> The theory is based on a separation of the concepts of the gravitational field and matter. While this may be a valid approximation for weak fields, it may presumably be quite inadequate for very high densities of matter. One may not therefore assume the validity of the equations for very high densities and it is just possible that in a unified theory there would be no such singularity.

We clearly need new **laws of physics** to describe the behaviour of matter in the vicinity of the Big Bang, when the density and temperature were much higher than can be achieved in laboratory experiments. In particular, any theory of matter under such extreme conditions must take account of quantum effects on a cosmological scale. The name given to the theory of **gravity** that would replace general relativity at ultra-high energies by taking these effects into account is **quantum gravity**, but unfortunately such a theory has still to be constructed.

There are, however, ways of avoiding the initial singularity in general relativity without appealing to quantum effects. Firstly, we could try to avoid the singularity by proposing an equation of state for matter in the very early Universe that does not obey the conditions laid down by Hawking and Penrose. The most important of these conditions is called the *strong energy condition*: that $\rho + 3p/c^2 > 0$, where ρ is the matter density and p is the pressure. There are various ways in which this condition might indeed be violated. In particular, it is violated by a **scalar field** whose evolution is dominated by its vacuum energy, which is the condition necessary for driving **inflationary Universe** models into an accelerated expansion. The vacuum energy of the scalar field may be regarded as an effective **cosmological constant**; models in which the cosmological constant is included generally have a 'bounce' rather than a singularity: if we turn back the clock we find that the Universe reaches a minimum size and then expands again.

Whether the singularity is avoidable or not remains an open question, and the issue of whether we can describe the very earliest phases of the Big Bang, before the **Planck time**, will remain unresolved at least until a complete theory of quantum gravity is constructed.

SEE ALSO: Essay 3.

FURTHER READING: Penrose, R., 'Gravitational collapse and space-time singularities', *Physical Review Letters*, 1965, **14**, 57; Hawking, S. W. and Penrose, R., 'The singularities of gravitational collapse and cosmology', *Proceedings of the Royal Society*, 1970, **A314**, 529.

SLIPHER, VESTO MELVIN (1875–1969) US astronomer, who worked at the Lowell Observatory. In 1912 he obtained the first radial-velocity measurement from the spectrum of a 'spiral nebula', the Andromeda Galaxy, and in 1914 he announced his discovery of the **redshifts** in the spectra of other such objects. By 1925 he had radial velocities for over forty of them, all too high for the objects to belong to the Milky Way system. This demonstrated that they were indeed distant systems external to our own Galaxy, and provided the observational evidence for the **expansion of the Universe**.

SMOOT, GEORGE (1945–) US physicist. He led the team that oversaw the instrumentation and data analysis for the **Cosmic Background Explorer** (COBE) satellite that discovered temperature fluctuations ('**ripples**') in the **cosmic microwave background radiation**. He described this discovery as like 'looking at the face of God'.

SOURCE COUNTS The oldest and simplest of the tests of **classical cosmology**, which consists simply of counting the number of sources of a particular type as a function of their apparent brightness. In **optical astronomy**, this basically means counting **galaxies** as a function of their apparent **magnitude**, while in **radio astronomy** radio sources are counted as a function of their radio flux density

The idea of the test is that faint sources (those with low observed fluxes) will, on average, tend to be more distant than bright ones. By simply counting sources to lower and lower flux limits we are therefore effectively probing to greater and greater depths. We are not, however, attempting to measure a distance for each source, so difficult **spectroscopy** is not required. In a static Euclidean universe in which the properties of the sources are unchanging with time, the number N observed with fluxes greater than a given value S would have the form of a simple power law. In fact, if we were to plot a graph of $\log N$ against $\log S$ the slope should be –3/2, since the **luminosity distance** depends on the square root of the flux limit. Any departure from this behaviour is evidence that at least one of the assumptions made is incorrect: the

Universe may be non-Euclidean (see **curvature of spacetime**) or non-static (see **expansion of the Universe**), or the sources may be evolving (see **evolution of galaxies**). The net behaviour of the $\log N$ versus $\log S$ relation is, however, affected by all these factors so it is not easy to see which of them accounts for the observed behaviour.

It was established (by Martin **Ryle** and others in the early 1960s) that the **steady state theory** (which does not allow evolution) is ruled out by radio source counts which showed clear signs of evolution. Counts of optically identified galaxies are also known to be dominated by evolution. This means that it is difficult to use source counts to test the geometry and expansion rate of the Universe, which was the original goal of this idea, as any cosmological effects are now known to be completely swamped by evolution. Recent observations with the **Hubble Space Telescope** have allowed optical source counts to be obtained to a staggering faint limit of 29th magnitude in blue light, but even this tells astronomers much more about the evolution of galaxies than it does about **cosmological models**.

FURTHER READING: Metcalfe, N. *et al.*, 'Galaxy formation at high redshifts', *Nature*, 1996, **383**, 236.

* **SPACETIME** In the theories of **special relativity** and **general relativity** the three dimensions of space and the one dimension of **time** are handled not separately, but as parts of a four-dimensional structure called *spacetime*. This amalgamation of different concepts is required because, in relativistic theories, physical laws are expressed in terms of quantities that are the same for all observers regardless of how they are moving. For observers undergoing relative motion in special relativity, clocks need not beat at the same rate, and rulers need not appear to be the same length. Events apparently occurring simultaneously when observed by one observer may be separated in time when observed by another. Time must therefore be measured in a relative manner, just as spatial positions are. However, we can construct intervals between events (at different times and/or in different places) in the four-dimensional spacetime in such a way that these intervals do not depend on the state of motion of whoever is measuring them.

Although space and time are, in some senses, treated equivalently in relativistic theories, they are not exactly the same. The difference is described by the **metric** which, for special relativity, is written as

$$ds^2 = c^2\,dt^2 - (dx^2 + dy^2 + dz^2)$$

Ignoring the first term on the right-hand side and disregarding the minus sign, this expression simply describes Pythagoras's theorem in flat, *Euclidean* space, so the spatial interval between two points is as expected. The term in $c\,dt$ brings time into the picture, but note that it has a different sign to the spatial coordinates. This indicates the characteristic signature of the metric, which emphasises the special nature of time. Light rays travel along paths in spacetime defined by $ds = 0$. These paths are called *null geodesics*. It makes more sense to think of light rays as existing in time and space rather than travelling in them. Just as we think of two different spatial points existing at the same time, so in some sense does the past, present and future of a light ray exist as well.

In general relativity the situation is rather more complicated because the spatial part of the metric is curved (this is how this theory can describe both acceleration and **gravity**). In this theory the spacetime is described in mathematical terms as having a *Riemannian* geometry (formally, it is a 4-dimensional Riemannian manifold). In general terms the metric of this geometry is a **tensor** denoted by g_{ij}, and the interval between two events is written as

$$ds^2 = g_{ij}\, dx^i\, dx^j$$

where repeated suffixes imply summation, with i and j both running from 0 to 3; $x^0 = ct$ is the time coordinate, and (x^1, x^2, x^3) are the space coordinates. Particles acted on by no gravitational forces (i.e. free particles) move along paths which are no longer straight because of the effects of the curvature of spacetime contained in g_{ij}. Paths in this four-dimensional space are called *geodesics*, a generalisation of the concept of the shortest path between two points on a curved surface. Geodesics are defined in such a way that they minimise the integral of a quantity called the *action* along the path. The paths of light rays are still null geodesics (with $ds = 0$). The trajectory of a free particle is a geodesic with ds not necessarily equal to zero. The general trajectory of a particle moving through spacetime (not necessarily on a geodesic) is called a *world line*. An extended body traces out a more complicated region of spacetime. For example, a thin piece of string traces out a 'world sheet', and a sheet would trace out a kind of 'world volume'.

It may help to visualise spacetime by considering a smaller number of dimensions and ignoring the spatial curvature. If the Universe had only one spatial and one temporal dimension, the world line could be drawn on a graph with time plotted vertically and spatial distance plotted horizontally. A particle at rest with respect to the coordinate system has a world line that coincides with the vertical axis; moving particles have world lines which are curves or straight lines sloping upwards. World lines cannot go downwards in this diagram because of the special nature of time, which does not allow travel in the time-reverse direction. For two spatial dimensions we can picture a stack of movie frames. Imagine a particle moving around within the camera shot. If we stacked these frames on top of one another so that time is represented in the vertical direction, then the world line of the particle would be a more complicated line, possibly spiralling upwards.

SEE ALSO: **hypersurface.**

FURTHER READING: Narlikar, J. V., *Introduction to Cosmology*, 2nd edition (Cambridge University Press, Cambridge, 1993); Rindler, W., *Essential Relativity: Special, General, and Cosmological*, revised 2nd edition (Springer-Verlag, New York, 1979).

SPECIAL RELATIVITY Albert **Einstein**'s theory of special relativity, published in 1905, stands as one of the greatest intellectual achievements in the history of human thought. The reason why special relativity should be regarded so highly is that Einstein managed to break away completely from an idea that most of us regard as being obviously true: that time is an absolute phenomenon. Although **general relativity** is a more complete and mathematically challenging theory than its precursor, the deep insights required to make the initial step are perhaps even more impressive than the monumental work that became the later generalisation of it.

The idea of relativity itself goes back to Galileo, who was the first to claim that it is only relative motion that matters. Galileo argued that if one were travelling in a boat at constant speed on a smooth lake, there would be no experiment that one could perform in a sealed cabin on the boat that would indicate that there was any motion at all. Einstein's version of the principle of relativity simply turned it into the statement that all **laws of physics** have to be exactly the same for all observers in relative motion. In particular, Einstein decided that this principle must apply to the recently developed theory of electromagnetism (see **fundamental interactions**). It is a consequence of James Clerk **Maxwell**'s equations of electromagnetism that the speed of light in a vacuum appears as a universal constant (usually denoted by c). The principle of relativity implies that all observers have to measure the same value of c. This seems straightforward enough, but the consequences are nothing short of revolutionary.

Einstein decided to ask himself specific questions about what would be observed in particular kinds of experiment in which light signals are exchanged. There are scores of fascinating examples, but here we give just one. Imagine that there is a flash bulb in the centre of a railway carriage moving along a track. At each end of the carriage there is a clock, and when the flash illuminates it we can see the time. When the flash goes off, the light signal reaches both ends of the carriage simultaneously from the point of view of someone sitting in the carriage: the same time is seen on each clock.

Now picture what happens from the point of view of an observer at rest who is watching the train from the trackside. The light flash travels with the same speed in the trackside observer's reference frame as it does for the passengers. But the passengers at the back of the carriage are moving towards the signal, while those at the front are moving away from it. The trackside observer therefore sees the clock at the back of the train lit up before the clock at the front. But when the clock at the front is lit up, it shows the same time as the clock at the back!

This example demonstrates that the concept of simultaneity is relative: the arrivals of the two light flashes are simultaneous in the frame of the carriage, but occur at different times in the frame of the track. Other examples of the same phenomenon are *time dilation* (moving clocks appear to run slow) and the so-called *Lorentz–Fitzgerald contraction* (moving rulers appear shorter). These are all consequences of the assumption that the speed of light must be the same when it is measured by all observers. Of course, the examples given above are a little unrealistic: in order to show noticeable effects the velocities concerned must be a sizeable fraction of c. Such speeds are unlikely to be reached in railway carriages. Nevertheless, experiments have been done which show that time dilation effects are real: the decay rate of radioactive particles is much slower when they are moving at high velocities because their internal clocks run slowly.

Special relativity also spawned the most famous equation in all of physics: $E = mc^2$, expressing the equivalence between matter and energy. This prediction has also been tested, rather too often, because it embodies the principle behind the explosion of atomic bombs.

In 1908 Hermann **Minkowski** expressed special relativity in the form in which it is usually used today: in terms of a **spacetime** described by a particular **metric**. It was this formulation that allowed Einstein to generalise the theory to incorporate not just uniform motion in straight lines, but also acceleration and gravity (see **general relativity**).

FURTHER READING: Rindler, W., *Essential Relativity: Special, General, and Cosmological*, revised 2nd edition (Springer-Verlag, New York, 1979).

SPECTROSCOPY A battery of observational techniques used to study the spectra of **electromagnetic radiation** emitted by astronomical sources of various kinds. Most astronomical sources emit some form of continuum radiation, upon which are superimposed various **emission lines** or **absorption lines**. The shape of the observed spectrum can be used to determine the chemical composition, density and temperature of the source. Measurements of the **Doppler effect** can be used to determine the motion of the object along the line of sight to the observer. The birth of spectroscopic methods in the 19th century, which greatly increased our scope for understanding the physical properties of distant objects, led to the birth of a new branch of science: astrophysics.

Different instruments are used for spectroscopic studies of different kinds of radiation, but they all feature some form of spectrograph mounted on a telescope. A *spectrograph* is a device for dispersing light so that the amount of energy present at each wavelength can be measured. It usually contains a slit through which the light must pass, excluding stray light from other objects in the telescope's field of view. This light is then collimated into a narrow beam, and a grating or prism disperses the light onto a detector where its properties are measured. Modern detectors are usually electronic (i.e. CCDs) but in the past photographic plates were used for this task. Instruments have been devised for many possible wavebands, but most spectroscopy is carried out with infrared, optical or ultraviolet light.

Optical spectroscopy is the method usually used to determine the recessional velocities of **galaxies** according to their **redshifts**. The optical portion of the electromagnetic spectrum of galaxies contains a large number of emission lines that can be used to measure the Doppler shift. Spectroscopic methods in this region can also be used to measure the rotation curves of galaxies, and thus to seek evidence for **dark matter**, although

more recently infrared techniques have been deployed for this task. The **Hubble Space Telescope** also has the capability to perform spectroscopy in the ultraviolet region.
SEE ALSO: Essay 4.

FURTHER READING: Aller, L. H., *Atoms, Stars, and Nebulae* (Harvard University Press, Cambridge, MA, 1971); Graham-Smith, F. and Lovell, B., *Pathways to the Universe* (Cambridge University Press, Cambridge, 1988).

** **SPONTANEOUS SYMMETRY-BREAKING** In particular many-particle systems, **phase transitions** take place when processes occur that move the system between some disordered phase, characterised by a certain degree of symmetry, and an ordered phase with a smaller degree of symmetry. In this type of order–disorder transition, some macroscopic quantity called the *order parameter* (here denoted by the symbol Φ) grows from its original value of zero in the disordered phase. The simplest physical examples of materials exhibiting these transitions are ferromagnetic substances and crystalline matter. At temperatures above a critical temperature T_C (the *Curie temperature*), the stable state of a ferromagnet is disordered and has net magnetisation $M = 0$; the quantity M in this case plays the role of the order parameter. At temperatures below T_C a nonzero magnetisation appears in different domains (called *Weiss domains*) and its direction in each domain breaks the rotational symmetry possessed by the original disordered phase at $T > T_C$. In the crystalline phase of solids the order parameter represents the deviation of the spatial distribution of ions from the homogeneous distribution they have at their critical temperature, which is simply the freezing point of the crystal, T_f. At temperatures $T < T_f$ the ions are arranged on a regular lattice, which possesses a different symmetry to the original liquid phase.

The lowering of the degree of symmetry of the system takes place even though the **laws of physics** that govern its evolution maintain the same degree of symmetry, even after the phase transition. For example, the macroscopic equations of the theory of ferromagnetism and the equations in solid-state physics do not favour any particular spatial position or direction. The ordered states that emerge from such phase transitions do, however, have a degree of symmetry which is less than that governing the system. We can say that the solutions corresponding to the ordered state form a *degenerate set* of solutions with the same symmetry as that possessed by the laws of physics. Returning to the above examples, the magnetisation M can in theory assume any direction. Likewise, the ions in the crystalline lattice can be positioned in an infinite number of different ways. Taking into account all these possibilities, we again obtain a homogeneous and isotropic state. Any small fluctuation, in the magnetic field of the domain for a ferromagnet or in the local electric field for a crystal, will have the effect of selecting one preferred solution from this degenerate set, and the system will end up in the state corresponding to that fluctuation. Repeating the phase transition with random fluctuations will eventually produce randomly aligned final states.

This is a little like the case of a free particle with velocity $\boldsymbol{v}$ and position $\boldsymbol{r}$, described in Newtonian mechanics by the requirement that $d\boldsymbol{v}/dt = 0$. This (very simple) law has both translational and rotational symmetries. The solutions of the equation are of the form $\boldsymbol{r}(t) = \boldsymbol{r}_0 + \boldsymbol{v}_0 t$, with some arbitrary initial choice of $\boldsymbol{r} = \boldsymbol{r}_0$ and $\boldsymbol{v} = \boldsymbol{v}_0$ at $t = 0$. These solutions form a set which respects the symmetry of the original equation. But the initial conditions $\boldsymbol{r}_0$ and $\boldsymbol{v}_0$ select, for one particular time, a solution from this set which does not have the same degree of symmetry as that of the equations of motion.

A symmetry-breaking transition, during which the order parameter Φ grows significantly, can be caused by external influences of sufficient intensity. For example, a strong magnetic field can magnetise a ferromagnet, even above the Curie temperature. Such phenomena are called *induced symmetry-breaking* processes, in order to distinguish them from *spontaneous symmetry-breaking*. The spontaneous breaking of a symmetry comes from a gradual change of the parameters of a system, such as its temperature.

It is useful to consider the free energy F of a system, which is defined in **thermodynamics** to be $F = E - TS$, where E is the internal energy, T is the temperature and S is the entropy; the condition for the system to have a stable equilibrium state is that F must have a minimum. The free energy coincides with the internal energy only at $T = 0$. At higher temperatures, whatever the form of E, an increase in the entropy (i.e. disorder) generally leads to a decrease in the free energy F, and it is therefore favourable. For systems in which there is a phase transition, F is a function of the order parameter Φ. Under some circumstances the free energy F must have a minimum at $\Phi = 0$ (corresponding to the disordered state), while in others it must have a minimum for some $\Phi \neq 0$ (corresponding to the ordered state).

Let us now consider the simplest possible example. Suppose that the situation we are dealing with respects a symmetry between Φ and $-\Phi$. The free energy function F must also respect this symmetry, so that we can expand $F(\Phi)$ in a power series containing only even powers of Φ:

$$F(\Phi) \approx F_0 + \alpha(T)\Phi^2 + \beta(T)\Phi^4$$

where the coefficients α and β of this expansion depend explicitly on temperature T. If $\alpha > 0$ and $\beta > 0$, the curve has one minimum at the origin, so the equilibrium state lies at $\Phi = 0$. If the coefficients change in such a way that at lower temperatures $\alpha < 0$ and $\beta > 0$, the original minimum becomes a maximum and two new minima appear at $\Phi = \pm\sqrt{(\alpha/2\beta)}$; in this case the disordered state is unstable, while the minima correspond to degenerate ordered states with the same probability. Any small external perturbation which renders one of the two minima slightly deeper or nudges the system towards one of them can make the system evolve towards one rather than the other. This is how spontaneous symmetry-breaking occurs. If there is only one parameter describing the system, say the temperature, and the coefficient α is written as $\alpha = a(T - T_C)$ with $\alpha > 0$, this transition is called a *second-order phase transition*, and it proceeds by a process known as *spinodal decomposition*: the order parameter appears or disappears gradually, and the difference ΔF in free energy between the two states is infinitesimally small when T is close to T_C.

There are also *first-order phase transitions*, in which at $T \approx T_C$ the order parameter appears or disappears rapidly, and the difference ΔF between the old and new minima is finite. This difference is called the *latent heat* of the phase transition. In first-order phase transitions, when T decreases the phenomenon known as *supercooling* can occur: the system remains in the disordered state characterised by $\Phi = 0$ even when T has fallen below the critical temperature. This is because the original minimum represents a *metastable* equilibrium. A local minimum of the free energy still persists there, protected by a *potential barrier* that prevents the system from moving immediately into the global minimum. As T decreases further, or the system is perturbed by either internal or external fluctuations, the system finally evolves into the global minimum that represents the true energetically stable state. This liberates the latent heat that was stored in the metastable state. The system, still in the ordered state, is heated up again to a temperature of around T_C by the release of this latent heat, a phenomenon called *reheating*.

To apply these notions to cosmology we need also to take into account quantum and thermal fluctuations and other effects of non-equilibrium thermodynamics. Different symmetries in the past also involved different fields interacting according to different potentials. Nevertheless, the basic general properties of phase transitions lie behind many models of the **inflationary Universe**.

SEE ALSO: **fundamental interactions**, **gauge theory**, **grand unified theory.**

FURTHER READING: Kolb, E. W. and Turner, M. S., *The Early Universe* (Addison-Wesley, Redwood City, CA, 1990).

STAR see **stellar evolution**.

STEADY STATE THEORY A **cosmological model**, advanced in the late 1940s by Thomas **Gold**, Fred **Hoyle**, Hermann **Bondi** and Jayant Narlikar (among others), based around a universe which is expanding but has the same properties at all times. The principle behind it is called the *perfect cosmological principle*, a generalisation of the **cosmological principle** which says that the Universe is homogeneous and isotropic in space, so as to include homogeneity with respect to time. For two decades steady state cosmology was a serious rival to the **Big Bang theory**. The steady state theory fell out of favour mainly because it appears to be inconsistent with the observed **black-body** spectrum of the **cosmic microwave background radiation**, and it cannot explain the observed strong correlation between **evolution of galaxies** and **quasars**, and **redshift**.

Because all the properties of steady state cosmology have to be constant in time, the expansion rate of this model is also a constant. It is then easy to find a solution of the **Einstein equations** that corresponds to this: the result is the *de Sitter solution* (see **cosmological models**), corresponding to an exponential time-dependence of the scale factor that describes the **expansion of the Universe**. But if the Universe is expanding, the density of matter must decrease with time. This is avoided in the steady state theory by postulating the existence of a field, called the *C-field*, which produces a steady stream of matter to counteract the dilution caused by the cosmic expansion. This process, called *continuous creation*, has never been observed in the laboratory, but the rate of creation required is so small (about one atom of hydrogen per cubic metre over the entire **age of the Universe**) that it is difficult to rule out continuous creation as a possible physical process by direct observations.

Although it is firmly excluded by present-day observations, the model played a vital role in the development of cosmology and astrophysics in the 1940s and 1950s. For example, it acted as a spur to fundamental work on stellar evolution, including the first detailed calculations of the nucleosynthesis of helium in stars.

From a philosophical point of view the steady state theory has many appealing features. For example, unlike the Big Bang theory it contains no initial **singularity**. Moreover, it embodies many of the concepts now incorporated in the **inflationary Universe** scenario. In inflation, the expansion of the Universe is described by the de Sitter cosmological model, as in the steady state theory. The C-field responsible for continuous creation is also a forerunner of the idea of a **scalar field**, like the one that drives inflation. In some versions of the inflationary Universe the overall properties of spacetime resemble a steady state model very strongly. In the *eternal inflation* model, for example, Big Bang universes appear as small bubbles in an infinite and eternal de Sitter vacuum. The only significant difference between this version of the Big Bang and the original steady state cosmology is the size of the creation event. The inflationary model gives rise to entire universes, rather than single hydrogen atoms.

FURTHER READING: Bondi, H. and Gold, T., 'The steady state theory of the expanding Universe', *Monthly Notices of the Royal Astronomical Society*, 1948, **108**, 252; Hoyle, F., 'A new model for the expanding Universe', *Monthly Notices of the Royal Astronomical Society*, 1948, **108**, 372; Hoyle, F. and Narlikar, J. V., 'Mach's principle and the creation of matter', *Proceedings of the Royal Society*, 1963, **A273**, 1; Linde, A. D., Linde, D. and Mezhlumian, A., 'From the Big Bang theory to the theory of a stationary Universe', *Physical Review* D, 1994, **49**, 1783; Krach, H., *Cosmology and Controversy: The Historical Development of Two Theories of the Universe* (Princeton University Press, Princeton, NJ, 1996).

STELLAR EVOLUTION The changes in luminosity, size and chemical composition that stars undergo during their lifetime. These changes are commonly represented on the fundamental diagram of stellar astrophysics, the *Hertzsprung–Russell* (HR) *diagram*, which is shown schematically in the Figure (overleaf) with a certain amount of artistic licence. What this diagram reveals is that stars are not spread out randomly when their temperature is plotted against their absolute luminosity. Many stars lie on a relatively narrow band on the HR diagram called the *main sequence*.

Other special kinds of star appear in particular regions of the diagram. The theory of stellar evolution explains both the main sequence and these extra features of the diagram.

In a nutshell, the life-history of a star according to the standard picture of stellar evolution is as follows. A star is born from the material contained in a cloud of gas when a fluctuation in the density of the gas becomes sufficiently large for a gravitational **Jeans instability** to develop and form what is called a *protostar*. For a static (non-expanding) gas cloud, this collapse can be very rapid. The protostar loses gravitational potential energy as it collapses, and this is turned into heat. Eventually the star becomes so hot that nuclear reactions begin and the protostar becomes a star. This sequence of events is represented on the HR diagram as a *Hayashi track*, a line moving from right to left on the diagram, and terminating on the main sequence. (The Figure shows Hayashi tracks for stars of 1.0, 0.01 and 0.001 solar masses; the smaller objects fail to reach the 'ignition temperature' for nuclear reactions, and do not make it to the main sequence.)

When the star is hot enough to allow hydrogen to be converted to helium (a process known as *burning*), the collapse is halted and the star enters a state of equilibrium in which the energy released by nuclear reactions counteracts the tendency to collapse: the radiation pressure from within holds up the outer layers of the star against gravitational collapse. This equilibrium state, which varies according to the mass of the star, represents the main sequence. How long the hydrogen-burning phase lasts depends on the star's mass. Although more massive stars have more hydrogen to burn, they also burn their fuel more rapidly and at a much higher temperature. The net effect of this is that very massive stars remain on the main sequence for a shorter time than less massive ones. Stars of lower mass than the Sun, for example, can remain on the main sequence for longer than the present **age of the Universe**.

Once the hydrogen in the core of a star has been exhausted, the core begins to contract. If the star's mass is more than about 40% of the mass of the Sun, the core then becomes hot enough to begin burning helium into carbon. What happens then depends sensitively on the mass. Stars whose masses are greater than the Sun's continue to burn hydrogen in a shell outside the core, while core itself burns helium. In this phase the star is cooler (and therefore redder), larger and brighter than it was on the main sequence. Such stars are called *giants*, and lie on the *giant branch*, coming off the main sequence towards the top of the HR diagram. More massive stars still become *supergiants*, at the top of the HR diagram. Once the helium in the core is exhausted, the core contracts again and other nuclear reactions fire up. This can happen several times, leading to a complicated sequence of events which may lead to variability and/or a very stratified chemical composition of the star as different nuclear reactions take place in different shells. Eventually, however, either the core cannot contract sufficiently to ignite any more nuclear reactions, or the core ends up being

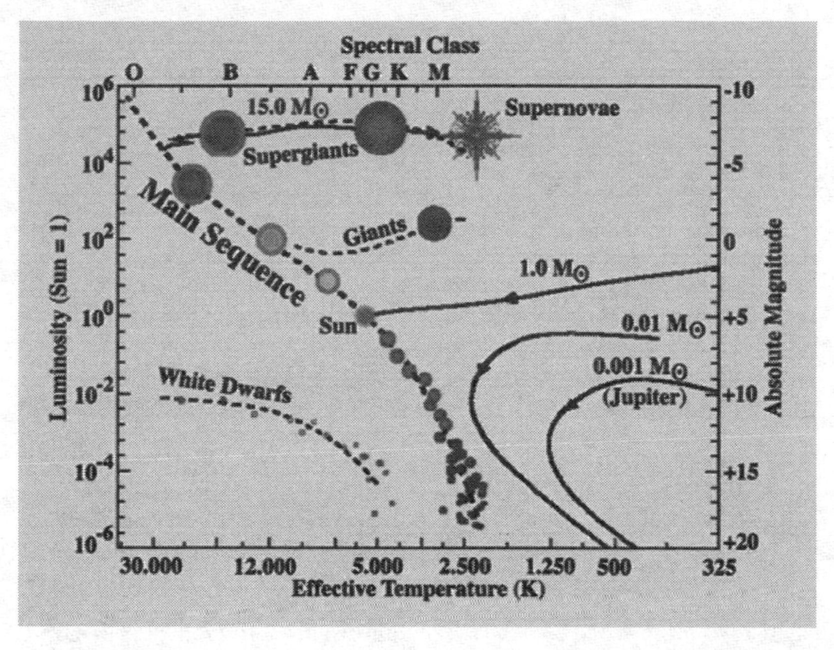

Stellar evolution Schematic illustration of the Hertzsprung–Russell diagram. The main sequence is the band where most stars spend most of their life: off this sequence are various possible evolutionary paths which stars of different masses might follow.

filled with iron. The latter eventuality awaits supergiant stars. Since iron is the most stable atomic nucleus known, it is not a nuclear fuel. At this point the star switches off. There being no production of energy left to support the star against its gravity, it begins a rapid and catastrophic collapse to a *neutron star* or even a **black hole**. The outer layers are blasted off during this process, in a **supernova** explosion.

In stars which are less massive than supergiants, the post-main-sequence evolution is different. These stars have denser cores, in which *degeneracy effects* can be significant. A degenerate core is one where the dominant source of pressure arises from the Pauli exclusion principle of **quantum theory**. This is a non-thermal source of pressure that depends only on the density. When the helium-burning phase begins for a star which has a degenerate core, it produces a *helium flash* and the core expands. The star then moves onto a region of the HR diagram called the *horizontal branch*. Helium burns in the core in this phase, while the outer shells burn hydrogen. Once the core helium is exhausted, the star continues to burn hydrogen outside the core and enters what is called the *asymptotic giant branch*. What happens then is rather uncertain: probably the outer layers are expelled to form a *planetary nebula*, while the degenerate core remains in the form of a *white dwarf*.

In any case, the final state is a compact object which gradually cools, much of the material of the star having been recycled into the **interstellar medium**, as in a supernova.
SEE ALSO: **globular clusters**.

FURTHER READING: Phillips, A. C., *The Physics of Stars* (John Wiley, Chichester, 1994); Tayler, R. J., *The Stars: Their Structure and Evolution* (Cambridge University Press, Cambridge, 1994).

STRING THEORY During the 1980s, mathematical physicists, including Michael Green (of Queen Mary College, University of London), became interested in a class of theories of the **fundamental interactions** that departed radically from the format of **gauge theories** that had been so successful in unified models of the physics of **elementary particles**. In these theories, known as *string theories*, the fundamental objects are not point-like objects (particles) but one-dimensional objects called *strings*. These strings exist only in spaces with a particular number of dimensions (either 10 or 26).

The equations that describe the motions of these strings in the space they inhabit are very complicated, but it was realised that certain kinds of vibration of the strings could be treated as representing discrete particle states. Amazingly, a feature emerged from these calculations that had not been predicted by any other forms of **grand unified theory**: there were closed loops of string corresponding to massless bosons that behaved exactly like *gravitons* – hypothetical bosons which are believed to mediate the gravitational interaction. A particular class of string theories was found that also produced the properties of **supersymmetry**: these are called *superstrings*. Many physicists at the time became very excited about superstring theory because it suggested that a **theory of everything** might well be within reach.

The fact that these strings exist in spaces of much higher dimensionality than our own is not a fundamental problem. A much older class of theories, called **Kaluza–Klein theories**, had shown that spaces with a very high dimensionality were possible if extra dimensions, over and above the four we usually experience, are wound up (*compactified*) on a very small length scale. It is possible, therefore, to construct a string theory in 26 dimensions, but wrap 22 of them up into such a tight bundle (with a scale of order the **Planck length**) that they are impossible for us to perceive.

Unfortunately there has been relatively little progress with superstring theory, chiefly because the mathematical formalism required to treat their complicated multidimensional motions is so difficult. Nevertheless, hope still remains that string theories, or generalisations of them such as membranes or M-theory, will pave the way for an eventual theory of everything.
SEE ALSO: Essay 3.

FURTHER READING: Barrow, J. D., *Theories of Everything* (Oxford University Press, Oxford, 1991).

STRUCTURE FORMATION From observations of the **cosmic microwave background radiation** (see also Essay 5), we know that the Universe was almost (but not quite) homogeneous when this radiation was last in contact with matter, about 300,000 years after the initial Big Bang. But we also know that the **Universe** around us today (perhaps 15 billion years later) is extremely inhomogeneous: matter is organised into **galaxies**, clusters of galaxies, superclusters, and so on in a complex hierarchy of **large-scale structure**. On sufficiently large scales the Universe does indeed begin to look homogeneous (as required by the **cosmological principle**), but there is clearly a great deal of structure around us that was not present at the stage of the **thermal history of the Universe** probed by the microwave background.

The need to explain how the lumpy Universe we see today emerged from the relatively featureless initial state of the early Universe calls for a theory of *structure formation*. There is a standard picture of how this might have happened, and it is based on the relatively simple physics of the gravitational **Jeans instability**. Since **gravity** is an attractive force, small fluctuations in the density from place to place get progressively amplified as the Universe evolves, eventually turning into the large structures we observe at the present time. Constructing a complete theory based on this idea is, however, far from straightforward, and no completely successful theory has yet emerged. To see how this has happened, it is instructive to consider the history of structure formation based on gravitational instability.

The first to tackle the problem of gravitational instability in an expanding **cosmological model** within the framework of general relativity was Evgeny Lifshitz in 1946. He studied the evolution of small fluctuations in the density of a **Friedmann model** with **perturbation theory**, using techniques similar to those still used today. The relativistic setting produces entirely similar results to the standard Jeans theory (which was obtained using only Newtonian gravity, and in a static background). Curiously, it was not until 1957 that the evolution of perturbations in a matter-dominated Friedmann model was investigated in Newtonian theory, by William Bonnor. In some ways the relativistic cosmological theory is more simple that the Newtonian analogue, which requires considerable mathematical subtlety.

These foundational studies were made at a time when the existence of the cosmic microwave background radiation was not known. There was no generally accepted cosmological model within which to frame the problem of structure formation, and there was no way to test the gravitational instability hypothesis for the origin of structure. Nevertheless, it was still clear that if the Universe was evolving with time (as **Hubble's law** indicated), then it was possible, in principle, for structure to have evolved by some mechanism similar to the Jeans process. The discovery of the microwave background in the 1960s at last gave theorists a favoured model in which to study this problem: the **Big Bang theory**. The existence of the microwave background in the present implied that there must have been a period in which the Universe consisted of a plasma of matter and radiation in **thermal equilibrium**. Under these physical conditions there are a number of processes, due to viscosity and thermal conduction in the radiative plasma, which could have influenced the evolution of a perturbation with a wavelength less than the usual Jeans length. The pioneering work by Joseph **Silk**, Jim **Peebles**, Yakov **Zel'dovich** and others between 1967 and 1972 represented the first attempts to derive a theory of galaxy and structure formation within the framework of modern cosmology.

At this time there was a rival theory in which galaxies were supposed to have formed as a result of primordial cosmic turbulence: that is, by large-scale vortical motions rather than the longitudinal adiabatic waves that appear in gravitational instability models. The vortical theory, however, rapidly fell from fashion when it was realised that it should lead to large fluctuations in the temperature of the microwave background on the sky. In fact, this point about the microwave background was then, and is now, important in all theories of galaxy formation. If structure grows by gravitational instability, it is, in principle, possible to reconcile the present highly inhomogeneous Universe with a past Universe which was much smoother. The microwave background seemed to be at the same temperature in all directions to within about one part in a hundred thousand, indicating a comparable lack of inhomogeneity in the early Universe. If gravitational instability were the correct explanation for the origin of structure, however, there should be some fluctuations in the microwave background temperature. This initiated a search, which met with success in the 1990s, for fluctuations in the cosmic microwave background.

In the 1970s, the origin of cosmic protostructure was modelled as two-component systems containing baryonic material and **radiation**. Two fundamental modes of perturbations can exist in such a two-component system: *adiabatic perturbations*, in which the matter fluctuations and radiation fluctuations are coupled together, and *isothermal perturbations*, in which the matter component is disturbed but the radiation component is uniform. These two kinds of perturbation evolve in a very different way, and this led to two distinct scenarios for structure formation:

1. The adiabatic scenario, in which the first structures to form are extremely massive, in the range 10^{12} to 10^{14} solar masses (the sizes of clusters of galaxies). This is because small-scale fluctuations in this model are eliminated in the early stages of their evolution by a process known as **Silk damping**. Galaxies then form by successive processes of fragmentation of these large objects. For this reason, the adiabatic scenario is also called a *top-down* model.
2. The isothermal scenario, in which the first structures, *protoclouds* or *protogalaxies*, are formed on a much smaller mass scale, of around 10^5 or 10^6 solar masses (similar to the mass of a **globular cluster**). Structures on larger scales than this are formed by successive mergers of these smaller objects in a process known as *hierarchical clustering*. For this reason, the isothermal scenario is described as a *bottom-up* model.

During the 1970s there was a vigorous debate between the adherents of these two pictures, roughly divided between the Soviet school led by Zel'dovich which favoured the adiabatic scenario, and the American school which favoured hierarchical clustering. Both these models were eventually abandoned: the former because it predicted larger fluctuations in the cosmic microwave background radiation than were observed, and the latter because no reasonable mechanism could be found for generating the required isothermal fluctuations.

These difficulties opened the way for the theories of the 1980s. These were built around the hypothesis that the Universe is dominated by (non-baryonic) **dark matter** in the form of **weakly interacting massive particles** (WIMPs). The WIMPs are collisionless **elementary particles**, perhaps massive neutrinos with a mass of around 10 eV, or some other more exotic species produced presumably at higher energies – perhaps the *photino* predicted by **supersymmetry** theory. These models had three components: baryonic material, non-baryonic material made of a single type of WIMP particle, and radiation. Again, as in the two-component system, there were two fundamental perturbation modes: these were *curvature perturbations* (essentially the same as the previous adiabatic modes) and *isocurvature perturbations*. In the first mode, all three components are perturbed together, and there is therefore a net perturbation in the energy density and hence a perturbation in the **curvature of spacetime**. In the second type of perturbation, however, the net energy density is constant, so there is no perturbation to the spatial curvature.

The fashionable models of the 1980s can also be divided into two categories along the lines of the top-down/bottom-up categories mentioned on the previous page. Here the important factor is not the type of initial perturbation, because no satisfactory way has been constructed for generating isocurvature fluctuations, just as was the case for the isothermal model. What counts in these models is the form of the WIMP. The two competing models were:

1. The *hot dark matter* (HDM) model, which is in some sense similar to the old adiabatic baryon model. This model starts with the assumption that the Universe is dominated by collisionless particles with a very large velocity dispersion (hence the 'hot dark matter'). The best candidate for an HDM particle would be the massive 10 eV neutrino mentioned above. In this model a process of **free streaming** occurs in the early Universe that erases structure on scales all the way up to the scale of a supercluster (greater than 10^{15} solar masses). Small-scale structure thus forms by the fragmentation of much larger-scale objects, as before.
2. The *cold dark matter* (CDM) model, which has certain similarities to the old isothermal picture. This model starts with the assumption that the Universe is, again, dominated by collisionless WIMPs, but this time with a very small velocity dispersion (hence

the 'cold'). This can occur if the particles decouple when they are no longer relativistic (typical examples are supersymmetric particles such as gravitinos and photinos) or if they have never been in thermal equilibrium with the other components (e.g. the *axion*). Small-scale structure survives in this model, but it is slightly suppressed, relative to the large-scale fluctuations, by the **Meszaros effect**. Structure formation basically proceeds hierarchically, but the hierarchy develops extremely rapidly.

Detailed calculations have shown that the fluctuations in the cosmic microwave background produced by these models are significantly lower than in the old baryonic models. This is essentially because the WIMP particles do not couple to the radiation field directly via scattering, so it is possible for there to be fluctuations in the WIMP density that are not accompanied by fluctuations in the temperature of the radiation. The HDM model fell out of favour in the early 1980s because it proved difficult to form objects early enough. The presence of **quasars** at redshifts greater than 4 requires superclusters to have already been formed by that epoch if small-scale structures are to form by fragmentation. The CDM model then emerged as the front-runner for most of the 1980s.

So far in this discussion we have concentrated only on certain aspects of the Jeans instability phenomenon, but a complete model that puts this into a cosmological context requires a number of different ingredients to be specified. Firstly, the parameters of the underlying Friedmann model (the **density parameter**, **Hubble parameter** and **cosmological constant**) need to be fixed. Secondly, the relative amounts of baryons and WIMPs need to be decided. And thirdly, the form and statistical properties of the **primordial density fluctuations** that start the whole instability process off need to be specified.

The standard CDM model that emerged in the mid-1980s served a very useful purpose because it established that the underlying cosmology was a flat Friedmann model with a **Hubble constant** of 50 kilometres per second per megaparsec, and no cosmological constant. The density of CDM particles was assumed to dominate all other species (so that $\Omega = 1$, due entirely to WIMPs). Finally, in accord with developments in the theory of the **inflationary Universe** that were happening at the same time, the initial fluctuations were assumed to be adiabatic fluctuations with the characteristic scale-free **power spectrum** predicted by most models of inflation.

Unfortunately, subsequent measurements of large-scale structure in the galaxy distribution from **redshift surveys** and, perhaps most importantly, the **ripples** seen by the COBE satellite, have effectively ruled out the CDM model. However, the early successes of CDM, and the fact that it fits all the data to within a factor of two or so, suggest that its basic premises may be correct. Various possible variations on the CDM theme have been suggested that might reconcile the basic picture with observations. One idea, which produces a hybrid scenario, is to have a mixture of hot and cold particles (this is often called the CHDM model). The most popular version of this model has a density parameter in CDM particles of about 0.7, and an HDM particle density of around 0.3. This model is reasonably successful in accounting for large-scale structure, but it still has problems on the scales of individual galaxies. It also requires an awkward fine-tuning to produce two particles of very different masses with a similar cosmological density. Unless a definite physical model can be advanced to account for this coincidence, the CHDM model must be regarded as rather unlikely.

It has also generally been assumed for most of the 1980s and 1990s that the Universe has to be very nearly flat, as suggested by the inflationary Universe models. This appeared to be a good idea before the COBE discovery of the ripples because the larger the total density of the Universe, the faster the fluctuations would grow. This means that a given level of structure now produces lower-temperature fluctuations on the microwave sky in a high-density Universe than in a low-density one. A low-density CDM model with a density parameter of around 0.3 to 0.4 actually matches all available data fairly well, and this may also be consistent with the measured amounts of cosmological dark matter. If we really want a low density and a **flat universe**, we can also add a cosmological constant. An alternative is to tinker with the initial fluctuation spectrum so that it is no longer scale-free, but tilted (i.e. the index of the power spectrum is no longer 1).

Whether one of these variants eventually emerges as a clear winner remains to be seen. All the variations on the CDM theme actually produce large-scale structure that looks qualitatively similar: detailed testing of the various models involves running complicated ***N*-body simulations** on supercomputers and comparing the results with statistical analyses of large-scale galaxy surveys (see the Figure). One of the problems with this approach is that, while the large-scale structure is relatively easy to predict, the same is not true for individual galaxies. On large scales the gravitational behaviour of the WIMPs completely dominates, and this is quite easy to compute, but on small scales the baryonic material comes into its own, introducing hydrodynamical, radiative and dissipative effects into the picture. In particular,

Structure formation Results of an *N*-body computation of the clustering of material expected in a Universe dominated by cold dark matter. The characteristic pattern of sheets, filaments and knots is in qualitative agreement with observations of large-scale structure.

it is difficult to predict the distribution of stars. Despite the enormous progress that has been made on the theoretical side of the structure formation problem, a complete theory for the formation of individual galaxies is still far away.

SEE ALSO: Essays 2, 3 and 5.

FURTHER READING: Lifshitz, E. M., 'On the gravitational instability of the expanding Universe', *Soviet Physics JETP*, 1946, **10**, 116; Bonnor, W. B., 'Jeans' formula for gravitational instability', *Monthly Notices of the Royal Astronomical Society*, 1957, **117**, 104; Silk, J., 'Fluctuations in the primordial fireball', *Nature*, 1967, **215**, 1155; Blumenthal, G. R., Faber, S. M., Primack, J. R. and Rees, M. J., 'Formation of galaxies and large-scale structure with cold dark matter', *Nature*, 1984, **311**, 517; Coles, P., 'The large-scale structure of the Universe', *Contemporary Physics*, 1996, **37**, 429.

SUNYAEV–ZEL'DOVICH EFFECT When photons from the **cosmic microwave background radiation** travel through a hot plasma (with a temperature of, say, around 10^8 K) they collide with energetic electrons and get scattered up to X-ray energies. If we look at the cosmic microwave background radiation through such a plasma cloud, we therefore see fewer microwave photons than we would if the cloud were not there. Paradoxically, this means that the plasma cloud looks like a cool patch on the microwave sky. This photon deficiency is the essence of the *Sunyaev–Zel'dovich effect*, named after Rashid Alievich Sunyaev (1943–) and Yakov **Zel'dovich**.

Quantitatively, the relative temperature dip $\Delta T/T$ depends on the temperature and number density of the scattering electrons (T_e and n_e) according to the formula

$$\Delta T/T = -2 \int (n_e k T_e \sigma / m_e c^2)\, \mathrm{d}l$$

where the integral is taken along the line of sight through the cloud; m_e is the mass of the electron, and σ is the Thomson scattering cross-section. This effect has been detected in observations of rich clusters of galaxies: the size of the temperature decrement $\Delta T/T$ is around 10^{-4}. Future fine-scale experiments designed to map the fluctuations in the cosmic microwave background radiation with an angular resolution of a few arc minutes are expected to detect large numbers of Sunyaev–Zel'dovich contributions from individual clusters.

A particularly interesting aspect of this method is that it is possible, at least in principle, to use it to obtain measurements of the distance to a cluster of galaxies in a manner that is independent of the cluster's **redshift**. To do this we need X-ray measurements of the cluster (see **X-ray astronomy**) which give information about n_e and T_e. Comparing these with the measured $\Delta T/T$ yields an estimate of the total path length ($L = \int \mathrm{d}l$) traversed by the photons on their way through the cluster. Assuming the cluster to be spherical, or by using a sample of clusters with random orientations, we can use L to estimate the physical size of the cluster. Knowing its apparent angular size on the sky then leads to an estimate of its distance; knowing its redshift then leads to a value of the **Hubble constant** H_0.

Attempts to apply this idea in practice have not been overwhelmingly successful, rather low values being obtained for H_0. On the other hand, it is a potentially important method because it does not rely on the complicated overlapping series of calibrations from which the **extragalactic distance scale** is usually constructed.

SEE ALSO: **intergalactic medium**.

FURTHER READING: Jones, M. *et al.*, 'An image of the Sunyaev–Zel'dovich effect', *Nature*, 1993, **365**, 320.

SUPERNOVA A catastrophic explosion which represents the end point of stellar evolution for massive stars. Supernovae are probably the most dramatic phenomena known to astronomy: they are more than a billion times brighter than the Sun and can outshine an entire galaxy for several weeks.

Supernovae have been observed throughout recorded history. The supernova seen in 1054 gave rise to the Crab Nebula, which now contains a rotating *neutron star*, known as a *pulsar*. Tycho Brahe observed a supernova in 1572. The last such event to be seen in our Galaxy was recorded in 1604, and was known as *Kepler's star*. Although ancient records suggest that the average rate of these explosions in the Milky Way appears to be one or two every century or so, none have been observed for nearly four hundred years. In 1987, however, a supernova did explode in the Large Magellanic Cloud, and was visible to the naked eye. The two categories of supernova labelled Type I and Type II are defined according to whether hydrogen is present in the spectrum: it is present for Type II supernovae, but not for Type I. Type I is subdivided into Types Ia, Ib and Ic, depending on further details of the shape of the spectra. Of particular interest for cosmology are the Type Ia supernovae; the Figure shows an example of a Type Ia light curve. These supernovae have very uniform peak luminosities, for the reason that they are all thought to be the result of the same kind of explosion. The usual model for these events is that a *white dwarf* (see **stellar evolution**) is gaining mass by accretion from a companion star. When the mass of the white dwarf exceeds a critical mass called the *Chandrasekhar limit* it explodes. Since the mass is always very close to this critical value, these objects are expected always to liberate the same amount of energy, and therefore provide us with a form of 'standard candle'. Type Ia supernovae are very promising objects with which to perform tests of the **curvature of space-time** using **classical cosmology**. They are also used in the construction of the **extragalactic distance scale**.

Type II supernovae are thought to originate directly from the explosions of massive stars, as described in the theory of **stellar evolution**. The final state of this explosion would be a

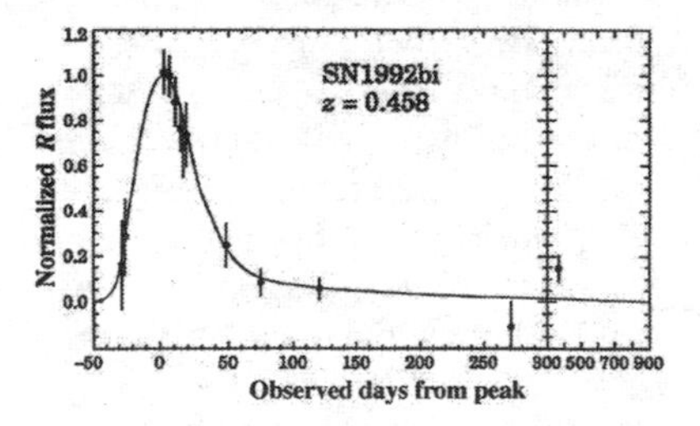

Supernova The light curve of a typical Type Ia supernova. The shape of this curve can be used to calibrate the distance of the exploding star, so these objects can be used as cosmological probes.

neutron star or **black hole**. The recent supernova 1987A was an example of this kind of event.

FURTHER READING: Phillips, A. C., *The Physics of Stars* (John Wiley, Chichester, 1994); Tayler, R. J., *The Stars: Their Structure and Evolution* (Cambridge University Press, Cambridge, 1994).

SUPERSYMMETRY Supersymmetry is an idea incorporated in certain **grand unified theories**. In the standard **gauge theories** of the **fundamental interactions** there is a fundamental difference between **elementary particles** that are *bosons* and those that are *fermions*. In particular, the particles that are responsible for mediating the various interactions are invariably bosons: the *gauge bosons*. For example, in quantum electrodynamics (QED) the gauge boson is simply the photon. In quantum chromodynamics (QCD), there are eight gauge bosons called *gluons*; the fermionic particles are the *quarks*.

In supersymmetric theories, every bosonic particle has a fermionic partner (and vice versa). The fermionic partner of the photon is called the *photino*; the hypothetical partner of the Higgs boson is the *Higgsino*, and so on. Quarks also have supersymmetric partners called *squarks*; leptons have partners called *sleptons*.

This idea has also been extended to a theory called *supergravity*, which seemed for a while to be a promising way of unifying the fundamental interactions with **general relativity**, thus forming **a theory of everything**. Supergravity theory, like general relativity, is independent of the choice of coordinate system. By including gravity, the conclusion is reached that the *graviton* (see **gravitational waves**) should have a fermionic partner called the *gravitino*.

None of the hypothetical supersymmetric particles have yet been detected; they must be extremely massive in order to have escaped detection so far, and are probably heavier than 100 GeV or so. This makes any stable supersymmetric particles promising candidates for non-baryonic **dark matter** in the form **of weakly interacting massive particles** (WIMPs). In particular, the lightest neutral sparticle, which is usually called the *neutralino*, has to be stable in any supersymmetric theory and should therefore have survived to the present time from the early stages of the **thermal history of the Universe**. Being so massive, these would constitute cold dark matter.

Although it is undoubtedly an elegant idea, there is as yet no compelling experimental evidence in favour of this theory.

FURTHER READING: Pagels, H. R., *Perfect Symmetry* (Penguin, London, 1992); Kolb, E. W. and Turner, M. S., *The Early Universe* (Addison-Wesley, Redwood City, CA, 1990).

T

Tachyon It is commonly asserted that **special relativity** forbids any particle to travel with a velocity greater than that of light. In fact this is not strictly true: special relativity merely rules out the possibility of accelerating any particle to a velocity greater than that of light. In **general relativity**, faster-than-light (*superluminal*) travel is mathematically possible, and some exact solutions of the **Einstein equations**, such as the Gödel universe (see **cosmological models**), have been found that permit this (see also **wormhole**). Some of the more speculative models of the **fundamental interactions**, including various superstring theories (see **string theory**), also allow the existence of particles that can travel at superluminal speeds. The generic name given to such particles is *tachyons*.

The attitude of most physicists to tachyons is that if they occur in a given theory then the theory must be wrong. This is essentially because tachyons would be able to perform time travel, and thus open up the possibility of logical paradoxes like travelling back in time and killing your parents before you were conceived. Although their existence would surprise many, no direct evidence either for or against the existence of tachyons has yet been obtained.

SEE ALSO: **light cone**.

****Tensor** In the theory of **general relativity**, the force of **gravity** is described in terms of transformations between accelerated **frames of reference**. The mathematical formulation of this theory depends on the inclusion of quantities that have manageable properties under transformations between different systems of coordinates. The formalism that is usually used for the task is a tensor formalism, and the **Einstein equations** are then written in the form of tensor equations.

A *tensor*, in the most general sense, is quantity with several components indicated by indices that can be written either as superscripts or as subscripts or as both. The number of indices required is called the *rank* of the tensor. A vector is a tensor of rank 1; tensors of rank 2 can be written as matrices, and so on. A general tensor (A) is an entity whose components transform in a particular way when the system of coordinates is changed from, say, x^i to x'^i. If the tensor in the original coordinate system is A becomes A' new system, then

$$A = A'^{kl...}{}_{pq...} = (\partial x'^k/\partial x^m)(\partial x'^l/\partial x^n) \ldots (\partial x^r/\partial x'^p)(\partial x^s/\partial x'^q) \ldots A^{mn...}{}_{rs...}$$

where we have allowed A to have an arbitrary rank. The upper indices are called *contravariant*, and the lower are *covariant*. The difference between these types of index can be illustrated by considering a tensor of rank 1 which, as mentioned above, is simply a vector. A vector will undergo a transformation according to certain rules when the coordinate system in which it is expressed is changed. Suppose we have an original coordinate system x^i, and we transform it to a new system x'^k. If the vector $\boldsymbol{A}$ transforms in such a way that $\boldsymbol{A}' = (\partial x'^k/\partial x^i)\boldsymbol{A}$, then it is a *contravariant vector* and is written with index 'upstairs': $\boldsymbol{A} = A^i$. If, however, $\boldsymbol{A}' = (\partial x^i/\partial x'^k)\boldsymbol{A}$, then it is a *covariant vector* and is written with the index 'downstairs': $\boldsymbol{A} = A_i$. The tangent vector to a curve is an example of a covariant vector; the normal to a surface is a covariant vector. The general rule given above is the generalisation of the concepts of covariant and contravariant vectors to tensors of arbitrary rank and to *mixed tensors* (with upstairs and downstairs indices).

A particularly important tensor in general relativity is the **metric** tensor g_{ij}, which has a property that describes the geometric structure of **spacetime**. This property is that

$$g^{im}g_{mk} = \delta^i{}_k$$

where $\delta^i{}_k$ is called the *Kronecker delta*; $\delta^i{}_k = 0$ if $i \neq k$, and $\delta^i{}_k = 1$ if $i = k$. This equation also illustrates an important convention that any tensor expression in which indices (in this case m) are repeated indicates that the repeated index is summed. For example, X_{ij} has no repeated indices, so it represents simply an arbitrary component of the tensor $\boldsymbol{X}$. On the other hand X_{ii} has a repeated index, so by the *summation convention* this is equal to $X_{11} + X_{22} + \ldots$ and so on, for whatever the range of the indices is, i.e. for however many components are needed to describe the tensor. In four-dimensional spacetime this number is usually four, but sometimes we need to deal only with the spatial behaviour of the metric, so the number of indices would be three. The metric tensor can also be used to raise or lower indices. For example, the 4-velocity of a fluid (which is used in the energy–momentum tensor of general relativity) can be written in covariant form as U_i or in contravariant form as U^k (the indices are free, since there is only one in each case). These forms are related by

$$U_i = g_{ik}U^k$$

The vector U^k is simply $\mathrm{d}x^k/\mathrm{d}s$, where s is the interval in the metric:

$$\mathrm{d}s^2 = g_{ij}\ \mathrm{d}x^i\ \mathrm{d}x^j$$

and $x^k(s)$ is called the *world line* of the particle; it is the trajectory followed in spacetime. In this equation the summation is over the index k.

Tensors are fairly complicated mathematical entities, but the formalism they provide is quite elegant. But defining what tensors are is only part of the story: general relativity constructs different equations for tensor variables, and this requires the construction of a special tensor calculus involving a particular kind of derivative called a *covariant derivative*. There is no space to provide an initiation to tensor calculus here; details are given in the textbooks listed below.

FURTHER READING: Berry, M. V., *Principles of Cosmology and Gravitation* (Adam Hilger, Bristol, 1989); Schutz, B.F., *A First Course in General Relativity*

(Cambridge University Press, Cambridge, 1985); Rindler, W., *Essential Relativity: Special, General, and Cosmological*, revised 2nd edition (Springer-Verlag, New York, 1979); Misner, C. W., Thorne, K. S. and Wheeler, J. A., *Gravitation* (W. H. Freeman, San Francisco, 1972); Weinberg, S., *Gravitation and Cosmology: Principles and Applications of the General Theory of Relativity* (John Wiley, New York, 1972).

Theory of everything (TOE) A theme that runs through the entire history of theoretical physics, going back at least to the days of Isaac **Newton**, is the continual process of unification of the **laws of physics**. For example, Newton unified the terrestrial phenomena of **gravity** (as caused the apocryphal apple to fall to the ground) with the motions of the celestial bodies around the Sun. He showed that these disparate effects could be represented in terms of a single unifying theory: the law of *universal gravitation*. James Clerk **Maxwell** in the 19th century similarly constructed a theory according to which the phenomena of electrostatics were unified with the properties of magnetism in a single theory of *electromagnetism*.

The advent of **gauge theories** made it possible for further unification of the **fundamental interactions** to be achieved: electromagnetism and the *weak nuclear interaction* were unified in a single *electroweak theory* in the 1960s. The fact that these forces are different in the low-energy world around us is simply a consequence of a broken symmetry (see **spontaneous symmetry-breaking**). Attempts to construct a theory that merges the *strong nuclear interaction* (described in **quantum field theory** terms by *quantum chromodynamics*) with the electroweak theory have proved somewhat less successful, mainly because the energy scale required to recover the more symmetric high-energy state of this unified theory is so large. Nevertheless, **grand unified theories** of these three interactions (electromagnetism, and the weak and strong nuclear interactions) appear to be feasible.

A *theory of everything* would take the unification of the laws of physics a stage further, to include also gravity. The main barrier to this final theory is the lack of any self-consistent theory of **quantum gravity**. Not until such a theory is constructed can gravity be unified with the other fundamental interactions. There have been many attempts to produce theories of everything, involving such exotic ideas as **supersymmetry** and **string theory** (or even a combination of the two, known as *superstring theory*). It remains to be seen whether such a grander-than-grand unification is possible.

However, the search for a theory of everything also raises interesting philosophical questions. Some physicists would regard the construction of a theory of everything as being, in some sense, like reading the mind of God, or at least unravelling the inner secrets of physical reality. Others simply argue that a physical theory is just a *description* of reality, rather like a map. A theory might be good for making predictions and understanding the outcomes of observation or experiment, but it is no more than that. At the moment, the map we use for gravity is different from the one we use for electromagnetism or for the weak nuclear interaction. This may be cumbersome, but it is not disastrous. A theory of everything would simply be a single map, rather than a set of different ones that we use in different circumstances. This latter philosophy is pragmatic: we use theories for the same reasons that we use maps – because they are useful. The famous London Underground map is certainly useful, but it is not a particularly accurate representation of physical reality. Nor does it need to be.

In any case, perhaps we should worry about the nature of explanation afforded by a theory of everything. How will it explain, for example, why the theory of everything is what it is, and not some other theory?

FURTHER READING: Weinberg, S., *Dreams of a Final Theory* (Vintage Books, London, 1993); Barrow, J.D., *Theories of Everything* (Oxford University Press, Oxford, 1991).

THERMAL EQUILIBRIUM, THERMODYNAMIC EQUILIBRIUM The condition that pertains when processes such as scattering, collisions or other physical interactions occur sufficiently quickly that they distribute all the energy available in the system uniformly among the allowed energy states. The concept of thermal equilibrium (or thermodynamic equilibrium – the two terms are virtually synonymous) is extremely important in the standard **Big Bang theory**. The assumption of thermal equilibrium, which can be strongly justified, makes it possible to perform relatively straightforward calculations of the **thermal history of the Universe**, as is illustrated here by a few examples.

In a state of thermal equilibrium, the rate at which energy is absorbed by a body is also equal to the rate at which it is emitted, so the body radiates as a **black body**. The very accurate observed black-body form of the spectrum of the **cosmic microwave background radiation** provides overwhelming evidence that it was produced in conditions of near-perfect thermal equilibrium. In other words, there was a time when processes in which both matter and **radiation** participated quickly enough for energy to be distributed in the required way among the various allowed energy states. Of course, the black-body radiation no longer interacts very effectively with matter, so this radiation is not in thermal equilibrium now, but it was when it was last scattered (see **last scattering surface**). When the Universe was so hot that most of the atomic matter was fully ionised, scattering processes (mainly *Thomson scattering*) were very efficient indeed, and they held the matter and radiation fluids at the same temperature. When the temperature fell to a few thousand degrees, a process of *decoupling* took place. As the number of free electrons fell, the time between successive scatterings of a photon became comparable to the characteristic time for the expansion of the Universe (essentially the inverse of the **Hubble parameter**). The radiation which had until then been held at the same temperature as the matter was then released and propagated; we see it now, with its shape preserved but redshifted to the much lower temperature of 2.73 K.

But the cosmic microwave background radiation is not the only possible repository for energy in the hot Big Bang, and is not the only thing to have been first held in thermal equilibrium, only later to undergo a kind of decoupling. All kinds of elementary particles can be held in thermal equilibrium as long as they scatter sufficiently rapidly to maintain thermal contact. Any kind of scattering process is characterised by a timescale τ which can be expressed in the form

$$\tau = 1/(n\sigma v)$$

where n is the number density of scatterers, σ (which describes the physical scattering process and which may be a function of energy) is called the *scattering cross-section*, and v is the relative velocity of the scattered particle and the scatterer. In the early Universe the relevant velocities were usually highly relativistic, so we can put $v = c$ for most applications. For the decoupling of the cosmic microwave background radiation the relevant cross-section is the Thomson scattering cross-section, which has changed only a little as the Universe has cooled. What changed to cause the decoupling was that the number density of free electrons (the scatterers) fell, so that τ became very large.

In other situations the process of decoupling can involve more subtle changes. For example, at temperatures of around 10^{10} K or higher, neutrinos would have been held in thermal equilibrium with radiation via weak the nuclear interaction. As the temperature fell, so did the cross-section for these processes. The result was the production of a cosmic *neutrino background* that decoupled at about this temperature. If we could develop a neutrino astronomy, it would be possible to probe the epoch of neutrino coupling in much the same way that we can probe the recombination epoch using observations of the cosmic microwave background radiation. There are other important applications of thermodynamic arguments to the early Universe. According to Boltzmann's theory in statistical mechanics, the relative numbers of particles in two energy states A and B in a state of thermal equilibrium can be written as

$$N_B/N_A = (g_B/g_A)\exp(-\Delta E/kT)$$

where ΔE is the difference in energy between the two states, T is the temperature and k is the Boltzmann constant. The factors g_B and g_A (called the *statistical weights*) represent the number of states at the given energy; they take account, for example, of degenerate states of different spin.

Now, neutrons and protons can interconvert via weak nuclear processes as long as the interaction rate (determined by the cross-section) is sufficiently high. They are therefore held in thermal equilibrium, so their relative numbers at a temperature T are given by the Boltzmann formula, with $\Delta E = \Delta mc^2$ determined by the mass difference between the neutron and the proton. As the Universe cooled and T fell, the ratio of protons to neutrons adjusted itself through collisions so that it always matched the equilibrium value for that temperature. But when the rate of collisions fell to a certain level, the equilibrium was no longer maintained and the ratio ceased to adjust. It became frozen out at the value it had just before the weak interactions went out of equilibrium. The freezing out of the neutron/proton ratio is of vital importance for the process of cosmological **nucleosynthesis**.

As we turn back the clock to earlier and earlier times, we find more and more examples of reactions occurring sufficiently quickly to maintain equilibrium at high temperatures when they were strongly out of equilibrium at lower temperatures. It has been speculated that, at temperatures on the grand unification energy scale, of about 10^{28} K, processes might have occurred that produced an equilibrium abundance of a particle whose freeze-out density today is sufficient for it to form the **weakly interacting massive particles** (WIMPs) from which the bulk of the observed **dark matter** might well be made.

SEE ALSO: Essays 1, 3 and 5.

FURTHER READING: Mandl, F., *Statistical Physics* (John Wiley, Chichester, 1971); Silk, J., *The Big Bang*, revised and updated edition (W. H. Freeman, New York, 1989); Narlikar, J. V., *Introduction to Cosmology*, 2nd edition (Cambridge University Press, Cambridge, 1993), Chapter 5; Weinberg, S., *Gravitation and Cosmology: Principles and Applications of the General Theory of Relativity* (John Wiley, New York, 1972); Kolb, E. W. and Turner, M. S., *The Early Universe* (Addison-Wesley, Redwood City, CA, 1990).

THERMAL HISTORY OF THE UNIVERSE
One of the main achievements of modern cosmology is to have reconstructed the past evolution of the cosmos by using the standard **Friedmann models** to describe the **expansion of the Universe**. The extremely accurate **black-body** spectrum of the **cosmic microwave background radiation** simplifies this task, as it strongly argues for the application of relatively simple laws of equilibrium **thermodynamics** at virtually all stages of its evolution. The reconstruction of the thermal history of the Universe within the framework of the **Big Bang theory** simply requires the repeated application of the idea of **thermal equilibrium** at different stages. The only exceptions to the rule of thermal equilibrium are the various non-equilibrium processes that occur at the cosmological **phase transitions** that are predicted to have taken place in the very early stages. The thermal history is reconstructed by taking present-day observations and **laws of physics** tested in the laboratory, and progressively turning back the clock to earlier and earlier times of higher and higher energy, density and temperature. This brief overview summarises the main stages and gives a qualitative description of what is thought to have gone on. Although this history has been constructed by working backwards, it makes sense to present it in chronological order.

The *Planck era* started with the initial Big Bang (at $t = 0$), when the temperature and density were both infinite. Since we have no adequate theory of **quantum gravity**, we have very little idea of what happened in this period, which lasted until the **Planck time** (10^{-43} seconds), at which the temperature was higher than 10^{32} K).

The *phase transition era* represents the interval that begins at the Planck time and takes in:

the epoch during which **grand unified theories** held sway (when the temperature was around 10^{28} K, and the time about 10^{-37} seconds);

the stage of electroweak symmetry-breaking (at around 10^{15} K, and after about 10^{-11} seconds);

the epoch when quarks were confined in hadrons (around 10^{11} K and 10^{-5} seconds).

This era is characterised by non-equilibrium processes (see **phase transitions**).

The *hadron era* followed shortly after the era of phase transitions, but was very brief: it lasted for only a few microseconds. Before the start of the hadron era, quarks behaved as free **elementary particles**. The process by which hadrons were formed from these quarks is called the *quark–hadron phase transition*. By the end of the hadron era, most hadron species had either decayed or been annihilated, leaving only the nucleons (protons and neutrons).

The *lepton era* followed immediately after the hadron era. During this epoch, the dominant contribution to the density of the Universe was from the various kinds of lepton (such as electrons, and their associated neutrinos). Pairs of leptons and antileptons were created in large numbers in the early Universe (see **antimatter**) but, as the Universe continued to cool, most lepton species annihilated. The end of the lepton era is usually taken to be the time by which most electron–positron pairs had annihilated, at a temperature of around 10^9 K, and about 1 second after the initial singularity. Cosmological **nucleosynthesis** began around this time.

The *radiation era* was the period in which the Universe was dominated by the effects of **radiation** or relativistic particles. This lasted from about the first second until about 10,000 years after the origin. At the end of the radiation era began the epoch of matter–radiation equivalence, when radiation and non-relativistic particles contributed equally to the dynamics of the Universe. The synthesis of the light elements was completed in this period, resulting in the formation of hydrogen, deuterium, helium-3, helium-4 and lithium-7.

The *plasma era*: after the epoch of matter–radiation equivalence, the Universe behaved essentially like a classical plasma. The temperature was high enough for all the hydrogen and helium to be fully ionised and, though the radiation no longer dominated the energy density, there was a strong residual effect of radiation drag on the matter. Any significant concentrations of baryonic material could have formed only after the Universe became neutral.

The *recombination era* was reached when the Universe had cooled to a temperature of a few thousand degrees, allowing electrons and protons to combine to form hydrogen atoms. In standard theories, this was about 300,000 years after the beginning. The onset of recombination was closely associated with the phenomenon of matter–radiation decoupling, which means that observations of the cosmic microwave background radiation provide a means of studying the Universe when it was roughly this age (see **thermal equilibrium**).

The *matter era* began after recombination. Some cosmologists prefer to take the attainment of matter–radiation equivalence as marking the beginning of the matter era, but this tends to diminish the importance of the plasma epoch as a distinct era in its own right. The matter era at last saw material freed from the influence of radiation; it was in this period that **structure formation** was initiated by the small **primordial density fluctuations** that were perhaps laid down during the phase-transition era.

SEE ALSO: Essays 1, 3 and 5.

FURTHER READING: Silk, J., *The Big Bang*, revised and updated edition (W. H. Freeman, New York, 1989); Narlikar, J. V., *Introduction to Cosmology*, 2nd edition (Cambridge University Press, Cambridge, 1993), Chapter 5; Weinberg, S., *Gravitation and Cosmology: Principles and Applications of the General Theory of Relativity* (John Wiley, New York, 1972); Kolb, E. W. and Turner, M. S., *The Early Universe* (Addison-Wesley, Redwood City, CA, 1990).

* **THERMODYNAMICS** The study of how the properties of matter change with temperature. It is one of the senior branches of modern physics, having begun in the late 18th century with a number of investigations into the nature of heat. In the early days, thermodynamics was conceived in terms of macroscopic quantities such as density, pressure and temperature. General rules, which later became the laws of thermodynamics, were devised to describe how such quantities varied in different conditions. One of the great achievements of 19th-century science was the work of physicists of the calibre of Ludwig **Boltzmann** and Josiah Willard Gibbs, who showed that these macroscopic laws could be derived from a microscopic description of the individual particles from which macroscopic bodies are constructed, beginning with the *kinetic theory of gases* and developing into the wide field of *statistical mechanics*. Later on, with the discovery of **quantum theory** and the properties of the **elementary particles**, the simple Boltzmann theory was modified to take account, for example, of the *Pauli exclusion principle*. Modern statistical mechanics recognises distinct statistical distributions applying to fermions and bosons: the so-called *Fermi–Dirac statistics* and *Bose–Einstein statistics*, respectively.

A microscopic description of thermodynamics is required for performing accurate calculations of the abundances of elementary particles produced in the early Universe. Some aspects of this problem are discussed under **thermal equilibrium**, and in the more specialised texts listed at the end of the entry.

The older macroscopic description of thermodynamics also leads to some important cosmological insights. The famous *laws of thermodynamics* can be written in the following (macroscopic) forms:

Zeroth law If two systems are each in thermal equilibrium with a third system, then they must be in thermal equilibrium with each other.

First law The equilibrium state of a system can be characterised by a quantity called the *internal energy*, *E*, which has the property that it is constant for an isolated system. If the system is allowed to interact with another system, then the change in its internal energy is $\Delta E = -W + Q$, where W is the work done by the system and Q is the heat absorbed by the system.

Second law An equilibrium state of a system can be characterised by a quantity called the *entropy*, S, which has the property that it never decreases for an isolated system. Moreover, if the system absorbs heat slowly in infinitesimal amounts $\mathrm{d}Q$, while its temperature is roughly constant, then the change in entropy is $\mathrm{d}S = \mathrm{d}Q/T$. This second statement refers to what are usually called *reversible changes*.

Third law The entropy of a system has the property that, as the temperature of the system tends to zero, the entropy tends to a constant value, independently of all other parameters that describe the system.

These three laws can be reproduced by the microscopic theory of statistical mechanics, but as stated in the forms presented above they involve only macroscopic quantities.

We can combine the first and second laws in a particularly useful form for small reversible changes:

$$\mathrm{d}E = \mathrm{d}Q - \mathrm{d}W$$

If the pressure of the system is p and the change in volume while it does the work is $\mathrm{d}V$, then $\mathrm{d}W = p\,\mathrm{d}V$ so that, by including the second law, we get

$$\mathrm{d}E = T\,\mathrm{d}S - p\,\mathrm{d}V$$

We can regard the early stages of the **thermal history of the Universe** as involving an *adiabatic expansion* which corresponds to no net change in entropy, $\mathrm{d}S = 0$, so the net result is $\mathrm{d}E = -p\,\mathrm{d}V$. The **Einstein equations** in the standard **Friedmann models** lead to a very similar expression, of the form

$$\mathrm{d}(\rho c^2 a^3) = -p\,\mathrm{d}a^3$$

where a is the *cosmic scale factor* describing the **expansion of the Universe**. Since the energy per unit volume can be written as ρc^2 and ρ is the matter density, this expression is entirely equivalent to the simple macroscopic law given above, even though it is derived from the Einstein field equations in which only the gravitational effect of pressure is taken into account. In this approximation, all non-relativistic gases behave as if they had no pressure at all. In fact, gases do exert a pressure according to the usual pV^{γ} law. But the gravitational effect of this pressure is much smaller than the gravitational effect of the mass, so it can be neglected for cosmological purposes. If the pressure is taken to be zero, then the density of matter must fall off as $1/a^3$, according to this equation. For radiation or ultra-relativistic particles the pressure is precisely one-third of the energy density, in which case the solution of the above equation is that the density ρ falls off as $1/a^4$; the same result was obtained using completely different arguments in the discussion of **radiation**.

SEE ALSO: **heat death of the Universe**.

FURTHER READING: Mandl, F., *Statistical Physics* (John Wiley, Chichester, 1971); Kolb, E. W. and Turner, M. S., *The Early Universe* (Addison-Wesley, Redwood City, CA, 1990).

TIME Ideally, of course, this entry should start with a clear definition of what time actually is. Everyone is familiar with what it does, and how events tend to be ordered in sequences. We are used to describing events that invariably follow other events in terms of a chain of cause and effect. But we cannot get much further than these simple ideas. In the end, probably the best we can do is to say that time is whatever it is that is measured by clocks.

Einstein's theories of **special relativity** and **general relativity** effectively destroyed for ever the Newtonian concepts of absolute space and absolute time. Instead of having three spatial dimensions and one time dimension which are absolute and unchanging regardless of the motions of particles or of experimenters, relativistic physics merges them together in a single four-dimensional entity called **spacetime**. For many purposes, time and space can be treated as mathematically equivalent in these theories: different observers measure different time intervals between the same two events, but the four-dimensional spacetime interval is always the same.

However, the successes of Einstein's theoretical breakthroughs tend to mask the fact that we all know from everyday experience that time and space are essentially different. We can travel north or south, east and west, but we can only go forwards in time to the future, not backwards to the past. And we are quite happy with the idea that both London and New York exist at a given time at different spatial locations. But nobody would say that the year 5001 exists in the same way that we think the present exists. We are also happy with the idea that what we do now causes things to happen in the future, but not with the idea that two different places at the same time can cause each other. Space and time really are quite different.

In cosmology, the **Friedmann models** have a clearly preferred time coordinate called cosmological **proper time** (see also **Robertson–Walker metric**). But the Friedmann equation is again time-symmetric. Our Universe happens to be expanding rather than contracting, but could it be that the directionality of time that we observe is somehow singled out by the large-scale **expansion of the Universe**? It has been speculated, by Stephen **Hawking** among others, that if we lived in a **closed universe** that eventually stopped expanding and began to contract, then time would effectively run backwards for all observers. In fact, if this happened we would not be able to tell the difference between a contracting universe with time running backwards and an expanding universe with time running forwards.

Another, more abstract problem stems from the fact that general relativity is fully four-dimensional: the entire *world line* of a particle, charting the whole history of its motions in spacetime, can be calculated from the theory. A particle exists at different times in the same way that two particles might exist at the same time in different places. This is strongly at odds with our ideas of free will. Does our future really exist already? Are things really predetermined in this way?

These questions are not restricted to relativity theory and cosmology. Many physical theories are symmetric between past and future in the same way that they are symmetric between different spatial locations. The question of how the perceived asymmetry of time can be reconciled with these theories is a deep philosophical puzzle. There are at least two other branches of physical theory in which there arises the question of the *arrow of time*, as it is sometimes called.

One emerges directly from the second law of **thermodynamics**. The entropy of a closed system never decreases; the degree of disorder of such a system always tends to increase. This is a statement cast in macroscopic terms, but it arises from the microscopic description of atoms and energy states provided by statistical mechanics. The laws governing these microstates are all entirely time-reversible. So how can an arrow of time emerge? Laws similar to the classical laws of thermodynamics have also been constructed to describe the properties of **black holes** and of gravitational fields in general. Although the entropy associated with gravitational fields is difficult to define, these laws seem to indicate that the arrow of time persists, even in a collapsing Universe.

Another arrow-of-time problem emerges from **quantum mechanics**, which is again time-symmetric, but in which weird phenomena occur such as the collapse of the wavefunction when an experiment is performed. Wavefunctions appear only to do this in one direction of time, and not the other.

FURTHER READING: Hawking, S. W., *A Brief History of Time* (Bantam, New York, 1988); Davies, P. C. W., *About Time: Einstein's Unfinished Revolution* (Penguin, London, 1995).

TIME DILATION see **general relativity**, **special relativity**.

TOPOLOGICAL DEFECTS Topological defects of various kinds are predicted to have occurred during **phase transitions** in the early Universe. The exact character of the defect that would have been produced depends exactly on the nature of the phase transition and the configuration of the fields involved in **spontaneous symmetry-breaking**. Their existence can be argued on general grounds because of the existence of **horizons** in cosmology. If a phase transition happens more or less simultaneously in all regions of the Universe, then there is no possibility that regions separated by more than the scale of the cosmological horizon at the time can exchange light signals. Whatever the configuration of the vacuum state of whatever field is undergoing the transition in one region, the state in a different, causally disconnected part of the Universe would be expected to be independent. This incoherence of the field would have resulted in defects, much like the defects that appear when liquids are rapidly cooled into a solid phase. Solids formed like this tend to have only short-range order within domains separated by defects in the form of walls. Other types of defect are possible, depending on the type of phase transition. Some of the cosmological defects that have been suggested are described overleaf.

Monopoles (sometimes called *magnetic monopoles*) are hypothetical point-like defects in the fabric of **spacetime**, produced in the early Universe according to some **grand unified theories** (GUTs). No monopoles have yet been detected in the laboratory. These objects are historically important (even though their existence is entirely speculative), because the inflationary Universe model was originally suggested as a means of reconciling the present lack of observed monopoles with GUT theories. The rapid expansion of the Universe associated with inflation simply dilutes the number of monopoles produced in the phase transition to an unobservably small value.

Cosmic strings are one-dimensional (line-like) defects, slightly similar to the vortex tubes that can be produced in liquid helium phase transitions. If produced in the framework of a GUT, such a string would be about 10^{-31} metres thick, and have a mass of about ten million solar masses per light year. Because of their strong gravitational effect on nearby matter, it has been suggested that cosmic strings might play a significant role in cosmological **structure formation** by generating large enough **primordial density fluctuations**. It is now generally accepted, however, that this is not the case because observations of **large-scale structure** and the fluctuations in the **cosmic microwave background radiation** disagree with the predictions of cosmic-string theory.

Domain walls would be two-dimensional (sheet-like) defects. In essence, they are wall-like structures in which energy is trapped, rather like the Bloch wall formed between the Weiss domains in a ferromagnet. Any theory of the fundamental interactions that predicts large numbers of domain walls would predict a highly inhomogeneous universe, contrary to observations, so these particular defects are to be avoided at all costs.

Cosmic textures are by far the hardest kind of defect to visualise; they involve a kind of twisting of the fabric of spacetime. Like cosmic strings, these entities have been suggested as possible sources for the primordial density fluctuations, but they have fallen out of favour because they fail to reproduce the so-called *Doppler peak* seen in observations of the **cosmic microwave background radiation** (see **Sakharov oscillations**).

SEE ALSO: Essays 3 and 5.

FURTHER READING: Kolb, E. W. and Turner, M. S., *The Early Universe* (Addison-Wesley, Redwood City, CA, 1990).

U

Ultraviolet astronomy The branch of observational astronomy that deals with the part of the spectrum of **electromagnetic radiation** between wavelengths of about 90 and 350 nm. These wavelengths are mostly blocked by the Earth's atmosphere (which is just as well, for they are harmful to life), so the field of ultraviolet astronomy only really started with the upsurge in rocket technology after the Second World War.

In more recent times this branch of observational astronomy has mainly been carried out from space. A series of ultraviolet space missions called the Orbiting Astronomical Observatories (OAO) began in 1968. The third in this series of satellites (OAO-3, also called Copernicus), began to map the distribution of matter in our own Galaxy as revealed by measurements in the ultraviolet region of the spectrum. Ultraviolet observations were also carried out in the 1970s from the Skylab space station.

The modern era of ultraviolet astronomy dawned in 1978 with the launch of the International Ultraviolet Explorer (IUE), which until 1996 performed ultraviolet **spectroscopy** on tens of thousands of objects, both galactic and extragalactic. One of the most important discoveries made by IUE was the presence of hot gaseous haloes around some galaxies. The Hubble Space Telescope (HST) has carried on where IUE left off; with its much higher spectral resolution it can observe much fainter objects. Extreme ultraviolet observations have also been performed by the X-ray satellite ROSAT (see **X-ray astronomy**).

SEE ALSO: Essay 4.

Universe The entirety of all that exists. The Greek word *cosmos*, the root of **cosmology**, means the same; cosmology is the study of the Universe. This definition seems relatively straightforward, but there are some confusing subtleties which often make for semantic and linguistic confusion. For example, what do we mean exactly by *exist*?

Modern scientific cosmology assumes the existence of a physical world that we can probe by experiment and observation. This is what many scientists mean by 'the Universe': the set of all physical things. But this raises problems of its own. How do we decide whether something exists or does not? We probably all accept that the planets go round the Sun. But in the 16th century there was simply a mathematical theory that explained the observed positions of planets on the sky in terms of a model in which idealised bodies (representing the planets) travel on elliptical paths around another idealised body (representing the Sun). Astronomers of the time did not immediately accept that the elements of this model actually represented (in mathematical terms, were in one-to-one correspondence with) elements in the real Universe. In other words, they were by no means sure that the real planets actually travelled on real elliptical orbits round the real Sun, but they knew that the model based on the work of Nicolaus **Copernicus** and Johannes **Kepler** enabled them to calculate the positions in which the planets would appear in the night sky. Now that the Solar System has been observed so many times, and explored by so many space probes whose trajectories rely on the same laws by which the planets move, we accept that things really are like that.

In the 16th century the Solar System was the frontier of science. This is no longer the case: modern science encompasses **elementary particles**, **fundamental interactions**, and a description of **gravity** in terms of the **curvature of spacetime**. But can we say that these phenomena exist in the same confident way that we say the Solar System exists? Do we *know* that electrons orbit around protons, that protons are made of quarks? Are we sure that primordial **nucleosynthesis** created the **light element abundances**? Clearly we have a model, the **Big Bang theory**, that represents the evolution of the Universe in terms of these concepts, but we would hesitate to argue that we are confident that this model is in exact correspondence with reality. This view of the limitations of cosmology leads us to adopt a conservative stance in which we are not so much concerned with what exists and what does not, but with seeking to explain the empirical properties of the world in terms of models. For this reason, it has become usual to distinguish between 'Universe' (the perhaps unknowable entirely of all existing things) and 'universe' (a **cosmological model** of the Universe).

Some adopt a different philosophical view. For some scientists what really exists is the **laws of physics**: our Universe is merely a consequence, or an outcome, of those laws. This approach more resembles a Platonist philosophy in which what exists is the idealised world of mathematics within which the laws of physics are framed. But do these laws exist, or do we invent them? Is mathematics an intrinsic property of the world, or is it simply a human invention that helps us to describe that world, in much the same way as a language? Is the Universe mathematical, or did we invent mathematics in order to create universes?

This may seem like an irrelevant philosophical detour, but it is of central importance: the way we see the nature of the Universe, what we actually believe it to be, defines the limits of what cosmology can hope to do. For example, if we accept that **time** is simply a property that the physical world possesses, then it makes little sense to 'explain' the beginning of time – the birth of the Universe – by invoking laws that existed, as it were, before the Universe came into being. Alternatively, if we think that the laws of physics existed before our Universe came into being, then it would make sense (perhaps) to consider constructing a theory of the actual Creation. This philosophical schism has led to (at least) two quite distinct approaches to **quantum cosmology.** On the one hand are ideas that describe the Creation of the Universe out of a 'vacuum state' by quantum tunnelling; on the other hand are ideas that require the Universe to have no real beginning at all.

The question of time also raises another linguistic problem. In a cosmological model based on **general relativity**, we have a description of all of **spacetime**: here, there and everywhere as well as past, present and future. Does the future exist? It must if it is part of the Universe. But does this not mean that everything is preordained? If it exists already, then surely it cannot be changed?

Even if we can settle on the meaning of the word 'Universe', there are still problems with the use of language in cosmology. For example, according to the standard **Friedmann models** the Universe can be either finite (**closed universe**) or infinite (**open universe**). But if the Universe is finite, what is outside it? If it is expanding, what is it expanding into? If the Universe is the entirety of all that exists, then our model universe cannot be embedded in anything. What is outside the Universe must be something that does not exist. It does not therefore make any sense to think of there being anything outside the Universe.

Although we do not know whether the Universe is finite or infinite we do know that, if the Big Bang theory is correct, the extent of all the parts of it that we shall ever be able to observe is finite. If the Big Bang happened 15 billion years ago then, roughly speaking, we cannot possibly see any farther than the distance light can travel in 15 billion years, i.e. 15 billion light years (for a more rigorous discussion, see **horizons**). Cosmologists therefore use the term *observable Universe* to indicate the part of the (possibly infinite) Universe that is amenable to astronomical investigation. But in some versions of the **inflationary Universe** model, the part of the Universe we can observe might be just one of a potentially infinite number of bubbles we can never observe. In this case a model universe (such as Friedmann model) is a model of the observable Universe, but not for the entire Universe (which may be extremely complicated and chaotic). Cosmologists often use the term *mini-universe* to describe any one of the small bubbles, part of which is our observable Universe (see also **baby universe**).

More confusing still is the problem introduced by adopting the *many-worlds* interpretation of **quantum physics**. In this interpretation, every time an experiment or an observation is performed the Universe splits into two. There is therefore an ensemble of *parallel universes*, in each of which all possible experiments have different outcomes. Do these parallel worlds exist, or is the ensemble simply a construction that allows us to calculate probabilities? Are they universes or parts of the Universe?

SEE ALSO: Essay 1.

FURTHER READING: Barrow, J. D., *Pi in the Sky* (Oxford University Press, Oxford, 1992); Barrow, J. D., *The World Within the World* (Oxford University Press, Oxford, 1988); Hawking, S. W. and Penrose, R., *The Nature of Space and Time* (Princeton University Press, Princeton, 1996); Deutsch, D., *The Fabric of Reality* (Allen Lane, London, 1997).

* **VIRIAL THEOREM** An important result from the field of statistical mechanics that deals with the properties of self-gravitating systems in equilibrium. According to the theory of the **Jeans instability**, small initial fluctuations grow by virtue of the attractive nature of **gravity** until they become sufficiently dense to collapse. When such a structure collapses it undergoes what is sometimes called *violent relaxation*: the material that makes up the structure rapidly adjusts itself so that it reaches a kind of pressure balance with the gravitational forces. The velocities of particles inside the structure become randomised, and the structure settles down into an equilibrium configuration whose properties do not undergo any further change. This process is sometimes called *virialisation*.

The *virial theorem*, which applies to gravitationally bound objects of this kind, states that the total kinetic energy T contained in the structure is related to the total gravitational potential energy V by the equation

$$2T + V = 0$$

This theorem can be applied to gravitationally bound objects such as some kinds of **galaxy** and clusters of galaxies, and its importance lies in the fact that it can be used to estimate the mass of the object in question.

Because the motions of matter within a virialised structure are random, they are characterised by some dispersion (or *variance*) around the mean velocity. If the object is a galaxy, we can estimate the variance of stellar motions within it by using **spectroscopy** to measure the widths of spectral lines affected by the **Doppler shift**. If the object is a galaxy cluster, we have to measure the **redshifts** of all the galaxies in the cluster. The mean redshift corresponds to the mean motion of the cluster caused by the **expansion of the Universe**; the variance around this mean represents the **peculiar motions** of the galaxies caused by the self-gravity of the material in the cluster. If the variance of the velocities is written as v^2, then the total kinetic energy of the object is simply $\frac{1}{2}Mv^2$, where M is the total mass.

If the object is spherical and has the physical dimension R, then the total gravitational potential energy will be of the form $-\alpha GM^2/R$, where α is a numerical factor that measures how strongly the object's mass is concentrated towards its centre. Note that V is negative because the object is gravitationally bound. We can therefore make use of the virial theorem to derive an expression for the mass M of the object in terms of quantities which are all measurable: $M = Rv^2/\alpha G$.

This illustration is very simplified, but illustrates the basic point. More detailed analyses do not assume spherical symmetry, and can also take into account forms of energy other than the kinetic and gravitational energy discussed here, such as the energy associated with gas pressure and **magnetic fields**. In rich clusters of galaxies, for example, the galaxies are moving through a very hot gas which emits X-rays: the high temperature of the gas reflects the fact that it too is in equilibrium with the gravitational field of the cluster. Viralisation can produce gas temperatures of hundreds of millions of degrees in this way, and this can also be used to measure the mass of clusters (see **X-ray astronomy**). A virial analysis by Fritz **Zwicky** of the dynamics of the relatively nearby Coma Cluster provided the first evidence that these objects contain significant amounts of **dark matter**. The material in these clusters is sufficient to allow a value of the **density parameter** of around $\Omega_0 \approx 0.2$.

A modified version of the virial theorem, called the *cosmic virial theorem*, applies on scales larger than individual gravitationally bound objects like galaxy clusters: it allows us to relate the statistics of galaxies' peculiar motions to the density parameter. This method also usually produces an estimated value of $\Omega_0 \approx 0.2$, indicating that dark matter exists on cosmological scales, but not enough to reach the critical density required for a **flat universe**.

FURTHER READING: Tayler, R. J., *The Hidden Universe* (Wiley-Praxis, Chichester, 1995); Coles, P. and Ellis, G. F. R., *Is the Universe Open or Closed?* (Cambridge University Press, Cambridge, 1997).

Weakly interacting massive particle (WIMP) The name given to the **elementary particles** of hypothetical non-baryonic **dark matter** that, in some theories, are assumed to pervade the cosmos. Such a particle could account for the dark matter seen in galaxies and in clusters of galaxies (see **large-scale structure**), and may assist in the problem of cosmological **structure formation**. At least part of the dark matter could be in some baryonic form such as **massive compact halo objects** (MACHOs), but if the theory of cosmological **nucleosynthesis** of the **light element abundances** is correct then there cannot be enough baryonic material to provide a critical-density **flat universe** (see also **gravitational lensing**). There are many possible candidates for the hypothetical WIMPs. These are usually divided into two classes: hot dark matter (HDM) and cold dark matter (CDM).

Any relic non-baryonic particle species which has an appreciable cosmological abundance at the present epoch, and which had a thermal velocity close to the velocity of light when it was produced in the early Universe, is called *hot dark matter*. If a particle produced in **thermal equilibrium** is to have such a large velocity, it has to be extremely light. The favoured candidate for such a particle is a neutrino with a rest mass of around 10 eV (electronvolts) which is 1/500,000 of the mass of the electron (see **elementary particles**). It is not known whether any of the known neutrino species actually has a nonzero rest mass. But if any do, and their mass is around 10 eV, then the standard **Big Bang theory** of the **thermal history of the Universe** predicts a present-day density of relic particles close to the critical density required to make the Universe recollapse. The Universe would then be expected to have a value of the **density parameter** close to 1. However, HDM does not seem to be a good candidate from the point of view of structure formation theories, because the extremely high velocities of the neutrinos tend to erase structure on scales up to and including those of superclusters of galaxies. It is unlikely, therefore, that the **Jeans instability** of HDM can on its own be responsible for the formation of **galaxies** and large-scale structure.

The alternative, *cold dark matter*, is a more promising candidate for the cosmological dark matter. Any relic non-baryonic particle species which has an appreciable cosmological abundance at the present epoch, and which had a thermal velocity much less than the velocity of light when it was produced, would be cold dark matter. In order to be moving slowly in a state of thermal equilibrium, a

CDM particle is normally (though not always) expected to be very massive. There are many possible candidates for CDM, suggested by various theories of the **fundamental interactions** and the physics of elementary particles (see e.g. **grand unified theory**). In some such theories, incorporating the idea of **supersymmetry**, all bosonic particles should have fermionic partners. Promising candidates for a CDM particle are therefore such objects as the *photino*, the supersymmetric partner of the photon. Another possible CDM candidate is the *axion*, which appears in certain grand unified theories. The axion actually has a very tiny mass (a mere one-hundred billionth of the mass of the electron) but interacts so weakly with **electromagnetic radiation** that it is never held in thermal equilibrium and therefore, paradoxically, has a very small velocity. It is even possible that primordial **black holes** with very small mass could behave like CDM particles. This form of dark matter has, until recently, been strongly favoured on theoretical grounds because it appears to assist in solving the problem of structure formation. Recent observational data, however, seem to suggest that the simplest versions of this picture of structure formation are not correct and some other ingredient is necessary, perhaps a smattering of HDM.

Experiments are under way to detect WIMPs experimentally using sensitive underground detectors. Hunting for particles of dark matter in this way is like looking for the proverbial needle in a haystack, because neither the mass nor the interaction rate is known.

SEE ALSO: Essay 2.

FURTHER READING: Riordan, M. and Schramm, D., *The Shadows of Creation: Dark Matter and the Structure of the Universe* (Oxford University Press, Oxford, 1993).

WEINBERG, STEVEN (1933–) US physicist. He originated the electroweak **gauge theory** of the **fundamental interactions**, and also worked on theoretical issues arising from the **cosmological constant** in **general relativity** and on the problem of cosmological **structure formation**.

WHITE HOLE see **black hole**.

WILSON, ROBERT WOODROW (1936–) US physicist and radio astronomer. With Arno **Penzias** he discovered the **cosmic microwave background radiation** while working with an antenna designed for use with communications satellites. The two men shared the 1978 Nobel Prize for Physics.

WORLD-LINE see **spacetime**.

WORMHOLE A hypothetical hole or tunnel in the fabric of **spacetime**. The standard cosmological models are based on the assumption that the **Universe** has a simple topological structure, like that of a sphere. However, some solutions of the equations of **general relativity** have been found that correspond to wormholes connecting regions of spacetime in a kind of short-cut. Intriguingly, the two ends of a wormhole might be located at different points in both space and time, and it appears to be possible (at least mathematically) to achieve time travel by moving along the tunnel. It is as if we could travel from London to Sydney through the Earth rather than along its surface.

The mathematical existence of wormholes was discovered by Albert **Einstein** and Nathan Rosen in 1916. They found that the Schwarzschild solution for a black hole could be represented by a bridge between two flat regions of spacetime. This bridge became known as the *Einstein–Rosen bridge*, but it was found to be mathematically impossible for anything to travel between the two regions, so the bridge was widely regarded as being unphysical and a mere mathematical curiosity. Later relativity theorists tried to show that wormholes were impossible, but found that they could not. What is more, new solutions have been obtained (by Kip Thorne and others) that represent *traversable* wormholes. It does indeed look as if general relativity permits time travel to occur through wormholes. On the other hand, it is true to say that the creation of a traversable wormhole requires the existence of a peculiar form of matter which acts as if it has negative mass. It may be, therefore, that while a wormhole is compatible with Einstein's theory of spacetime, it is not compatible with the properties of matter: no one yet knows how to create material with the required properties.

Although the idea of time travel through wormholes may remain the intellectual property of science fiction writers, there is another context in which wormholes might appear with potentially important consequences. Some **quantum cosmology** theorists have argued that on scales of order the **Planck length** (10^{-35} metres), the topology of spacetime might resemble a 'foam', with tubes like wormholes connecting apparently disconnected regions. Such wormholes would be continually opening and closing on a timescale of the order of the **Planck time** (10^{-43} seconds), but it is possible that this quantum turmoil might lead to the formation of **baby universes**. This would have a profound effect on our understanding of the initial stages of the **Big Bang theory**.

FURTHER READING: Thorne, K. S., *Black Holes and Time Warps* (Norton, New York, 1994); Hawking, S. W., *Black Holes and Baby Universes and Other Essays* (Bantam, New York, 1993).

WRINKLES see **ripples**.

X-RAY ASTRONOMY The branch of observational astronomy that is concerned with the X-ray region of the spectrum of **electromagnetic radiation**, between the ultraviolet and gamma-ray regions, and dealing with photons with energies ranging from around a hundred eV to hundreds of MeV. These energies are typically observed in **active galaxies** and in the hot **intergalactic medium** seen in rich galaxy clusters. The first astronomical X-ray observations were made in 1949 from a rocket-borne experiment that detected X-ray emission from the Sun. During the 1960s, rockets and high-altitude balloons revealed a large variety of X-ray sources on the sky, as well as the existence of the diffuse background radiation now known as the extragalactic **X-ray background**.

The first satellite mission dedicated to X-ray astronomy was Uhuru, launched in 1970, which carried out an all-sky survey. Other missions were subsequently flown, but activity in the 1970s was still largely confined to rocket and balloon-borne experiments. In 1977, however, NASA launched the first High Energy Astrophysical Observatory (HEAO-1), which was much larger than any X-ray mission that had been flown before. This satellite compiled a sky survey over the energy band between 0.1 keV and 10 MeV. The second satellite in this series (HEAO-2), later renamed the Einstein Observatory, carried a grazing-incidence telescope which enabled it to record detailed images of X-ray sources. This kind of telescope is now the standard instrument for X-ray astronomy. The imaging capability, together with the Einstein Observatory's excellent sensitivity, put X-ray astronomy on a par with other branches of observational astronomy.

Progress in X-ray astronomy has since then been rapid. The European Space Agency launched Exosat in 1983. Starting in 1979, the Japanese launched three X-ray missions of increasing size and complexity called Hakucho, Tenma and Ginga. The last of these, Ginga, was equipped with a large array of proportional counters that allowed it to perform detailed **spectroscopy** of X-ray sources. The 1990s have seen equally rapid progress. ROSAT (the name comes from the German Röntgen Satellit), launched in 1990, undertook the first imaging survey of the X-ray sky, cataloguing more than 60,000 sources (see the Figure overleaf). An X-ray telescope carried on the space shuttle (the Astro-1 mission) and a recent Japanese mission, the Advanced Satellite for Cosmology and Astrophysics (ASCA), have deployed new CCD detectors that enable them to perform more detailed spectral

X-ray astronomy The Coma Cluster seen in X-rays. The strong X-ray emission is produced in hot gas in the cluster by a process known as thermal bremsstrahlung.

measurements than has hitherto been possible. BeppoSAX was launched in 1996, and another relevant experiment, Spectrum-X, is scheduled for launch before the millennium. Two further missions planned for early in the 21st century — the Advanced X-ray Astrophysics Facility (AXAF) and the X-ray Multi-Mirror Mission (XMM) — are expected to have mission lifetimes of around 10 years.

X-RAY BACKGROUND The existence of the **cosmic microwave background radiation** (CMB) is one of the pieces of evidence in favour of the standard **Big Bang theory**, but this is not the only form of diffuse emission known to astronomy. There are also extragalactic backgrounds in other parts of the spectrum of **electromagnetic radiation**, one of the most important of which is the *X-ray background*.

Like the CMB, the X-ray background is quite smoothly distributed on the sky, indicating that it is not produced by local sources within our Galaxy. Although the isotropy of this radiation is not as extreme as that of the CMB (which is uniform to one part in a hundred thousand), it does furnish further evidence in favour of the large-scale homogeneity and isotropy described by the **cosmological principle**. In fact, the observed X-ray flux is isotropic on the sky to about one part in a thousand for energies between 2 and 20 keV. Since the bulk of the emission of this radiation is thought to arise from discrete sources (unlike the cosmic microwave background radiation), this observation itself places strong constraints on how inhomogeneous the distribution of these sources can be.

It is not known at present precisely what is responsible for the X-ray background, but many classes of object can, in principle, contribute. Individual **galaxies**, **quasars** and **active galaxies** all produce X-rays at some level, as do rich galaxy clusters and some of the superclusters that make up the **large-scale structure** of the galaxy distribution. Indeed, the bulk of the radiation may be produced by objects, such as **quasars**, at quite high cosmological **redshifts**. Disentangling the various components is difficult, and is the subject of considerable controversy; even so, it does seem that all these sources make significant contributions to the net X-ray flux we observe. Unlike the CMB, therefore, which has a relatively simple origin, the production of the X-ray background involves very messy processes occurring in a variety of astronomical sources with different physical conditions. When the nature and origin of the background is clarified, its properties may well shed some light on the origin of the objects responsible for its generation (see **structure formation**).

The properties of the X-ray background are also strongly related to the properties of the **intergalactic medium** (IGM). For many years, astronomers believed that a possible hiding-place for **dark matter** in the form of baryons might be a very hot, fully ionised IGM consisting mainly of a hydrogen plasma. This plasma would have to have a temperature as high as 100 million degrees, for at this temperature the plasma radiates in the X-ray region, which might account for at least some of the observed X-ray background.

Full ionisation of gas at this temperature makes it difficult to observe such a plasma directly in wavebands other than X-rays, for example by using **absorption lines**, as can be done with colder, neutral gas. However, photons from the CMB would scatter off energetic free electrons in such a plasma. The result of this scattering would be that photons got boosted to X-ray energies, thus creating a deficit in the original microwave region. This effect is known as the **Sunyaev–Zel'dovich effect**. Since the scattering medium is, by hypothesis, uniform everywhere, the characteristic **black-body** spectrum of the CMB should be distorted. Such distortions are not observed. We can therefore conclude that a plasma sufficiently hot and dense to contribute significantly (i.e. at more than the level of a few per cent) to the X-ray background would create distortions of the black-body spectrum which are larger than those observed. The model of dark matter in the form of a hot IGM is therefore excluded by observations.

SEE ALSO: **infrared background**.

FURTHER READING: Boldt, E., 'The cosmic X-ray background', *Physics Reports*, 1987, **146**, 215.

Z

Zel'dovich, Yakov Borisovich (1914–87) Soviet physicist. He began by working in nuclear physics and rocket science, but then turned to cosmology and relativity theory. He carried out important work on **black holes**, **active galaxies** and **quasars**, and then developed early ideas of cosmological **structure formation**. Together with Rashid Sunyaev, he predicted the **Sunyaev–Zel'dovich effect**, which has now been measured observationally.

Zwicky, Fritz (1898–1974) Bulgarian-Swiss-US astronomer. He worked mainly on **stellar evolution** and observations of **supernovae**, but also made pioneering studies of the **large-scale structure** of the Universe and obtained the first compelling evidence for significant quantities of **dark matter**.

Zel'dovich–Sunyaev effect see **Sunyaev–Zel'dovich effect**.

PHYSICAL CONSTANTS AND COMMON UNITS

	Symbol	Numerical value	SI unit
Boltzmann constant	k	1.38×10^{-23}	J K
Newtonian gravitational constant	G	6.67×10^{-11}	$\mathrm{N\ m^2\ kg^{-2}}$
Planck constant	h	6.63×10^{-34}	J s
Velocity of light	c	3.0×10^{8}	$\mathrm{m\ s^{-1}}$
Stefan–Bolzmann constant	s	5.67×10^{-8}	$\mathrm{W\ m^{-2}\ K^{4}}$
Mass of Sun	M	1.99×10^{30}	kg
Mass of proton	m_p	1.67×10^{-27}	kg
Mass of electron	m_e	9.11×10^{-31}	kg
Charge on electron	e	1.602×10^{-19}	C
Electronvolt	eV	1.60×10^{-19}	J
Year	*	3.156×10^{7}	s
Light year	†	9.46×10^{15}	m
Parsec	pc	3.09×10^{16}	m
Kiloparsec	kpc	3.09×10^{19}	m
Megaparsec	Mpc	3.09×10^{22}	m

* The recommended but rarely used symbol is ‘a’; ‘y’ and ‘yr’ are the usual alternatives.

† There is no recommended symbol; ‘l.y’ is often used.

BIBLIOGRAPHY

Popular books, requiring little or no knowledge of mathematics or physics:

Barbree, J. and Caidin, M., *A Journey Through Time: Exploring the Universe with the Hubble Space Telescope* (Penguin, London, 1995).

Barrow, J. D., *The World Within the World* (Oxford University Press, Oxford, 1988).

Barrow, J. D., *Theories of Everything* (Oxford University Press, Oxford, 1991).

Barrow, J. D., *Pi in the Sky* (Oxford University Press, Oxford, 1992).

Barrow, J. D., *The Origin of the Universe* (Orion, London, 1995).

Barrow, J. D. and Silk, J., *The Left Hand of Creation: The Origin and Evolution of the Expanding Universe* (Basic Books, New York, 1983).

Chown, M., *The Afterglow of Creation: From the Fireball to the Discovery of Cosmic Ripples* (Arrow, London, 1993).

Close, F., *The Cosmic Onion* (Heinemann, London, 1983).

Crowe, M. J., *Modern Theories of the Universe from Herschel to Hubble* (Dover, New York, 1994).

Davies, P. C. W., *The Forces of Nature* (Cambridge University Press, Cambridge, 1979).

Davies, P. C. W., *About Time: Einstein's Unfinished Revolution* (Penguin, London, 1995).

Deutsch, D., *The Fabric of Reality* (Allen Lane, London, 1997).

Eddington, A. S., *The Nature of the Physical World* (Cambridge University Press, Cambridge, 1928).

Fischer, D. and Duerbeck, H., *Hubble: A New Window to the Universe* (Springer-Verlag, New York, 1996).

Florence, R., *The Perfect Machine: Building the Palomar Telescope* (Harper Collins, New York, 1994).

Goldsmith, D. and Owen, T., *The Search for Life in the Universe*, revised 2nd edition (Addison-Wesley, Reading, MA, 1993).

Goldsmith, D., *Einstein's Greatest Blunder? The Cosmological Constant and Other Fudge Factors in the Physics of the Universe* (Harvard University Press, Cambridge, MA, 1995).

Graham-Smith, F. and Lovell, B., *Pathways to the Universe* (Cambridge University Press, Cambridge, 1988).

Gribbin, J., *In Search of the Big Bang* (Corgi, London, 1986).

Gribbin, J., *Companion to the Cosmos* (Orion, London, 1997).

Gribbin, J. and Goodwin, S., *Origins: Our Place in Hubble's Universe* (Constable, London, 1997).

Gribbin, J. and Rees, M. J., *The Stuff of the Universe* (Penguin, London, 1995).

Guth, A. H., *The Inflationary Universe* (Jonathan Cape, New York, 1996).

Harrison, E., *Darkness at Night: A Riddle of the Universe* (Harvard University Press, Cambridge, MA, 1987).

Hawking, S. W., *A Brief History of Time* (Bantam, New York, 1988).

Hawking, S. W., *Black Holes and Baby Universes and Other Essays* (Bantam, New York, 1993).

Hetherington, N. S., *Encyclopedia of Cosmology* (Garland, New York, 1993).

Hey, J. S., *The Radio Universe*, 3rd edition (Pergamon Press, Oxford, 1983).

Hoskin, M., (editor) *The Cambridge Illustrated History of Astronomy* (Cambridge University Press, Cambridge, 1997).

Hubble, E., *The Realm of the Nebulae* (Yale University Press, Newhaven, CT, 1936).

Kauffman, S., *At Home in the Universe* (Penguin, London, 1996).

Kline, M., *Mathematics in Western Culture* (Penguin, London, 1987).

Krauss, L. M., *The Fifth Essence: The Search for Dark Matter in the Universe* (Basic Books, New York, 1989).

Lightman, A., *Ancient Light: Our Changing View of the Universe* (Harvard University Press, Cambridge, MA, 1991).

North, J., *The Fontana History of Astronomy and Cosmology* (Fontana, London, 1994).

Overbye, D., *Lonely Hearts of the Cosmos: The Story of the Scientific Quest for the Secret of the Universe* (HarperCollins, New York, 1991).

Pagels, H. R., *Perfect Symmetry* (Penguin, London, 1992).

Petersen, C. C. and Brandt, J. C., *Hubble Vision: Astronomy with the Hubble Space Telescope* (Cambridge University Press, 1995).

Preston, R., *First Light: The Search for the Edge of the Universe* (Random House, New York, 1996).

Riordan, M. and Schramm, D., *The Shadows of Creation: Dark Matter and the Structure of the Universe* (Oxford University Press, Oxford, 1993).

Silk, J., *The Big Bang*, revised and updated edition (W.H. Freeman, New York, 1989).

Smoot, G. F. and Davidson, K., *Wrinkles in Time* (William Morrow, New York, 1993).

Thorne, K. S., *Black Holes and Time Warps* (Norton, New York, 1994).

Tipler, F. J., *The Physics of Immortality* (Macmillan, London, 1995).

Tucker, W. and Tucker, K., *The Cosmic Inquirers: Modern Telescopes and Their Makers* (Harvard University Press, Harvard, 1986).

Weinberg, S., *The First Three Minutes: A Modern View of the Origin of the Universe* (Fontana, London, 1983).

Weinberg, S., *Dreams of a Final Theory* (Vintage Books, London, 1993).

Undergraduate books; some background in mathematics and/or physics needed:

Aller, L. H., *Atoms, Stars, and Nebulae* (Harvard University Press, Cambridge, MA, 1971).

Arp, H. C., *Quasars, Redshifts, and Controversies* (Interstellar Media, Berkeley, CA, 1987).

Barrow, J. D. and Tipler, F. J., *The Anthropic Cosmological Principle* (Oxford University Press, Oxford, 1986).

Bernstein, J. and Feinberg, G., (editors) *Cosmological Constants* (Columbia University Press, New York, 1986).

Berry, M. V., *Principles of Cosmology and Gravitation* (Adam Hilger, Bristol, 1989).

Clark, S., *Towards the Edge of the Universe* (Wiley-Praxis, Chichester, 1997).

Coles, P. and Ellis, G. F. R., *Is the Universe Open or Closed?* (Cambridge University Press, Cambridge, 1997).

Dick, S. J., *The Biological Universe* (Cambridge University Press, Cambridge, 1996).

Harrison, E., *Cosmology: The Science of the Universe* (Cambridge University Press, Cambridge, 1981).

Hawking, S. W. and Israel, W. (editors), *300 Years of Gravitation* (Cambridge University Press, Cambridge, 1987).

Hockney, R. W. and Eastwood, J. W., *Computer Simulation Using Particles* (Adam Hilger, Bristol, 1988).

Islam, J. N., *An Introduction to Mathematical Cosmology* (Cambridge University Press, Cambridge, 1992).

Krach, H., *Cosmology and Controversy: The Historical Development of Two Theories of the Universe* (Princeton University Press, Princeton, NJ, 1996).

McDonough, T. R., *The Search for Extraterrestrial Intelligence* (John Wiley, Chichester, 1987).

Mandelbrot, B. B., *The Fractal Geometry of Nature* (W.H. Freeman, San Francisco, 1982).

Mandl, F., *Statistical Physics* (John Wiley, Chichester, 1971).

Mihalas, D. and Binney, J., *Galactic Astronomy* (W.H. Freeeman, New York, 1981).

Narlikar, J. V., *Introduction to Cosmology*, 2nd edition (Cambridge University Press, Cambridge, 1993).

Phillips, A. C., *The Physics of Stars* (John Wiley, Chichester, 1994).

Rindler, W., *Essential Relativity: Special, General, and Cosmological*, revised 2nd edition (Springer-Verlag, New York, 1979).

Robson, I., *Active Galactic Nuclei* (John Wiley, Chichester, 1996).

Roos, M., *Introduction to Cosmology*, 2nd edition (John Wiley, Chichester, 1997).

Rowan-Robinson, M. G., *The Cosmological Distance Ladder* (W.H. Freeman, New York, 1985).

Rybicki, G. and Lightman, A. P., *Radiative Processes in Astrophysics* (John Wiley, New York, 1979).

Schutz, B. F., *A First Course in General Relativity* (Cambridge University Press, Cambridge, 1985).

Shklovskii, I. S. and Sagan, C., *Intelligent Life in the Universe* (Holden-Day, New York, 1966).

Squires, E., *The Mystery of the Quantum World* (Adam Hilger, Bristol, 1986).

Symon, K. R., *Mechanics*, 3rd edition (Addison-Wesley, Reading MA, 1980).

Tayler, R. J., *The Stars: Their Structure and Evolution* (Cambridge University Press, Cambridge, 1994).

Tayler, R. J., *The Hidden Universe* (Wiley-Praxis, Chichester, 1995).

Tayler, R. J., *Galaxies: Structure and Evolution* (Wiley-Praxis, Chichester, 1997).

Advanced texts, suitable for specialists only:

Binney, J. and Tremaine, S., *Galactic Dynamics* (Princeton University Press, Princeton, 1987).

Coles, P. and Lucchin, F., *Cosmology: The Origin and Evolution of Cosmic Structure* (John Wiley, Chichester, 1995).

Collins, P. D. B., Martin, A. D. and Squires, E. J., *Particle Physics and Cosmology* (John Wiley, New York, 1989).

Hawking, S. W. and Ellis, G. F. R., *The Large-Scale Structure of Space-Time* (Cambridge University Press, Cambridge, 1973).

Hawking, S. W. and Penrose, R., *The Nature of Space and Time* (Princeton University Press, Princeton, 1996).

Heck, A. and Perdang, J. M. (editors), *Applying Fractals in Astronomy* (Springer-Verlag, Berlin, 1991).

Kolb, E. W. and Turner, M. S., *The Early Universe* (Addison-Wesley, Redwood City, CA, 1990).

Linde, A. D., *Particle Physics and Inflationary Cosmology* (Harwood Academic, London, 1990).

Makino, J. and Taiji, M., *Scientific Simulations with Special-Purpose Computers* (John Wiley, Chichester, 1998).

Misner, C. W., Thorne, K. S. and Wheeler, J. A., *Gravitation* (W. H. Freeman, San Francisco, 1972).

Padmanabhan, T., *Structure Formation in the Universe* (Cambridge University Press, Cambridge, 1993).

Parker, E. N., *Cosmical Magnetic Fields* (Clarendon Press, Oxford, 1979).

Peebles, P. J. E., *The Large-Scale Structure of the Universe* (Princeton University Press, Princeton, 1980).

Peebles, P. J. E., *Principles of Physical Cosmology* (Princeton University Press, Princeton, 1993).

Rees, M. J., *Perspectives in Astrophysical Cosmology* (Cambridge University Press, Cambridge, 1995).

Weinberg, S., *Gravitation and Cosmology: Principles and Applications of the General Theory of Relativity* (John Wiley, New York, 1972).

Weinberg, S., *The Quantum Theory of Fields* (2 vols) (Cambridge University Press, Cambridge, 1995).

Selected technical articles:

Abell, G. O., 'The distribution of rich clusters of galaxies', *Astrophysical Journal Supplement Series*, 1958, **3**, 211.

Albrecht, A. and Steinhardt, P. J., 'Cosmology for grand unified theories with radiatively induced symmetry breaking', *Physical Review Letters*, 1982, **48**, 1220.

Alcock, C., *et al*. 'Possible gravitational microlensing of a star in the Large Magellanic Cloud', *Nature*, 1993, **365**, 621.

Arp, H. C., Burbidge, G., Hoyle, F., Narlikar, J. V. and Wickramasinghe, N. C., 'The extragalactic Universe: An alternative view', *Nature*, 1990, **346**, 807.

Bertschinger, E. and Gelb, J. M., 'Cosmological *N*-body simulations', *Computers in Physics*, 1991, **5**, 164.

Blandford, R. and Narayan, R., 'Cosmological applications of gravitational lensing', *Annual Reviews of Astronomy and Astrophysics*, 1992, **30**, 311.

Blumenthal, G. R., Faber, S. M., Primack, J. R. and Rees, M. J., 'Formation of galaxies and large-scale structure with cold dark matter', *Nature*, 1984, **311**, 517

Boldt, E., 'The cosmic X-ray background', *Physics Reports*, 1987, **146**, 215.

Bond, J. R., Carr, B. J. and Hogan, C. J., 'The spectrum and anisotropy of the cosmic infrared background', *Astrophysical Journal*, 1986, **306**, 428.

Bondi, H. and Gold, T., 'The steady state theory of the expanding Universe', *Monthly Notices of the Royal Astronomical Society*, 1948, **108**, 252.

Bonnor, W. B., 'Jeans' formula for gravitational instability', *Monthly Notices of the Royal Astronomical Society*, 1957, **117**, 104.

Brans, C. and Dicke, R. H., 'Mach's principle and a relativistic theory of gravitation', *Physical Review Letters*, 1961, **124**, 125.

Burstein, D., 'Large-scale motions in the Universe: A review', *Reports on Progress in Physics*, 1990, **53**, 421.

Coles, P., 'The large-scale structure of the Universe', *Contemporary Physics*, 1996, **37**, 429.

Coles, P. and Ellis, G. F. R., 'The case for an open Universe', *Nature*, 1994, **370**, 609.

de Lapparent, V., Geller, M. J. and Huchra, J. P., 'The large-scale structure of the Universe', *Astrophysical Journal*, 1986, 302, L1.

De Sitter, W., 'On Einstein's theory of gravitation and its astronomical consequences: Third paper', *Monthly Notices of the Royal Astronomical Society*, 1917, **78**, 3.

Dicke, R. H., Peebles, P. J. E, Roll, P. G. and Wilkinson, D. T., 'Cosmic black-body radiation', *Astrophysical Journal*, 1965, **142**, 414.

Dirac, P. A. M., 'The cosmological constants', *Nature*, 1937, **139**, 323.

Dyson, F. J., 'Time without end: Physics and biology in an open universe', *Reviews of Modern Physics*, 1979, **51**, 447.

Einstein, A., 'Cosmological considerations on the general theory of relativity', 1917, reprinted in *The Principle of Relativity*, edited by H.A. Lorentz *et al.* (Dover, New York, 1950).

Ellis, G. F. R., 'Alternatives to the Big Bang', *Annual Reviews of Astronomy and Astrophysics*, 1987, **22**, 157.

Fort, B. and Mellier, Y., 'Arc(let)s in clusters of galaxies', *Astronomy and Astrophysics Review*, 1994, **5**, 239.

Franx, N. *et al.*, 'A pair of lensed galaxies at $z = 4.92$ in the field of CL1358+62', *Astrophysical Journal*, 1997, **486**, L75.

Freeman, W. *et al.*, 'Distance to the Virgo Cluster galaxy M100 from Hubble Space Telescope observations of Cepheids', *Nature*, 1994, **371**, 757.

Friedmann, A., 'Über die Krummung des Raumes', *Zeitschrift für Physik*, 1922, **10**, 377. (English translation: 'On the curvature of space' in *Cosmological Constants*, edited by J. Bernstein and G. Feinberg (Columbia University Press, New York, 1986), p. 49.

Guth, A. H., 'Inflationary Universe: A possible solution to the horizon and flatness problems', *Physical Review* D, 1981, **23**, 347.

Harrison, E. R., 'Fluctuations at the threshold of classical cosmology', *Physical Review* D, 1970, **1**, 2726.

Hartle, J. B. and Hawking, S. W., 'The wave function of the Universe', *Physical Review* D, 1983, **28**, 2960.

Hawking, S. W., 'Black hole explosions?', *Nature*, 1974, **248**, 30.

Hawking, S. W. and Penrose, R., 'The singularities of gravitational collapse and cosmology', *Proceedings of the Royal Society*, 1970, **A314**, 529.

Hoyle, F., 'A new model for the expanding Universe', *Monthly Notices of the Royal Astronomical Society*, 1948, **108**, 372.

Hoyle, F. and Narlikar, J. V., 'Mach's principle and the creation of matter', *Proceedings of the Royal Society*, 1963, **A273**, 1.

Hubble, E., 'A relation between distance and radial velocity among extra-galactic nebulae', *Proceedings of the National Academy of Sciences*, 1929, **15**, 168.

Jones, M. *et al.*, 'An image of the Sunyaev–Zel'dovich effect', *Nature*, 1993, **365**, 320.

Kellerman, K. I., 'The cosmological deceleration parameter estimated from the angular size/redshift relation for compact radio sources', *Nature*, 1993, **361**, 134.

Kneib, J.-P. and Ellis, R. S., 'Einstein applied', *Astronomy Now*, May 1996, p. 43.

Lemaître, G., 'A homogeneous universe of constant mass and increasing radius accounting for the radial velocity of the extragalactic nebulae', *Monthly Notices of the Royal Astronomical Society*, 1931, **91**, 483.

Lifshitz, E. M., 'On the gravitational instability of the expanding Universe', *Soviet Physics JETP*, 1946, **10**, 116.

Linde, A. D., 'Scalar field fluctuations in the expanding Universe and the new inflationary Universe scenario', *Physics Letters* B, 1982, **116**, 335.

Linde, A. D., Linde, D. and Mezhlumian, A., 'From the Big Bang theory to the theory of a stationary Universe', *Physical Review* D, 1994, **49**, 1783.

Lineweaver, C. H. and Barbosa, D., 'What can cosmic microwave background observations already say about cosmological parameters in critical-density and open CDM models?', *Astrophysical Journal*, 1998, **496**, in press.

MacCallum, M. A. H., 'Anisotropic and inhomogeneous cosmologies' in *The Renaissance of General Relativity and Cosmology*, edited by G.F.R. Ellis *et al.* (Cambridge University Press, Cambridge, 1993), p. 213.

Meszaros, P., 'The behaviour of point masses in an expanding cosmological substratum', *Astronomy and Astrophysics*, 1974, **37**, 225.

Metcalfe, N. *et al.*, 'Galaxy formation at high redshifts', *Nature*, 1996, **383**, 236.

Misner, C. W., 'The isotropy of the Universe', *Astrophysical Journal*, 1968, **151**, 431.

Narlikar, J. V. and Padmanabhan, T., 'Inflation for astronomers', *Annual Reviews of Astronomy and Astrophysics*, 1991, **29**, 325.

Newman, W. I. and Sagan, C., 'Galactic civilizations: Population dynamics and interstellar diffusion', *Icarus*, 1981, **46**, 293.

Peebles, P. J. E., Schramm, D. N., Turner, E. L. and Kron, R. G., 'The case for the hot relativistic Big Bang cosmology', *Nature*, 1991, **353**, 769.

Penrose, R., 'Gravitational collapse and space-time singularities', *Physical Review Letters*, 1965, **14**, 57.

Penzias, A. A. and Wilson, R. W., 'A measurement of excess antenna temperature at 4080 Mc/s', *Astrophysical Journal*, 1965, **142**, 419.

Perlmutter, S. *et al.*, 'Measurements of the cosmological parameters omega and lambda from the first seven supernovae at $z > 0.35$', *Astrophysical Journal*, 1997, **483**, 565.

Perlmutter, S. *et al.*, 'Discovery of a supernova explosion at half of the age of the Universe', *Nature*, 1998, **391**, 51.

Rowan-Robinson, M. *et al.*, 'A sparse-sampled redshift survey of IRAS galaxies: I. The convergence of the IRAS dipole and the origin of our motion with respect to the microwave background', *Monthly Notices of the Royal Astronomical Society*, 1990, **247**, 1.

Sachs, R. K. and Wolfe, A. M., 'Perturbations of a cosmological model and angular variations of the cosmic microwave background', *Astrophysical Journal*, 1967, **147**, 73.

Sandage, A. R., 'Cosmology: The search for two numbers', *Physics Today*, 1970, **23**, 34.

Sandage, A. R., 'Distances to galaxies: The Hubble constant, the Friedmann time and the edge of the Universe', *Quarterly Journal of the Royal Astronomical Society*, 1972, **13**, 282.

Sandage, A. R., 'Observational tests of world models', *Annual Reviews of Astronomy and Astrophysics*, 1988, **26**, 561.

Saunders, W. *et al.*, 'The density field of the local Universe', *Nature*, 1991, **349**, 32.

Shechtman, S. *et al.*, 'The Las Campanas Redshift Survey', *Astrophysical Journal*, 1996, **470**, 172.

Silk, J., 'Fluctuations in the primordial fireball', *Nature*, 1967, **215**, 1155.

Slipher, V. M., 'Spectrographic observations of nebulae', *Popular Astronomy*, 1915, **23**, 21.

Smoot, G. F. *et al.*, 'Structure in the COBE Differential Microwave Radiometer first-year maps', *Astrophysical Journal Letters*, 1992, **396**, L1.

Vilenkin, A. 'Boundary conditions in quantum cosmology', *Physical Review* D, 1986, **33**, 3560.

Walsh, D., Carswell, R. F. and Weymann, R. J., '0957+561A,B: Twin quasistellar objects or gravitational lens?', *Nature*, 1979, **279**, 381.

Weinberg, S., 'The cosmological constant problem', *Reviews of Modern Physics*, 1989, **68**, 1.

White, S. D. M. *et al.*, 'The baryon content of galaxy clusters: A challenge to cosmological orthodoxy', *Nature*, 1993, **366**, 429.

Zel'dovich, Ya. B., 'A hypothesis unifying the structure and entropy of the Universe', *Monthly Notices of the Royal Astronomical Society*, 1972, **160**, 1P.

Index

Bold is used for terms which have entries in the dictionary section, and for the corresponding page numbers.